Materials

MITCHELL'S BUILDING SERIES

Materials

Alan Everett *RIBA*

The Mitchell Publishing Company Limited
London

© Alan Everett 1970, 1978, 1981 and 1986
Reprinted 1972
Reprinted 1975
New edition 1978
New edition 1981
Revised 1983
New edition 1986
Reprinted 1989, 1991

Typeset by Latimer Trend & Company Ltd
Plymouth, Devon

Printed by Courier International Ltd
East Kilbride, Scotland

Published by
The Mitchell Publishing Company Limited
a subsidiary of BT Batsford Limited
4 Fitzhardinge Street, London W1H 0AH

British Library Cataloguing in Publication Data

Everett, Alan
  Materials. – New ed. –
  (Mitchell's building series)
  1. Building materials
  I. Title   II. Series
  691   TA403

ISBN 0–7134–5442–3

Contents

CONTENTS

Plasters are dealt with in Mitchell's *Finishes*, chapter 13

British Standards Institution
BSI Enquiry Service and publications referred to in this volume are obtainable by post from:
The British Standards Institution
Linford Wood
Milton Keynes
MK14 6LE

Counter purchases can be made at
192 Pentonville Road
London N1 9ND
and at eight regional addresses.

Acknowledgment

I am grateful to the Director and Clerk of the Greater London Council for permission to quote from the London Building Acts 1930–39 and the Constructional Bylaws made under these Acts, and to the Controller of Her Majesty's stationery Office for permission to quote from the Building Regulations 1976, and to the Directors of the following organizations for their kind permission for quotations to be made from their publications:

The Building Research Establishment
The British Standards Institution
The Fire Research Station
The Princes Risborough Laboratory
Research and Development Associations of the construction industry.

I am also grateful to many firms for permission to quote from their publications.

Every effort has been made to acknowledge each quotation individually in the text, but I trust that any omission to do so will be forgiven.

Thanks are due to Sir Isaac Pitman and Sons Ltd for permission to print figure 25 which is based on a table in *Properties of Concrete* by A M Neville MC, Msc (Eng), PhD, MICE, AMIstruct E; also to CR Books Ltd for permission to print figure 21 from *Concrete Practice* Vol. II by R H Elvery Bsc (Eng), MICE, to Heinemann Ltd for the information in table 40 taken from *Stone in Building* by Hugh O'Neill, and to Butterworths Ltd for values derived from *The Metals Reference Book* by CJ Smithells MC, DSc, FIM, an invaluable source of information.

I have endeavoured to condense information without distorting the meaning, but no matter how carefully done, summaries, and quotations isolated from their context, can be misleading and the reader is referred to the original document in each case.

Sixteen chapters on diverse subjects can only be written with the help of many authorities and I have been extremely fortunate in this respect. I am particularly grateful to the undernamed experts for the time and patience they expended in reading and correcting my manuscripts, typescripts and/or proofs. The first named under each subject heading were my principal advisers who, in the case of the longer chapters, devoted a very considerable amount of time and energy to the task. I am very happy to have many new friends and hasten to state that they are not responsible for any errors which may have crept into the text.

Introduction and properties generally D E Hull ARIBA and B R Currell PhD, BSc (London), FRIC.

Fire R W Fisher MRSH of The Fire Protection Association and H L Malhotra BSc, CEng, MICE, MIFire E of the Fire Research Station.

Costs J Nisbet FRICS

Acoustics J E Moore FRIBA of The Polytechnic of North London

Timber R P Wood BA For (Cantab) FIWSc, of the Timber Research and Development Association; R L Dawson, AIWSc; V Serry (Metrication) of The Phoenix Timber Co Ltd and the late E Joyce MSIA of Brighton College of Art

Boards and slabs E A Raynham BSc, AInst P of The Fibre Building Board Development Organization Ltd (FIDOR) and R S Vickerstaff

Stones R J H Dix of The Stone Firms Ltd; B L Clarke BSc and D B Honeyborne BSc (Hons) of the BRE and K W Young FGS, MGLC of the Polytechnic of North London

Ceramics I L Freeman BSc (Hons), FICeram of the BRE and J Shore of Carters Tiles Ltd

Bricks and blocks K Thomas CEng, MIStruct E, ARTC, L Bevis and J R Harding TD, CEng, FInst F, FICeram of the Brick Development Association (BDA) and Welwyn Hall Research Association

Limes and cements R A Keen of The Cement and Concrete Association (CCA); D E Shirley

BSc, AMIChem E, AM Inst F, AICeram of Lafarge Aluminous Cement Co Ltd and AT Corish BSc Eng, ACGI, MICE, AMInst HE of The Cement Marketing Co Ltd

Concretes BW Shacklock MSc, FICE, MInst Struct E, MInst HE, CD Pomeroy, DSc FInst P and D Palmer BSc Eng, DIC, MICE of The Cement and Concrete Association (CCS) and W Kinniburgh FRIC (lightweight concretes) of the BRE

Metals – generally GME Cooke BSc (Mech Eng) of the Corporate Laboratories of the British Steel Corporation (BISRA)

Corrosion of metals KA Chandler BSc, ARSM, FICorr T of The Corporate Laboratories of the British Steel Corporation (BISRA)

Ferrous metals GME Cooke BSc (Mech Eng) of the Corporate Laboratories of the British Steel Corporation (BISRA); DT Williams AMI Struct E, MSoc CE (France) of the British Steel Corporation; HV Hill MSc, AMICE, AMIStruct E of The Sheet Steel Industries Development Association (SSIDA); P Shaw AIM of The Stainless Steel Development Association (SSDA)

Copper E Carr PhD, BSc of The Copper Development Association (CDS) and DR Harrington BSc of The Delta Metal Company Ltd

Lead and zinc JF Quinn ARIBA of The Lead and Zinc Development Associations (LDA and ZDA)

Aluminium FC Clarke BSc (Eng) of The Aluminium Federation

Asbestos products PW Steggals of Cape Universal Building Products Ltd and DI Jones of Turners Asbestos Cement Co Ltd

Bituminous products Shell International Petroleum Co Ltd

Flat glass DW Armstrong BSc, ARCS, AINS, and Brian Waldron, of Pilkington Brothers Ltd

Decorative glass AR Fisher MGP of Whitefriars Glass Works Ltd

Plastics and rubbers WF Ratcliffe BSc (Eng), C Eng, API of Shell International Chemical Co Ltd; JR Crowder BSc, PhD of the BRS; P Fordham AA Dip, ARIBA of Imperial Chemicals Industries Ltd; JA Brydson FPI and M Kaufman PhD, FRIC of The Polytechnic of North London and EA Everest of the British Plastics Federation

Adhesives E Van der Straeten of CIBA (ARL) Ltd and Dr Aubrey of the Polytechnic of North London

Mortars JF Ryder BSc of the BRE and P Robinson of the Welwyn Hall Research Association

Sealants Bostik Ltd; RG Groeger BSc, ARPS of John Laing Research and Development Ltd; TA Baker of the BRE and RAV Meikle AIMar E, of Expandite Ltd

My sincere thanks are due no less to any whom I have inadvertently omitted to mention and to eleven experts in one organization who must remain anonymous. I am also indebted to many firms for providing information about their products.

My thanks are due also to RE Sowden BSc (Hons) for metrication, to GW Dilks for the drawings from my originals and to Ernest Joyce for figures 2, 3, 6, 7, 9 10 and 11 concerned with timber and boards, to Margaret Bird for skilled typewritten transcriptions from my hieroglyphics and to Mr and Mrs TM Doust for compiling the index.

I am extremely grateful to Thelma M Nye for her patience, encouragement and expert editorial advice.

Infinite thanks are extended to my wife for uncomplainingly suffering the work to be done.

The author also thanks anyone who very kindly brings mistakes to his attention or who, in other ways, enables him to improve future editions of this volume.

London 1970 AE

The author continues to be grateful to those who contributed in many ways to the first edition of this book. I am grateful, also, to the readers who kindly responded to my request for comments and suggestions, and venture to hope that others will do so in the future.

I thank many friends who have provided information and advice since 1970, in particular:
DW Aubrey, BSc, PhD, FRIC, CChem, (adhesives) Polytechnic of North London; VA Callcut, CEng, FIM, MIQA, Copper Development Association; DW Clegg, ALA, SfB Agency UK Ltd; J Harding, TD, F Inst, F, CEng, FICeram, Brick Development Association; IDG Lee BSc (Eng), FIWSc, Chipboard Promotion Association; M Wragg, Pilkington Brothers Ltd.

My thanks are again extended to the editor, Thelma M Nye, for her painstaking processing of the text. Special thanks are always due to my ever patient wife.

Oxford 1981   AE

I am grateful to those listed above and also to many who have enabled me to 'keep up with later developments' in particular to the always helpful staffs of the Oxford libraries.

Oxford 1986   AE

13

Introduction

Since I first wrote this volume in 1970, considerable changes have taken place in building, and there has been greater appreciation of the need for knowledge about materials and for skill in specifying and supervising work.

Beneficial changes have followed a greater awareness of, and demands for, safety from fire and from hazardous materials such as asbestos and lead. Problems associated with the use of calcium chloride, alkali-reactive aggregates and high alumina cement in concrete, have become better known.

It is heartening to note greater concern for safety and, in particular, concern for the comfort of physically handicapped users of buildings. Sadly, however, changes are often slow and occur only belatedly in response to serious failures which happen in spite of adequate warnings.

It is sad also that buildings must now be vandal-resistant and 'fortified' – materials and products are now being advertised as not being readily disposed of by thieves!

In materials, steel frames are again economical for multi-storey buildings, and glass fibre reinforced polyester is well established. Visually, brass is back. Fashionable timbers have been, in turn, mahogany, oak, teak and now naked knotty pine has been brought from the humble kitchen into the boudoir.

Design

Satisfactory design of safe and economical building structures, fabrics, components, services and finishes, dealt with in other volumes of *Mitchell's Building Series*, demands knowledge of the properties and behaviour of materials and combinations of materials. The designer should also have an understanding of problems in the manufacture and assembly of materials and components, in factories and on building sites, and of their initial and long-term costs. Equally important, the successful designer must be meticulous in detailing his intentions in drawings and specifications, and in checking the compliance of products and their incorporation in the work.

The 1985 *Building Regulations* apply throughout England and Wales.

With the exception of thermal insulation, the new Regulations are in the form of performance requirements expressed as 'reasonable', adequate, safe, etc. *Approved Documents* give guidance as to acceptable ways of satisfying the Regulations.

Regulation 7 refers to *Materials and Workmanship*.

Sources of information

These have increased substantially since 1970. The problems and risks involved in summarising information are considerable and the reader should consult the authorities on the respective subjects.

Information is provided by government and trade research and development organisations. *Building Centres* exist in some towns.

Publications concerning building materials include:

Building Research Establishment (BRE) publications, in particular:

Principles of Modern Buildings Volumes I and II, HMSO

BRE Digests, HMSO

British Standards Institution (BSI) publications, in particular, *British Standard Specifications* and *Codes of Practice*. The full documents can be consulted in major libraries.

British Standard Handbook no. 3 (four volumes) comprises summaries of 1500 British Standards (BS) relating to building, including: glossaries, specifications, guides, codes of practice and methods of sampling and testing.

BSI Catalogue is annual, and *BSI News* describes new publications, amendments, etc, each month.

Manual of British Standards in building, Hutchinson Education – broadly matches BS Handbook no. 3, but is much briefer.

British Board of Agrément Certificates – see page 18.
Building Regulations and other *statutory requirements*.
The *National Building Specification* (NBS), RIBA.
Office *libraries* and *Microfiche systems* are provided and updated by the *RIBA* and the *Barbour Index*
Materials for Building, Lyall Addleston
  Volume 1 *Physical and chemical aspects of matter and the strength of materials*, Illiffe
  Volumes 2 and 3 *Water and its effects*, Illiffe
  Volume 4 *Heat and fire and their effects*, Butterworth
Properties of building materials, HJ Eldridge, MTP Construction
Specification (annual, six volumes), Architectural Press
The Chemistry of Building Materials, RME Diamant, MSc, DipChemE, AMInstF, Business Books
Professional and trade journals

Resources

Materials consume varying amounts of energy in primary manufacture. Thus, metals, cement, clay bricks, plastics and glass are made mainly from ores, clay, limestone, petroleum and sand. In all cases, further energy must be used in shaping materials into components, in transport and in fixing in buildings. Although there are still vast reserves of the basic materials, their costs increase as they are exhausted locally, or their availability is reduced by other land uses including building itself, or as further exploitation, even of timber, becomes unacceptable to landscape conservationists. A conflict between needs to conserve two resources has been seen in the construction of the Kielder reservoir at the expense of 1.5 million trees. It is interesting to note that vegetable products, including timber, need not run out if they are systematically replanted and cultivated.

Economies can be effected by minimising waste in manufacture and on the building site. BRE Digest 247 deals with *Waste of building materials* and BRE digest 259 with *Materials control to avoid waste*.

Clearly, it is sensible to use industrial waste at least where the cost of its recovery does not exceed that of 'new' materials. BS 6543:1985 a *Guide to the use of industrial by-products and waste materials* encourages the use of materials such as wood particles, blast furnace slag and pulverized fuel ash (PFA), as fills and aggregates and in the manufacture of boards, bricks, blocks and cements.

Economics

The cost of building materials is an important factor in the initial capital cost of buildings. In houses, the ratio of costs of materials to site labour is about 53:47, and in office buildings about 60:40, while in buildings such as factories the relative cost of materials is higher. However, cost decisions should take into account not only initial costs, but equally, running and maintenance costs including the costs of replacements during the whole life of the building.

Just as the extra cost of additional thermal insulation of a building may reduce the required capacity of a heating installation, the fact that over a period of sixty years the initial cost of materials is only about 10 per cent of the total cost of running and maintaining a typical building, suggests that extra costs of maintenance-free materials may be justified. While the use of less durable materials soon leads to incipient decay and maintenance work causes frequent inconvenience and disruption, durable materials generally 'mellow' with age and provide continuing aesthetic satisfaction leading to better building-user and staff relations. Clearly, if savings have to be made initially, they should be obtained from finishes which can be replaced with little disturbance, rather than from, say, basement tanking. Although often costing more than short-lived materials, durable materials are justified even for 'temporary' buildings which can be used for other purposes later.

However, other factors which must be taken into consideration in calculating an overall cost-in-use, include the interest rate on capital, inflation and taxation and any grants which favour either greater capital or running and maintenance expenditures.

References include:

Building Economy: A synoptic approach, PA Stone, Pergamon Press.

Principles of modern building, Volume I Chapter 10, Building economics, HMSO.

Economic Theory and the Construction Industry, Patricia M Hillebrandt, Macmillan.

Spon's Architects' and Builders' Price Book, E and F N Spon.

Research, development and patents

The Building Research Establishment (BRE) incorporates the *Fire Research Station* (FRS), the Building Research Station (BRS) and the Princes Risborough Laboratory (PRL), formerly the *Forest Products Research Laboratory* (FPRL). Industrial organizations include: *The British Ceramic Research Association* (BCRA), *The Brick Development Association* (BDA), *The Laboratories of the British Steel Corporation, The Cement and Concrete Association* (CCA) and *The Timber Research and Development Association* (TRADA). Research and development is also undertaken by individual firms, eg Imperial Chemical Industries (ICI) and Pilkington Flat Glass Ltd, and by the Universities and Polytechnics.

Development is encouraged by, and often enabled to proceed, only by the protection afforded by *patents* which are granted in return for disclosure of inventions. Annual increases in fees payable by patentees reflect the increases in income they may enjoy up to twenty years.

The applications and procedures outlined in the Patent Office leaflet *Applying for a patent* are complex and require the expertise of a Patent Agent.

Notice of the existence of a patent for a product must be accompanied by its number.

Standardization

The benefits of standardization of products and technological procedures was recognised from 1910 by the establishment of the *Engineering Standards Committee* which preceeded the *British Standards Institution* the head office of which is 2 Park Street, London, W1A 2BS. Departments for information, technical help for exporters and marketing are at Linford Wood, Milton Keynes, MK14 6LE. The BSI publishes *glossaries* of terms, *specifications* describing sizes, tolerances and required properties of materials and components, and *codes of practice* describing recommended methods for incorporating them in buildings. Until recently, codes of practice (CP) were listed separately, but now they are being included in British Standards (BS).

Specifications and Codes are means of establishing safe and reasonable minimum levels of quality and performance, and in some cases, compliance with British Standards and Codes of Practice is 'deemed to satisfy' Building Regulations.

The standardisation of technical terms is of primary importance. BS 6100, the *Glossary of building and civil engineering terms* should be adhered to.

Part 0:1984 explains the objects of the Standard. Other parts deal with:

Part 1 Section 1.0:1984 *General construction terms.*

Subsection 1.5:1984 *Coordination of dimensions, tolerances, and accuracy*

Part 2 Section 2.1:1984 *Structural design and elements*

Part 4 *Forest products*

Part 5 *Masonry*

Part 6 *Concrete and plaster*

BS 4949:1973 is a *Glossary of terms relating to building performance*.

Costly mistakes can result from the use of incorrect terms, eg the word 'either' where 'both' is intended. Other commonly confused terms are:

additive – admixture
adsorption – absorption
asbestos – asbestos cement
a plastic – a plastics material
breeze blocks – lightweight aggregate blocks
cobble – sett
corrugated iron – corrugated steel
fire proof – fire resisting
floor – flooring
glass fibre – glass fibre reinforced plastics
hardwood – dense and dark coloured softwood
incombustible – noncombustible
inflammable – flammable

polystyrene – expanded polystyrene (EPS)
roof – roofing
setting – hardening
stress – strain
tank – cistern
wrought iron – bent mild steel

Registered trade names should not be used unless the specific proprietary product is required, eg *Fibreglass, Bitumastic, Perspex, Armourplate* and *Formica.*

The rationalization and standardization of product sizes reduces the multiplicity of stocks required to be held by suppliers and merchants, and importantly, it reduces the labour in cutting-to-fit on the building site.

Relevant *British Standards* include:
DD 22:1972
Recommendations for the co-ordination of dimensions in building. Tolerances and fits for building. The calculation of work sizes and joint clearances for building components.
BS 4011:1966
Recommendations for the co-ordination of dimensions in building. Co-ordinating sizes for building components and assemblies.
BS 4330:1968
Recommendations for the co-ordination of dimensions in building. Co-ordinating sizes for building, eg floor-floor and floor-ceiling dimensions, to assist dimensional coordination of components.

Other publications of the BSI include *Drafts for Development (DD)* which deal with topics where firm requirements cannot be established until comments have been considered by the drafting committee. *PD*s are *Published Documents* which do not fall under other headings.

BSI publications are prepared with due regard to the work of the International Organisation for Standardization (ISO).

British Standards Institution publications are obtainable from any of the addresses given on page 10.

Certification

New and untested materials should not be specified unless the building owner accepts the risk of possible failure. Accelerated ageing tests require expert devising and interpretation, which takes into account the likely standards of specification, workmanship and supervision as well as the conditions in service.

The testing and certification of new products for specific uses is undertaken by *The British Board of Agrément (BBA)*, Hemel Hempstead, Herts, is an independent body which employs the resources of BRE and other organizations. The *Agrément system* encourages the development of new products and processes by examining prototypes and by issuing certificates.

These define products and give basic design data. They state suitability for uses and how they satisfy various aspects of the Building Regulations. They describe how products should be associated with other parts and materials and how they should be stored and handled on sites. Requirements for maintenance are also given.

Quality control in production

Specifiers of building products and processes must be satisfied that their requirements can be, and are complied with. Manufacturers who apply statistical methods of *Quality control* may be granted the *BSI Kitemark* which involves regular and systematic checking of manufacture, testing procedures and records, by BSI inspectors.

Clearly, the costs of quality control must be balanced by savings in the costs of production and 'after-sales' costs. BS 5750 (many parts) *Quality systems* is for manufacturers.

BS 4778:1979 is a *Glossary of terms used in quality control.*

Selection of products

Reputable manufacturers provide information as to the properties and merits of their products and warn against their use in unsuitable circumstances. Products should be used strictly in accordance with manufacturers' recommendations. Collections of well presented trade literature are contained in the *RIBA Index* and the *Barbour Index* and *Compendium.*

In selecting materials or products, the degrees to which available materials or products satisfy each *performance requirement* should be evaluated. For economy in cost and space in buildings, wherever practicable materials should perform more than one function, eg, combinations of: thermal insulation and sound insulation and provide strength, weather resistance and good appearance. The final choice may be determined by considerations such as delivery dates, costs of incorporation in the building and *costs-in-use* see page 16.

Specification

A conventional specification describes materials and workmanship. To avoid misunderstandings, the specifier should use only terms which are clearly defined in BS 6100. BSPD 6112:1967 is a *Guide to the Preparation of Specifications*.

The National Building Specification (NBS), published by the RIBA in four volumes and a 'small jobs' version, and *Specification*, Architectural Press Ltd, provide standard clauses.

In requiring compliance with British Standards it is very important to know the grades, qualities and other alternatives they may contain.

Performance specifications enable manufacturers to employ their specialized expertise and resources in the best way to satisfy stated functional and aesthetic requirements. However, requirements must be precisely defined, and suitable testing facilities are essential in order to compare the merits of different solutions. A reference is *Performance specification writing for building components*, HMSO.

Workmanship and supervision on site

Rightly, British Standards, eg BS 1186 for joinery, and the Building Regulations, are now concerned with workmanship as well as materials and products.

Supervision is necessary to ensure compliance with instructions provided by designers, authorities and by manufacturers.

Increasing sophistication of products places additional responsibility on supervisors to ensure compliance with unfamiliar and specialised techniques. If this labour-intensive supervision and checking cannot be provided it is better to adhere to old fashioned materials and methods.

Supervision is dealt with in *The Supervision of Construction — a guide to site inspection*, John Watts, Batsford.

1 Properties generally

Some properties, which relate to certain materials only, are considered in other chapters.

Properties which relate to materials generally are dealt with in this chapter, ie:

(a) Density and specific gravity
(b) Strength
(c) Optical properties – page 23
(d) Electrical properties
(e) Thermal properties
(f) Acoustic properties
(g) Deformations – page 27
   1 Movements caused by applied loads
   2 Movements caused by changes in moisture content of materials
   3 Movements caused by changes in temperature
   4 Stresses due to thermal and moisture changes
   5 Common defects due to movements and their prevention
(h) Deterioration – page 32
   1 Corrosion of metals – see also page 193.
   2 Sunlight – page 32
   3 Biological agencies
   4 Water
   5 Crystallization of salts – page 33
   6 Frost
   7 Chemical action
   8 Loss of volatiles – page 34
   9 Abrasion and impact
  10 Vibration
  11 Fire
(i) Appearance – page 43

(a) Density and specific gravity

BS 648:1964 *Schedule of weights of building materials* is a useful reference for calculation purposes. Table 1 gives densities for many materials in kg/m^3, and others are given at the appropriate points in the text.

Specific gravity is a ratio of the density of a substance at a given temperature to the density of water at 4°C.

(b) Strength

The New Science of Strong Materials by J E Gordon, Pelican, is recommended reading.

Materials must be capable of safely supporting their own weight and any applied loads without distortion which would reduce the efficiency of a structure, or be unsightly. Strength properties are defined on this page and values are given for a range of materials on pages 196–197.

When a material is said to be 'strong' it is its stength in tension which is usually referred to, but it is often necessary to know its strength properties in compression, shear and torsion. Also, strength properties vary with the rate and frequency of loading and, in non-homogeneous materials, with the direction of loading. Strength properties are further influenced by the moisture content of materials such as timber and the temperature of materials such as plastics. Absorbent materials are usually tested when wet to allow for loss of strength in that state.

Materials which are subjected to a force are said to be *stressed* and the change in shape is known as *strain* (cause and effect). In *elastic* materials up to an *elastic limit* strain is proportional to the load applied and they recover their original shape and size when the load is removed.

Materials such as mild steel which suffer a relatively small amount of strain when subjected to a given load are said to be *stiff* or *rigid*, a property which must not be equated with breaking strength. Thus aluminium alloys which are as strong in tension as steel are far less rigid.

With increasing load, at a point which is not always clearly defined, materials cease to be *elastic* and become *plastic* and undergo permanent distortion. Materials which do this to a high degree are *ductile* as distinct from *brittle* materials. Ductility generally decreases with strength in tension.

In varying degrees, materials undergo slow plastic deformation or *creep*, under a constant stress. Steel has very small creep at normal working stresses and temperature, but in a year or two even a small load increases the deformation of a concrete member as much as three times the initial elastic movement.

(c)   Optical properties

Some reference is made to this subject under *Glass* chapter 12.

(d)   Electrical properties

The reader is referred to *MBS: E and S*, chapter 15.

(e)   Thermal properties and insulation

British Standards concerning thermal insulation include:

BS 8027:1985 *CP for energy efficiency in buildings* which sets out briefly principles and criteria for energy design.

BS 874 (confirmed 1980) *Methods of determining thermal properties, with definitions of thermal insulating terms.*

BS 3533:1981 *Glossary of thermal insulation terms.*

BS 3958:Parts 1–5, *Thermal insulating materials* describes various types and forms of preformed insulation.

BS 5803, Parts 1–5:1985 deal with *mineral and cellulose fibre insulation in pitched roof spaces in dwellings.*

Heat is lost from buildings by ventilation – both intentional, eg open windows and flues, and accidental, through gaps around doors and windows. Heat is also lost through floors, walls, roofs and closed doors and windows – see BRED 190.

Heat losses and gains are achieved by:

1   *Cavities in structures and components*

Insulation improves with sealed cavities up to an optimum width of around 20 mm.

Cavities in external walls can be filled with insulating materials.

BS 8208 is a *Guide to assessment of suitability of external walls for filling with thermal insulants,*

Part 1:1985 deals with walls of *Existing traditional cavity construction* up to 12 m high.

BS 6676, Parts 1 and 2, 1986 is *Thermal insulation of cavity walls using man-made mineral fibre batts (slabs).*

Cavities in walls were first introduced to stop water penetrating, and this benefit can be retained by leaving a 20 mm gap between batts and the external masonry leaf.

BRED 277 deals with *Built-in cavity wall insulation for housing.*

2   *Cellular materials*

All materials accord some resistance to heat transmission, but those with small and discontinous voids are most effective. Such insulating materials tend to be lightweight and weak, eg cork.

Specific *insulating materials* are listed in the index to this volume.

3   *Thermal conductivity*

Thermal conductivity (k) is a measure of the rate of heat transfer through a unit thickness and an area of a MATERIAL from FACE TO FACE (NOT from air to air). It is expressed as:

heat units transmitted in unit time Watts(J/s)
through unit thickness   m
of unit area m^2
for unit temperature difference
between the faces   deg C
ie   **W/m deg C**[1]

Table 1 lists some common building materials in order of increasing thermal conductivities (k) and decreasing resistivity (1/k). The *Chartered Institute of Building Services Engineers Guide* is a source.

The k values of materials vary with density, in the examples quoted from 0.029 to 400 W/m deg C with corresponding variations in density from 64 to 9000 kg/m^3. Conductivity values also vary with temperature, porosity and moisture content. With a moisture content of 20 per cent, volume by volume, most building materials transmit between two and three times as much heat as they do when they are dry. Hygroscopic materials such as timber vary in moisture content with the rela-

[1]This expression is derived from Wm/m^2 deg C by cancelling the metre thickness.

Bulk density kg/m³	Material	Thermal conductivity (k) W/m deg C	Thermal resistivity (1/k) m deg C/W
64	Expanded ebonite	0.029	34.48
16; 24	Expanded polystyrene	0.035; 0.033	28.6; 30.4
24; 40	Foamed polyurethane	0.024; 0.039	41.7; 25.6
16–48	Glass fibre quilt	0.032–0.04	31.3–25.0
48	Mineral and slag wools	0.03–0.04	33.3–25.0
120	Wool, hair and jute fibre felts	0.036	27.8
128	Corkboard (baked)	0.040	25.0
160	Balsa	0.045	22.2
80–240	Sprayed insulating coatings	0.043–0.058	23.25–17.3
240–350	Insulating fibre building boards	0.053–0.058	18.9–15.4
80–144	Exfoliated vermiculite (loose)	0.047–0.058	21.2–17.3
128–136	Rigid foamed glass slabs	0.050–0.052	20.0–19.2
350–800	Medium fibre building boards	0.072–0.101	13.9–9.9
320–700	Aerated concrete (low density)	0.084–0.18	11.9–5.55
365	Compressed straw slabs	0.101	9.9
450	Wood-wool slabs	0.093	10.75
400–800	Exfoliated vermiculite concrete	0.094–0.260	10.6–38.5
320–1040	Expanded clay – loose	0.12	8.83
800 and 961 (min)	Standard and tempered hardboards	0.125 and 0.180	8.0 and 5.6
513	Softwoods and plywoods	0.124	8.07
721	Diatomaceous earth brick	0.141	7.10
881 (max)	Asbestos-silica-lime insulating board (BS 3536)	0.144 (max)	6.95
449–800	Particle boards	0.101–0.158	9.92–6.34
961	Plasterboard	0.16	6.25
769	Hardwoods	0.16	6.25
641	Exfoliated vermiculite plaster	0.19	5.26
1190	Perspex (ICI)	0.21	4.8
960–2000	Foamed blastfurnace slag concrete	0.24–0.93	4.17–1.08
720–1760	Expanded clay and sintered PFA concretes	0.24–0.91	4.17–1.10
1620	Polyester glass fibre laminate (GRP)	0.35	2.86
1200	Asbestos cement (semi-compressed (BS690))	0.37	2.70
1041–1522	Clinker concrete	0.37–0.58	2.70–1.73
1442	Plaster (dense)	0.48	2.08
1142–1842	No-fines concrete	0.562–0.75	1.78–1.33
1600 (minimum)	Asbestos cement (fully compressed (BS 4036))	0.65	1.54
1602	Aerated concrete (high density)	0.65	1.54
1700	Brickwork	1.45–0.73	0.69–1.38
2100	Mastic asphalt	0.60	1.67
2306	Cement:sand	0.53	1.89
2520	Glass	1.05	0.98
1778	Rendering	1.15–1.21	0.87–0.83
2500	Sandstone	1.29	0.77
2260	Concrete 1 cement : 2 coarse aggregate : 4 sand	1.44	0.69
2310	Limestone	1.53	0.65
2590	Slate	1.88	0.53
2662	Granite	2.93	0.34
7140	Zinc	117.64	0.0085
7850	Steel	57	0.0176
2700	Aluminium and alloys	214	0.0047
9000	Copper	400	0.0025
11340	Lead	35.71	0.028

Table 1   Densities, thermal conductivities and thermal resistivities of common materials at normal moisture contents and at normal temperatures

tive humidity of the atmosphere. Common sources of dampness in materials are defective copings and damp proof courses and condensation.

Thermal resistivity of materials

Thermal resistivity (l/k) is a measure of the resistance to heat flow through unit thickness and area of *materials* from FACE to FACE being the reciprocal of *conductivity* values. Thus:

$$\frac{\text{thickness (m)}}{\text{conductivity (W/m deg C)}} = \textbf{m deg C/W}$$

Reflective materials

These materials have high reflectivity and low emissivity, eg aluminium foil, see page 24. It will be noted that a very thin film of materials with high conductivity can be a useful insulator.

Any two bodies facing each other exchange heat by radiation, the rate of emission and absorption depending upon the nature of their surfaces, and the temperature difference.

Figure 1 shows the mechanism of radiant heat transfer.

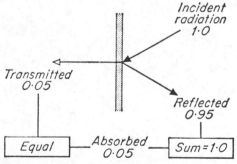

1 Radiant heat transfer of typical aluminium foil at normal temperature

Table 2 gives emissivities at normal temperatures and absorptivities of solar radiation of various surfaces.

It will be seen that highly polished aluminium is the most effective reflector of both normal temperature and solar radiation. Differences in colour are not important at normal temperatures but white is a good reflector of solar radiation and even better than aluminium which has become dulled by oxidation. Thus, the thermal resistance of an unventilated cavity 19 mm or more wide is nearly doubled if it is faced on one[1] or both sides with a surface of low emissivity such as shiney aluminium foil. Resistance is further increased if foil forms an airtight division between two cavities each at least 19 mm wide. White surfaces are particularly effective and in keeping mastic asphalt and bitumen felt roofings cool in summer.

Surface	Emissivity at 10–38°C	Absorptivity of solar radiation
Black non-metallic surfaces	0.90–0.98	0.85–0.98
Red brick, concrete and stone, dark paints	0.85–0.95	0.65–0.80
Yellow brick and stone	0.85–0.95	0.50–0.70
White brick, tile, paint, whitewash	0.85–0.95	0.30–0.50
Window glass	0.90–0.95	transparent
Bright aluminium, gilt or bronze paints	0.40–0.60	0.30–0.50
Dull copper, aluminium and galvanized steel	0.20–0.30	0.40–0.65
Polished copper	0.02–0.05	0.30–0.50
Highly polished aluminium	0.02–0.40	0.10–0.40

Table 2   Emissivity and absorptivity of surfaces

From *Heating, Ventilating and Air Conditioning Guide,* published by American Society of Heating and Ventilating Engineers

Thermal transmittance

Thermal transmittance (U) is the rate of heat transfer through a CONSTRUCTION from AIR TO AIR expressed as:

heat units in unit time	Watts (J/s)
over unit area	m²
with unit difference in temperature	
from air to air	deg C
ie	**W/m² deg C**

Thermal transmittance is the reciprocal of the sum of all the thermal resistances offered by a construction from air to air, ie:

---

[1]Bright aluminium has high reflectivity and low emissivity so it is equally effective on either side of a cavity:

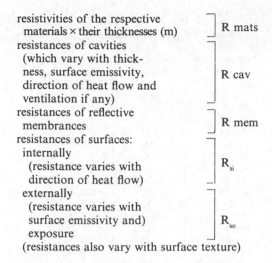

resistivities of the respective materials × their thicknesses (m)] R mats

resistances of cavities (which vary with thickness, surface emissivity, direction of heat flow and ventilation if any)] R cav

resistances of reflective membranes] R mem

resistances of surfaces: internally (resistance varies with direction of heat flow)] R_{si}

externally (resistance varies with surface emissivity and) exposure] R_{so}

(resistances also vary with surface texture)

BRE Digest 108 gives: *Standard U-values*

Building Regulations, Approved Document L, gives maximum permitted U-values for elements in various buildings, eg for dwellings: exposed walls and floors 0.6; roofs 0.35. The areas of single, double and triple glazing are limited.

The 0.6 W/m^2 deg C maximum value for walls is provided by a brick outer leaf, a 50 mm cavity and 100 mm blockwork, with insulation either as an inner lining or in the cavity.

Pattern staining

Dirt in air is deposited preferentially where heat loss is greater, ie on relatively cool surfaces and shows as a pattern which mirrors the variations in thermal conduction through different parts of walls or ceilings. Typically, wide dark bands show on plaster ceilings below the voids between timber ceiling joists, and narrow dark bands show on plaster ceilings below the solid concrete beams in *hollow tile* floors.

Thermal movements

Movements of materials arising from changes in temperature are dealt with on page 29.

Condensation

(see also *MBS: E and S* and Building Regulations 1985, Approved Document F2 *Condensation*.

Condensation occurs when air which holds water vapour is cooled below the *dew point*. Common sources of water vapour are damp walls, floors and ceilings, cooking and washing, flueless gas and oil heaters and the exhalation of occupants. At any given temperature air can support a limited amount of water as vapour, the quantity increasing with temperature. The amount held in air is usually expressed as *relative humidity*, ie the mass of water vapour in a unit volume of air as a percentage of the mass of water vapour in a unit volume of saturated air at the same temperature. The temperature at which air becomes saturated and vapour condenses is known as the *dew point*. Although the air in a room is usually well above its dew point, often the temperature of some surfaces is not, and typically condensation shows first on the insides of window panes. The vapour pressure in such conditions being higher inside a building than outside, water vapour passes through porous constructions and where this air falls to the dew point it condenses as *interstitial condensation*. In traditional construction water vapour was able to evaporate outwards but today *interstitial condensation* often tends to form on the inner surfaces of impervious membranes on the cold side of a construction, eg flat roof coverings, metal and glass wall claddings.

Interstitial condensation can be prevented by a *vapour barrier*, provided this itself is at a temperature above the dew point of the internal atmosphere and it must be continuous. Table 3 lists some materials in order of increasing resistance to the passage of water vapour. It is important to distinguish between vapour diffusivity, resistivity and diffusance.

Vapour diffusivity is the weight of water vapour which passes through unit thickness and area of a material per second under unit water vapour pressure gradient and at a given temperature (the permeability of films to water vapour approximately doubles for every 10°C rise in temperature). It is expressed as *g m/MNs*.

Vapour resistivity, the reciprocal of vapour diffusivity, is expressed as *MN/ s/g m*. It is the time taken for unit weight vapour to flow through unit area and thickness of a material when there is a unit difference of vapour pressure between the faces.

(It will be noted that these terms ending with

	Vapour resistivity of materials for unit thickness $MN\ s/g\ m$	Vapour resistance for stated thickness	
		thickness mm	$MN\ s/g$
Foamed urea formaldehyde	20–30	25	0.5–0.7
Wood-wool slab	15–40	50	0.7–2.0
Fibre building board	15–60	13	0.2–0.8
Brickwork	25–100	100	2.5–10.0
Concrete	30–100	100	3.0–10.0
Plasterboard	45–60	9.5	0.4–0.6
Timbers	45–75	50	2.2–3.7
Compressed straw slabs	45–75	50	2.2–3.7
Plaster	60	12	0.75
Rendering	100	12	1.25
Foamed polyurethane (open or closed cell)	30–1000	25	0.7–25.0
Expanded polystyrene	100–600	25	2.5–15.0
Hardboard	450–750	6	2.3–3.8
Plywoods	1500–6000	6	7.0–30.0
Expanded ebonite	11 000–60 000	25	250–1 500
Average gloss paint film	—	—	7.5–40
Polythene sheet	—	0.06	250
Aluminium foil	—	—	4 000

Based on data contained in BRE Digest 110, *Condensation*, HMSO

Table 3 *Vapour resistivities and resistances: typical values at normal temperatures*

ivity relate to unit thickness, as *conductivity* and *resistivity* do in respect of heat transfer.)

Vapour resistance The resistance of an homogeneous material is resistivity × thickness. Resistance values can be added together to find the overall resistance of a compound construction, which is expressed as $MN\ s/g$.

Vapour diffusance is the weight of water vapour which passes through unit area of a construction per hour under unit vapour pressure gradient and at a given temperature. It is the reciprocal of the total resistances, expressed as: $g/MN\ s$.

The BRE considers that membranes having a vapour resistance of $15\ MN\ s/g$ are suitable as vapour barriers in buildings, but because of the risk of interstitial condensation in cellular materials, additional vapour-sealing skins may be required.

Specific heat is the quantity of heat required to raise unit mass of a substance through 1°C. Examples are:

	J/kg deg C
Granite	330
Copper	376
Mild steel	502
Glass	830
Concrete	880–1 040
Aluminium	920
PVC	1 040
Polystyrene	1 250
Standard hardboard	1 250
Insulating fibreboard	1 400
Perspex (ICI)	1 460
Timber	1 500
Polyethylene	2 300
Water	4 187

Thermal capacity is the capacity of a body to store heat, expressed as the product of its *mass* (grammes) and its *specific heat*.

In steady temperature conditions the same U value can be provided either by a thick heavy wall or by a thin lightweight wall, but when sudden temperature changes occur the thick wall having high thermal capacity will be slow both to warm up and to cool down, and the converse is true of the second type of wall. Clearly, high thermal capacity is desirable in off-peak storage heaters, and in buildings with very high day temperatures and low night temperatures. In buildings which are heated intermittently high thermal capacity structures are not ideal, although lining interior surfaces with materials of low thermal capacity reduces the amount of heat required to warm a room. Thus, the choice between log cabin and stone castle walls, both of which could have the same U value, may depend upon the climate and the form of heating to be employed.

(f) Acoustic properties

Sound insulation is mainly achieved by mass and avoidance of direct paths – see *MBS: E and S*, chapter 6.

Sound absorption is considered here.

It is important to realise that sound absorbents are usually very poor sound insulators, in fact an open window which is a perfect absorbent, transmits virtually all sound. However, examples where sound absorbents can contribute to sound insulation are linings, in 'cut-off' lobbies, and in the reveals of double windows and baffles in ventilating ducts.

The amount of sound absorbed and the proportions absorbed at different frequencies (or pitch) vary with the nature of materials and forms of contruction. Thus:

1 *porous materials* absorb mainly at the higher frequencies

2 *resonant panels* absorb mainly at the lower frequencies

3 *cavity resonators* absorb mainly one frequency.

In practice a wider range of absorption can be obtained by combining two or more of the methods listed. For example, a perforated panel may be mounted over an air space containing porous absorbent.

Incidentally, porous treatments in particular, may also provide useful thermal insulation.

Unfortunately many sound absorbent treatments introduce combustible materials which spread flame rapidly and cavities behind such materials present a fire risk. Flame retardant impregnants and paints may be of value, although the latter reduce sound absorption.

We now consider the three types of absorbents:

1 Absorption of *porous materials* at low and middle frequencies increases with thickness. Decoration can substantially reduce sound absorption, but plastic films, provided they are not more than 0.051 mm thick and free to vibrate, allow the sound waves to pass into the absorbing material without interference. Porous materials may have more or less non-porous surfaces but with holes or slots which allow sound to enter and be absorbed within the thickness of the materials.

2 *Resonant panels* absorb mainly at particular resonant frequencies which become lower with greater weight of panel and with increased depth of the air space behind.

3 *Cavity* or *Helmholtz resonators* are usually employed in auditoria to correct a specific acoustic fault at a single frequency. The efficiency of a cavity resonator may be increased by introducing porous material into the neck.

The approximate sound absorptions of surfaces at various frequencies were given in BRS Digest 36 (first series), *Sound absorbent treatments*. The absorptions of some common surfaces at the middle frequency of 500 Hz are quoted in order of decreasing effectiveness in table 4 in which the more effective absorbents are broadly classified as follows:

A general wide band absorption

B preferential absorption at middle and high frequencies

C preferential absorption at low frequencies

(g) Deformations

Quite large movements may not be important where loss of strength, watertightness or good appearance will not result, but they should be foreseen and prevented, eg by keeping timber 'dry', minimized by appropriate choice of materials and structural forms, or allowed to occur in such a way that no damage results, eg by providing movement joints – see chapter 16.

Broadly, deformations may be caused by *deterioration of materials* (see page 32), eg expansion caused by sulphate attack on Portland cement, or

Finish	Absorption coefficient at 500 Hz	Category
Floorings		
Carpet, medium: on solid concrete floor	} 0.3	B
on boards on joists or battens		
Wood boards on joists or battens	0.1	
Linoleum and rubber		
Cork tiles	} 0.05	
Wood blocks		
Hard tiles	} 0.03	
Composition flooring		
Granolithic	0.02	
Windows		
Glass: up to 5 mm	0.1	C
6 mm or thicker in large sheets	0.04	
Walls and ceilings		
Perforated steel trays suspended below air space with: 25 mm rockwool	0.65	
51 mm rockwool	0.85	B
76 mm Wood-wool slabs, mounted solidly, unplastered	0.8	A
25 mm Glass wool or mineral wool mounted over air space on solid backing	} 0.8	
51 mm Glass wool or mineral wool on solid backing		B
25 mm Glass wool or mineral wool on solid backing	0.7	B
25 mm Hair felt covered by perforated membrane, eg muslin on solid backing	0.7	B
25 mm Wood-wool slabs unplastered mounted over 19 mm air space on solid backing	0.6	B
25 mm Clinker concrete unplastered	0.6	
Sprayed insulating coatings	0.55[1]	B
Curtains (medium weight): hung in folds or spaced away from wall	0.4	B
hung straight and close to wall	0.25	
25 mm Wood-wool slabs, unplastered:		B
mounted solidly	0.4	A
mounted over air space on solid backing or on joists or studs	0.3	C
ditto painted	0.15	
13 mm Insulating fibreboard on solid backing	0.15	
ditto painted	0.1	
Plywood:		
panels mounted over air space on solid backing or mounted on studs		
panels mounted on studs with porous material in air space	0.15	C
19 mm Matchboarding over air space on solid wall	0.10	C
Plaster (lime or gypsum) on lathing over air space on solid backing or on joists or studs	} 0.1	C
Plasterboards or fibrous plaster over air space on solid backing or on joists or studs		
Plywood mounted solid to wall	0.05	
Plaster (lime or gypsum) on solid backing	0.02	
Brickwork		
Concrete	} 0.02	
Tooled or polished stone including marble		
Glass bedded solid to wall		
Glazed tiles fixed direct to wall	} 0.01	
Marble fixed direct to wall		

[1]BS 3590 minimum   Data mainly from BRE Digest 36 (withdrawn)

Table 4   Approximate sound absorption coefficients of common finishes at 500 Hz

by frost, corrosion or fire. Here we consider deformations caused by:

1 applied loads, either *design loads*, or *accidental loads* arising from errors in structural design or from overloading
2 changes in moisture content of materials
3 changes in temperature of materials.

In practice, two or more causes of movement may occur together and where adjoining materials are mutually restrained the effective overall movement may be either greater or less than that of either of them.

1 Movements caused by applied loads

In small buildings deformations arising from applied loads are normally insignificant but in large structures they may require to be accommodated by pin joints and sliding bearings.

Most materials are elastic to some degree and many materials exhibit *plastic flow* or *creep* and are permanently distorted if a load is sustained. Steel has negligible creep at normal temperatures and stresses, but concrete creeps even with small loads notably during the first month or so, and then at a decreasing rate for several years, ultimate creep of gravel concrete cast in-situ being about twice the elastic strain. Creep is high in plastics, particularly in thermoplastics.

Mechanical properties are defined on page 191, and the properties of various materials are compared in tables 80 and 81 pages 197 and 198.

2 Movements caused by changes in moisture content of materials

Most materials expand to some extent when they are wetted and contract when they dry – see table 5. These dimensional changes are known as *moisture movements*, abbreviated in timber technology to *movements* (not to be confused with movement of moisture *through* materials). Initial movements of manufactured materials (eg shrinkage in drying of new concrete products, and expansion in absorption of water by ceramic products which begins immediately they are removed from the kiln) are partially *irreversible movements*. These tend to be greater than the *reversible movements* which accompany subsequent wetting or drying.

Clearly it is preferable not to use concrete and

	Movement per cent	
	Irreversible	*Reversible*
Metal, Glass	—	—
Limestone – Portland		0.004
Clay brick (typical good facing)	0.10–0.20 (expansion)	0.01
Calcium silicate bricks	0.001–0.05	0.001–0.05
Glass fibre polyester		0.02
Lightweight aggregate concretes		0.03–0.35
Dense concrete and mortars	0.02–0.08 drying (shrinkage)	0.01–0.06
Aerated concrete (autoclaved)		0.06–0.07
GRC		0.07
Timber – longitudinal		0.10
Harboards	slight	0.11–0.32
Sandstone–Darley Dale		0.15
Perspex (ICI)		0.35
Insulating fibreboard		0.20–0.37
Laminated plastics		0.10–0.50
Aerated concrete (air-cured)		0.17–0.22
Plywoods		0.15–0.30
Softwoods: radial	see	0.45–2.0
tangential	*Stress*	0.6–2.6
Hardwoods: radial	*setting*	0.5–2.5
tangential	*page 56*	0.8–4.0
Chipboard	large	0.1–12.0

Table 5   Moisture movements from dry to saturated condition

ceramic products until they have completed most, if not all, of their irreversible moisture movements.

Detailed information is given in BRE Digests *Estimation of thermal and moisture movements and stresses:* 227:Part 1, 228:Part 2, and 229:Part 3, 1979.

3 Movements caused by changes in temperature

Almost all materials expand when they are heated and contract when they cool. The movement of solids is expressed as the increase in length per unit length for one degree Celsius rise in temperature. Coefficients for typical building materials are given in table 6.

Changes in temperature of materials may result from atmospheric heat, solar radiation or from heating installations.

The *Principles of Modern Building*, Vol. 1, points out that air temperatures can vary up to 22°C between night and day in clear summer weather, while the extreme range of temperature

	Material	Coefficient of thermal expansion $\times 10^{-6}$ per deg C
High 26–200	Polythene, HD/LD	144/198
	Acrylics	72–90
	PVC	70
	Timber–across fibres	30–70
	Phenolics	15.3–45
	Zinc	31
	Lead	29
Medium 15–25	Aluminium	24
	Polyesters	18–25
	Brass	18
	Copper	17.3
	Stainless steel	17.3
	Gypsum plaster	16.6
	GRC	13–20
Low 1–14	Sandstones	7–16
	Concretes – various aggregates	10–14
	Mild steel	11–13
	Glass	6–9
	Granite	8–10
	Slates	6–10
	Marbles	1.4–11
	Limestones	2.4–9
	Aerated concretes	8
	Plywood	4–16
	Fired clay bricks– length	4–8
	width and height	8–12
	Mortars	11–13
	Asbestos-cement BS 690	12
	Asbestos-silica-lime insulating board	5
	Timber–longitudinal	3–6

Table 6   Thermal movements

between a hot summer day and a cold winter night can be as much as 50°C. Materials with small heat capacities will respond readily to changes in air temperature, more so if they are separated from massive structures by thermal insulation.

Normally the outer parts of buildings suffer the greatest changes in temperature, in particular those parts exposed to both cold winds and sunshine, such as roofs and parapets.

Table 2 shows that dark materials absorb more solar heat than light ones. Thus, in this country a black panel which is insulated at the back can rise as high as 71°C.

Detailed information is given in BRE Digests 227, 228 and 229: Parts 1–3, 1979.

4 Stresses due to thermal and moisture content changes

Movements[1] can be substantial and give rise to considerable stresses and if these exceed the strength of materials, cracks or buckling occur.

Calculations of stresses is complicated by a number of factors which cannot be accurately predicted, eg

(i) climatic variations
(ii) the ranges and rates of thermal and moisture content changes in materials in service
(iii) the magnitude of movements in the actual products to be used assuming them to be unrestrained. Published data for products of the same type is generally sufficiently accurate
(iv) the degree and type of restraint afforded by connections between materials.

Table 7 shows the stresses caused by a moderate rise in temperature in completely restrained walls. In Britain variation between night and day temperature may be as much as 22°C. Stress is directly proportional to the elastic moduli of materials. The latter being generally higher for stronger materials the stress given for strong brickwork is quite high. It will be seen that the stresses as a fraction of failing strengths in com-

[1] Movements include expansion with rise in temperature or with absorption of water by ceramic products and timber and contraction with fall in temperature and drying of cement, calcium silicate bricks and timber.

	Modulus of elasticity N/mm^2	Coefficient of thermal expansion per deg C	Failing stress in compression N/mm^2	Stress due to 15°C rise in temperature N/mm^2
Medium strength bricks in lime mortar	1 400	6×10^{-6}	71 000	2 800
Medium strength bricks in cement mortar	6 200	6×10^{-6}	180 000	12 000
Strong bricks in cement mortar	19 000	6×10^{-6}	400 000	33 000
Granite	48 000	10×10^{-6}	1 700 000	150 000

Table 7   Stresses caused by thermal changes

pression are about one twelfth in the strong brickwork compared with only about one twenty-fifth in the low strength brickwork.

In practice, thermal stresses may be modified by moisture movement stresses and because the larger temperature changes take place slowly the resulting stresses are reduced by one-half or more by the creep of the materials.

5   Common defects due to movements and their prevention

Some problems are now considered.

Dark coloured roof coverings often cause long, strong concrete slabs to expand and form horizontal cracks in top floor partitions just below the ceiling and to exert a thrust at their ends pushing the flank wall outwards. Coverings to flat roofs are always best finished with a solar heat reflecting surface, and in addition roof slabs require movement joints at not more than about 30 m intervals, and joints should be provided above partitions.

Copings are traditionally jointed in cement mortar which shrinks and allows the entry of water sometimes leading to sulphate attack on the mortar in clay brickwork[1] which then expands. This is a case where more than one mechanism may be at work, thus copings, particularly dark

[1] See chapter 6 Bricks, page 126.

coloured copings such as slate, are liable to substantial temperature change and resulting movement. Where there are abutments at the ends they will tend to be distorted or the coping may show signs of crushing. If the ends are unrestrained the coping may extend in its length in a series of 'pushes' caused by mortar settling in the joints between the slabs. A sealant topped movement joint at say 15 m intervals would usually prevent these defects. Partial restraint can lead to noisy stick-slip movements, eg in aluminium.

The Principles of Modern Building, Vol 1, draws particular attention to the fact that unrestrained aluminium cladding frames 12 m long increase in length nearly 10 mm in response to a 22°C rise in temperature. The structure behind will remain relatively static, and clearly such members must have fixings so they can move as a whole, or alternatively be fixed in small units so the relative movements will not be troublesome.

Coloured glass on a sunny facade with thermal insulation behind it poses a special problem. It must be free to move with a clearance of 6 mm on a dimension of 1 m and the avoidance of any possible contact between the edge of the glass and bead screws or other metal attachments are essential precautions. Differences in temperature due to shading can also lead to cracking, especially at the edges and it is important that these should be cut cleanly and that beads should not exceed 10 mm in width.

In buildings of rectangular plan where walls are free to expand in any direction movement joints should be provided at not more than 30 m intervals. Long walls which are restrained at their ends may distort their abutments or fail at 'weak links' in their length. PMB illustrates a 90 m long steel framed building which distorted the walls of brickwork buildings at its ends. It also illustrates a garden wall 45 m long restrained by buildings at its ends in which cracks in an arch opened at midday and closed again in the early morning.

BRE Digest 65 recommends that movement joints should be provided in long brick walls sufficient to accommodate an expansion of 9.5 mm in 12 m.

(h) Deterioration

This subject was dealt with by CP3, Code of basic design data for the design of buildings, Chapter IX: 1950 *Durability*.

All materials deteriorate and the function of the designer is to anticipate the changes which will occur in service.

Only wishful expectation of the most unlikely coincidence of good fortunes can explain the common disregard of the known causes and effects of deteriorating agencies on various materials. For example, the designer who expects that two coats of varnish will preserve the new appearance of timber cladding for very long, or that a building owner will take kindly to renewing the finish at frequent intervals, has not studied *MBS: Finishes*, chapter 6, or indeed, human behaviour.

With due regard to *cost-in-use* the designer must use materials in the correct context and inform the building-owner what after care will be needed to ensure a maximum life. He must also foresee not only common enventualities but the occasional events such as the ten year storm, attempts at burglary, and fire.

Differences in exposure have obvious effects upon durability, thus while a paint film may last a thousand years in an air-conditioned museum, on a south-facing slope in an industrial atmosphere it may fail in a year or so.

The lives of many materials which are exposed externally, or of those subject to abrasion, eg floorings (see *MBS: Finishes*, chapter 1), are often

dependent upon the removal of contaminants, the maintenance of a first line of defence in the form of protective coatings, or upon ancillary materials which have relatively short lives, such as putties and mastics.

Direct or indirect causes of deterioration include:

1 Corrosion of metals
2 Sunlight
3 Biological agencies
4 Water
5 Crystallization of salts
6 Frost
7 Chemical action
8 Loss of volatiles
9 Abrasion and impact
10 Vibration
11 Fire.

1 Corrosion of metals

This important cause of deterioration is discussed on page 193.

2 Sunlight

Sunlight causes degradation of clear finishes, paints, rubber bituminous products and some plastics and loss of colour in pigments.

3 Biological agencies

Insects and some animals attack organic materials, mainly timber. Importantly, sulphate-reducing bacteria produce sulphides from sulphates which corrode, iron, steel and lead pipes in the ground.

Plants such as creepers and trees, sometimes cause damage but in temperate climates fungal attack upon cellulosic materials which are persistently damp or wet is the most serious hazard under this heading – see page 72. It is preventable by the use of inherently resistant materials or by pretreatment with fungicide and by suitably designed construction,

4 Water

Some building materials, such as gypsum plasters and magnesium oxychloride flooring, are sufficiently soluble in water to preclude their use in damp situations. Limestones may be very slowly dissolved (and thereby kept clean where they are freely washed by rain). Timber, ordinary fibre-boards, and chip boards, wood-wool slabs and similar materials lose a proportion of their strength, and many flooring materials have less resistance to abrasion when they are wet.

Water also provides conditions which favour fungal attack, certain chemical reactions including electrolytic action, and frost damage. Alternate wetting and drying causes surface crazing and cracking of timber. Water also acts as a vehicle for the migration of salts and soluble substances.

It is important to distinguish between *adsorption* which is the bonding of water molecules to those of another material, from *absorption* which is the entry of water into the pores of a material.

5 Crystallization of salts

Soluble salts are derived from the ground or marine atmospheres or are formed by the constituents of building materials. Moisture evaporating from surfaces brings salts forward, giving rise to *efflorescence*. Where salts crystallize on the surface the effect may be merely disfiguring, but crystallization in the pores of the surface layer may cause gradual erosion or flaking as shown in table 8. Often parts which are most sheltered suffer more severely than where rain tends to wash out the salts. Thus, decay often occurs on the inside surfaces of stone mullions and transomes rather than outside. Calcium sulphate may be washed from limestone on to brickwork or sandstone and cause decay. Decay of poor brick or stone may be accelerated by the use of impervious mortar for pointing, which concentrates movement of water carrying salts in solution through the less dense materials.

6 Frost

Water expands when it freezes, at 4°C, and where it is contained in pores or other interstices it may cause decay. Materials with a laminar structure are more liable to deteriorate.

Microporous materials with small pores have greater capillary attraction and are less frost resistant than *macroporous* materials.

Class of material	How affected
Clay bricks	Good bricks rarely affected. Some underfired bricks may be badly attacked.
Clay tiles	Good tiles rarely affected. Some tiles suffer erosion at the top.
Terra-cotta	Rarely affected unless underfired.
Sand-lime bricks	Good bricks rarely affected.
Cast stone and Portland cement concrete	Rarely affected.
Sandstones	Vary greatly in resistance. The best stones are excellent.
Limestones	Vary greatly in resistance. This is the property which is most important in distinguishing a durable stone.

From *Principles of Modern Building*, Vol. 1, HMSO

Table 8 *Susceptibility of materials to deterioration, due to crystallization of soluble salts*

The ratio of the volume of the pore space filled by immersion in cold water to the total volume of the pore space of a specimen (as determined by prolonged immersion in boiling water followed by cooling while submerged), is called the *saturation coefficient*. As an indication of water's ability to freeze without bursting materials, the test applies to only a few materials – see BRED 269 *The selection of natural building stones*. Other laboratory tests, including cycles of freezing and thawing, may give some indication of frost resistance, but the only reliable guide is the behaviour of a material for a number of years in conditions resembling those in which it will be used.

Table 9 gives general information about the performance of some common materials.

7 Chemical action

Chemical action may affect volume (see *Deformations*, page 27), strength as in the corrosion of metals, appearance, eg acid attack on stones,

Class of material	How affected
Natural stone	Variable. Best stones unaffected. Some stones with pronounced cleavage along bedding planes are unsuitable for copings or cornices
Clay products	Best bricks and tiles unaffected but some insufficiently fired products and those having flaws of structure originating in the machine, may deteriorate, especially bricks in copings, and tiles on flat-pitched roofs.
Cast stone, concrete, asbestos cement	Rarely affected if of good quality.

From *Principles of Modern Building*, Vol. 1, HMSO
Table 9 Susceptibility of materials to deterioration as a result of frost action

Class of material	How affected
Clay products	*Rarely affected*, but may retain soot
Siliceous sandstones	*Rarely affected*, but retain soot
Slates	*Generally highly resistant*, but some suffer rapid decay – see BS 680
Cast stone and cement products generally	*Good quality products only slightly affected*. Dense mixes desirable for high degree of pollution.
Sand-lime bricks	*Very slightly* affected.
Limestones	*All attacked to some extent*. The more durable stones have a long life in the worst environment. Care needed in selection
Calcareous sandstones	Liable to be badly attacked.

From *Principles of Modern Building*, Vol. 1, HMSO
Table 10 Susceptibility of materials to attack by acid gases in polluted atmospheres

bricks and concrete and it causes deterioration of roof coverings. Contributory factors are heat and in nearly all chemical changes water enters into the chemical change or brings together reactive substances.

Aggressive gases increase the erosive effect of rainwater, and materials in industrial plants may require special protection – see table 10.

Sulphurous and sulphuric acids from burning fossil fuels, produce sulphur dioxide and sulphur trioxide in rain.

Ground waters, industrial wastes, wet clay bricks, soil and ashes often contain *soluble sulphates* which attack cement products and metals. The effect is expansion of Portland cement and hydraulic limes and their disintegration. Incompletely burnt colliery shales swell when they are wetted and should not be used as filling below ground level concrete floor slabs.

The carbonation of Portland cement in concretes leads to shrinkage and surface crazing, allowing moisture to enter to reinforcement.

8 Loss of volatiles

This is a cause of embrittlement and shrinkage of plastics, paints and sealants.

9 Abrasion and impact

In buildings these are mainly confined to floorings (see *MBS: Finishes*), but where abnormal hazards can be anticipated, suitably resistant surfaces should be provided, eg carborundum in nosings to stair treads and metal arrises in plasterwork and tiling.

10 Vibration

Vibration is most likely to be troublesome in light constructions and with brittle materials, and in such cases joints require special attention.

11 Fire

The behaviour of *materials* and *constructions* in building fires is considered here.

Relevant British Standards are:
BS 5588 *Fire precautions in the design and construction of buildings*

Part 1:1983 *CP for residential buildings*
Part 3:1983 *CP for office buildings*

Fire is a chemical reaction for which fuel, heat and a critical proportion of oxygen are all needed. Most solids and liquids give off vapour when they are heated and it is this which burns as flame. Some liquids give off flammable vapours even at normal atmospheric temperatures and are readily ignited when mixed with oxygen in critical proportions, ie they have low *flash points*. The flash point of petrol, for example, is 43°C below zero and butane boils at 0°C. Solids can burn only at or near their surfaces and are relatively difficult to ignite. Open textured materials burn more rapidly having a large 'surface area' throughout their mass, while finely divided dusts of materials such as coal, wood, flour, many plastics, magnesium and aluminium become explosive when they are suspended in air.

BRE Digest 288:1984, *Dust explosions*, discusses dust hazards and methods of protection in various industries.

In recent years, the need to save fuel has encouraged the adoption of thermal insulation which has often increased fire hazards in buildings. Relevant BRE Digests are: *Cavity barriers and fire stops*, 214:Part 1 and 215:Part 2 and BRE Digest 233 *Fire performance of walls and linings*.

BS 5803 *Thermal insulation for use in roof spaces in dwellings* recommends that if cables are covered with insulation, they should carry currents below their ratings to avoid overheating softening the sheathing, resulting in short circuits.

The calorific values of materials are given in *Postwar Building Study*, No 20, Appendix III (HMSO). The grading of building occupancies on the basis of *fire load* (expressed as kJ/m^2) is dealt with in *MBS: S and F*, Part 2.

In fires, materials may melt, burn, lose strength, expand, or shrink and crack.

Flame and collapse of buildings cause injury and loss of life but smoke and gases are even more dangerous causing loss of sensibility, panic, loss of vision and asphyxiation, often where no flame is present.

BRED 300 discusses *Toxic effects of fires* and BRED 260 deals with *Smoke control in buildings – design principles*.

The phenomenon of *flash over* deserves special mention. Solids can smoulder in a confined space for a long time, with only one third of the normal oxygen supply, and then suddenly and explosively burst into flame. This is explained by a gradual build-up of heat and vapour pressure causing sudden collapse, perhaps of enclosing panelling, and the entry of fresh air.

Terminology

It is most important that clearly defined standard terminology is used as given in BRE Digests 225, *Fire terminology* and in BS 4222 (five parts) *Glossary of terms associated with fire* which includes Part 1:1969 *The phenomenon of fire*, Part 2:1971 *Building materials and structures* and Part 5:1976 *Miscellaneous terms*. Examples of terms which should not be employed in respect of building materials are:'fire-proof', 'non-ignitable', 'self-extinguishing', 'incombustible' and 'flame-resistant'.

Advertisements for products should make it clear which tests, if any, have been carried out and the classifications achieved. This should be supported by *Test Report FROSI No – –* being the reference allocated by the Joint Fire Research Organization, Boreham Wood, Hertfordshire. If tests have not been performed a written assessment given by the JFRO on the basis of their accumulated experience may be accepted by the Building Regulations authorities.

BS 6336:1982 is a *Guide for the development and presentation of fire tests and their use in hazard behaviour and hazard assessment*. It deals with the theories of fire behaviour and hazard assessment.

BS 476

The main British Standard concerned with fire performance of buildings is BS 476 *Fire tests on building materials and structures*. The specification is in many parts. (There are no Parts 1 or 2.) Part 10:1983 is a *Guide to the principles and application of fire tests*. Tests are related to the factors which occur in building fires, but it is important to realise that the results of individual tests are of no value in isolation and must be related to the fire hazard as a whole.

The current Parts 3–8 inclusive, which are discussed here will later be replaced by revised and supplementary Parts 12–31. Parts 11–19 will deal with *Fire performances of products*, Parts 20–29 *Elements of building construction* and Parts 30 and 31 *Miscellaneous methods*.

It is important to note that the results of any test are of no value in isolation and must be related to the fire hazard as a whole.

The BS tests will now be considered in turn:

External fire exposure roof tests This is the title of BS 476: Part 3:1975. The purpose of the tests described is to provide information on the behaviour of roofs when there is a fire nearby but outside the building itself. It is important to note that the tests are neither capable of, nor intended to, predict the performance of a roof in the event of internal fire. Representative specimens of roof constructions 1.5 m × 1.2 m are exposed to heat conditions simulating a fire in a building of known size and a certain distance away. A test flame is used to simulate falling burning brands in a fire. To determine the extent of surface flaming, specimens are also subjected to a graduated intensity of radiant heat.

The 1975 specification classifies times of penetration as:

P60 ⎫
P30 ⎬ equivalent to the 1958 designations
P15 ⎭

AA, AB, AC BA, BB, BC

AD, BD, CA CB, CC, CD

P5   unclassifiable

Results of tests are to be described for:
(a) inclination of roof
(b) performance in preliminary ignition test
(c) extent of surface ignition in mm

Combustibility The term combustibility is defined by BS 476:Part 4:1970 (amended 1983) *Non-combustibility tests for materials*. If small samples plunged into a furnace maintained at 750° ignite, ie if any flaming occurs continuously for ten seconds or more, give off flammable gases, or if the temperature of the sample or furnace rises by 50°C or more they are deemed to be *combustible*. Otherwise they are *non-combustible*; there is no other grading.

Examples of *combustible* and *non-combustible* materials are given in table 11. The correct classification of mixtures of materials will only be known by subjecting them to the standard test.

Combustible	Non-combustible
Timber (even if impregnated with flame retardant)	Asbestos-cement products
Fibre building boards (even if impregnated with flame retardant)	Asbestos insulation board
Cork	Gypsum plaster
Wood-wool slabs	Glass
Compressed straw slabs	Glass wool (containing
Gypsum plasterboard (rendered combustible by the paper liner)	not more than 4–5 per cent bonding agent)
Bitumen felts (including asbestos fibre-based felt)	Bricks
	Stones
	Concretes
Glass wool or mineral wool with combustible bonding agent or covering	Metals
	Vermiculite
	Mineral wool
All plastics and rubbers	
Wood-cement chipboards	

Table 11   Combustible and non-combustible materials

Non-combustible materials add nothing to the fire load, flame does not spread over them, and they are essential for purposes such as flue linings.

Non-combustible materials may become combustible due to resinous binders or paper facings. Also, they do not necessarily contribute to fire resistance of structures. Thus, in fires steel expands and may push walls over, and loss of strength may cause collapse, while ordinary asbestos cement shrinks, cracks and allows the passage of flames and smoke. Glass is non-combustible but does not contribute much to fire resistance.

Combustible materials cannot be made non-combustible by coatings or impregnation. They support combustion and produce toxic gases and smoke. However materials such as timber can contribute to *fire resistance* – see page 39.

Ignitability BS 476: Part 5:1979 *Ignitability test for materials* is primarily intended for materials in slab or sheet form, but not for fabrics. The test identifies easily ignitable materials of low heat contribution, the full hazard of which is not necessarily shown by the *Fire propagation test*.

After being subjected to a small standard flame for 10 seconds 228 mm × 228 mm specimens of stated thicknesses are classified as either:

'*Easily ignitable*' ('*X*') if specimen flames for a further 10 seconds or if burning extends to an edge, or '*Not easily ignitable*' (*P*) if none of three specimens is classified as easily ignitable.

Fire propagation The surface spread of flame test does not measure all the properties which contribute to the fire hazard presented by lining materials and BS 476: Part 6:1981 *Fire propagation test for products* expresses as a numerical index the amount and rate of heat evolved by a specimen which is heated in an enclosed space.

Test results must be accompanied by a '*P*' or '*Y*' classification for *ignitability* (Part 5) and it is important to state the thickness of the specimen.

Surface spread of flame Spread of flame over combustible surfaces, more particularly walls and ceilings – can assist growth of fires. In cavities it is particularly dangerous and wherever possible they should be avoided or limited by the provision of *fire stops*.

The BS 476: Part 7:1971 test for surface spread of flame is carried out by subjecting the surface of samples mounted perpendicular to one edge of a three foot square radiant panel to a graduated intensity of heat. The distance and rate of flame spread along the surface is measured and the material is placed into one of four classifications:

Class 1 Surfaces of *very low flame spread*
Class 2 Surfaces of *low flame spread*
Class 3 Surfaces of *medium flame spread*
Class 4 Surfaces of *rapid flame spread*

In addition, the Building Regulations 1985 designate wholly non-combustible walls and ceilings as *Class O*. This 'highest class' for spread of flame over surfaces is also provided by a surface material which when tested in accordance with BS 476: Part 6:1981, either by themselves or where bonded throughout to a substrate, provides prescribed indices of *fire* propagation.[1]

However, the faces of plastics surface-materials which have softening points below 120°, (102°C test, BS 2782:1970) qualify as *Class O* only if they:

(a) are bonded throughout to a non-plastic substrate where the surface materials and substrate tested together satisfy the stipulated criteria for fire propagation[1], or
(b) where used as a lining to a wall, plastic surfaces with softening points below 120°C satisfy the fire propagation criteria[1] and, if the lining was not present, expose a surface, other than a plastics material with a softening point below 120°C, which satisfied the fire propagation criteria[1].

The Building Regulations 1985, Approved Document B, controls the surface spread of flame characteristics of ceilings, soffits and walls, (excluding doors, windows, skirtings, trim, fitted furniture etc), according to the purpose group of the building and the size of the room.

In rooms of two-storey houses *Class 4* wall and ceiling surfaces are not permitted. In rooms with more than $4 \, m^2$ floor area, walls exceeding half the floor area or $20 \, m^2$, and ceilings to three-storey houses must be *Class 1*. The requirements are stricter for circulation spaces, stairs, etc but floor surfaces are not controlled.

Table 12 gives the surface spread of flame classifications of some common building materials.

It will be seen that untreated plasterboards qualify for *Class surface spread of flame*, the paper liner being less than 0.8 mm thick.

Classifications can be improved by impregnation with flame retardant agents, usually based on ammonium phosphate. Their value is not reduced by surface abrasion or by cutting the material which has been treated, but their effective lives are not known, particularly under humid conditions. Moreover, salts in flame retardants may cause abnormal moisture movement, loss of strength and adhesion of decorative finishes.

The application of certain paints increases the rate of spread of flame of untreated surfaces – see page 38, but classifications can be improved by specialized surface applications, eg *intumescent* paints.

Reference: *Results of surface spread of flame tests on building products*, HMSO.

Fire resistance Fire resistance is a property of an element, eg walls, columns, floors, beams,

[1]An index of performance (I) not exceeding 12 and a sub-index (i_1) not exceeding 6. BS 476: Part 6, 1981.

	Buildings Regulations 1976 Class O	BS 476 Classes			
		1 Very low	2 Low	3 Medium	4 Rapid
Asbestos non-combustible boards	U/T				
Plasterboard	U/T				
Wood-wool slabs		U/T			
Synthetic resin-bonded paper laminates		Including some fire-retardant additives	U/T		
Hardboards —density more than 800 kg/m³		Treated with some flame-retardant treatments Impregnated with flame-retardant salts (not tempered hardboards)	Treated with – chlorinated rubber paint some stove enamels Faced with some plastics	U/T Treated with some stove enamels emulsion paints, oil, paints and enamels	Treated with some flammable paints, eg cellulose lacquers
Chipboard		Faced with exfoliated vermiculite or some flame-retardant treatments		U/T	
Timber and plywood —density more than 400 kg/m³		Treated with some flame-retardant treatments including some clear finishes		U/T	
Compressed straw slabs	Coated with 4.76 mm plaster	Faced with asbsestos paper or with certain flame retardant treatments		U/T	
Acrylic sheets at least 3.17 mm thick (polymethyl methacrylate)				U/T	
Glass reinforced polyester resin laminates (GRP)		Including flame-retardant additives and containing fillers	Including flame retardant additives	U/T	U/T
Timber and plywood density less than 400 kg/m³		Treated with – some flame-retardant treatments including some clear finishes			U/T
Insulating fibre building boards density not more than 35.0 kg/m	Coated with 4.76 mm plaster	Treated with – certain flame-retardant paints, three coats non-washable distemper,[1] one coat non-washable distemper on a sized board[1] Faced with – aluminium foil asbestos paper Impregnated with flame-retardant salts	Treated with one coat flat oil paint,[1] one coat washable or non-washable distemper,[1] chlorinated rubber paint,[1] aluminium paint,[1] some emulsion paints[1]	Treated with some emulsion paints	U/T

[1]Subject to confirmatory tests being carried out by the FRS.

U/T = untreated surface

Test results indicating the performance of specific products should be obtained from manufacturers.

Table 12   Spread of flame Classifications (BS 476:Part 7)

glazing and doors, and not of individual materials. It is expressed as the period of time in hours and minutes during which an element survives the test laid down in BS 476:Part 8:1972 while continuing to perform its normal structural or separating function. An element under test is deemed to have failed in the event of:

(a) collapse
(b) the formation of holes or orifices in a separating element through which flames can pass
(c) excessive heat transmission through a separating structure likely to lead to the ignition of combustible materials in contact with the outer face. The mean temperature of the unexposed surface must not rise more than 140°C, or the temperature at any point more than 180°C above the initial temperature. (In the case of doors and glazing it is sometimes assumed that combustible materials will not be placed against them and this requirement is waived.)

Load bearing elements subjected to load calculated to produce the maximum design stresses are required to resist collapse for the duration of the heating period while providing a margin of safety, and to withstand re-application of the test load two days after heating.

Points on the time-temperature curve operated in the furnace are:

> 583°C at 5 minutes
> 704°C at 10 minutes
> 843°C at 30 minutes
> 927°C at 1 hour
> 1 010°C at 2 hours
> 1 121°C at 4 hours
> 1 204°C at 6 hours.

Elements are graded in standard periods of $\frac{1}{2}$, 1, 2, 3, 4 and 6 hours and a further grading of $1\frac{1}{2}$ hours was recognized under the Building Regulations.

The period of fire resistance required by by-laws and regulations varies according to the purpose group of the building, its height, floor area and cubic capacity and whether the element is above or below ground.

Heat emission BS 476:1982 is a *Method for assessing the heat emission from building materials.*

Behaviour of materials in fire

This section discusses the general behaviour in fire of some common building materials used in their normal forms. See also the respective chapters.

Timber Moisture in timber absorbs some heat but it is easily ignited at about 220 to 300°C. Treatment with flame retardant chemicals by impregnation or by surface coatings reduce the rate of spread of flame, but the timber still carbonizes as if untreated.

Being organic, in burning timber produces highly toxic carbon monoxide and large quantities of smoke.

Most woods are in *Class 3 ('medium') spread of flame.* However, *Dricon* by Kopper Limited, when applied to wood by pressure impregnation, is claimed to have nil-flame spread, negligible heat release and will not support combustion. It does not 'wet' wood and is non-hydroscopic. It does not discolour applied finishes or corrode metals.

In fires, typical softwoods and hardwoods char at about 0.7 and 0.5 mm per minute, respectively. The charcoal insulates the interior and, unlike steel, there is no significant loss of strength even with a serious rise in temperature, moreover there is no increase in length which in the case of steel beams often causes walls to overturn. The thermal insulation of timber prevents a marked rise in the temperature of members on the side remote from a fire, so timber encasures can protect steelwork from the effects of fire.

The predictable rates of charring, and the dimensional stability and retention of strength by timber in fires, allows the performance of structural members to be calculated – see page 70.

Laminated timber members glued with typical synthetic resins have a fire resistance approximating to that of solid timber of the same species.

Fibre building boards These boards (BS 1142:Parts 1 and 2:1971 and Part 3:1972) are described on page 104. Ease of ignition and spread of flame vary with density and any impregnation or surface treatment, but no treatment makes them non-combustible.

Insulating fibre building boards (less than 400 kg/m³) are in *Class 4 ('rapid') spread of flame* (BS 476). This grading can be raised to *Class 1* by: impregnation with flame retardant salts, by aluminium foil or vermiculite surfaces or certain paint treatments. A 5 mm coat of gypsum plaster raises insulating fibreboards to *Class 0* of the Building Regulations 1985.

Hardboards Standard hardboard (more than 800 kg/m³) has a *Class 3 ('medium') spread of flame* BS 476.

However, although it is in a superior spread of flame class than untreated insulating board, it holds surface finishes less firmly under fire conditions and fire performance is less readily improved by flame retardant treatments.

Wood-wool slabs The wood fibres coated with cement are less easily ignited and the slabs have *Class 1 spread of flame* and a *Fire Propagation index* of about 11.

Stones Stone blocks and slabs are generally satisfactory in fires but overhanging features and lintels are liable to fail. Free quartz, eg in granite, disrupts suddenly at 575°C and should not be included in any stone where high fire resistance is required.

Sandstones behave better than granite, but in drying they shrink and may crack, with 30 to 50 per cent loss of strength.

Limestones give off CO_2 at about 800°C with loss of strength but little change in volume. Building fires normally exceed this temperature but the chemical dissociation absorbs a great deal of heat and is normally very slow. Limestones, with the exception of those which contain quartz crystals, do not spall and for that reason behave better in fires than granites or sandstones.

(Stones used as concrete aggregates – see pages 41 and 163.)

Plastics Although many are available in flame retardant grades, all plastics are combustible, and some of them generate large quantities of toxic smoke.

Cellulose nitrate (celluloid) is notorious, and expanded plastics such as foamed polystyrene

present serious fire hazards. They burn rapidly, and burning fragments drop from ceiling tiles.

The fire properties of thermoplastics vary widely according to their composition, form and other factors. For example, PVC melts at temperatures below that at which they can be tested for spread of flame, but certain wire mesh reinforced PVC sheets are in *Class 1 spread of flame* category.

Most thermosetting plastics char at temperatures above 400°C and burn at 700 to 900°C. It will be seen that while normal polyester glass-fibre reinforced laminates (GRP) are in *Class 3* or *4* spread of flame, modified products may be in *Class 1* or *2* – see table 12.

Clay products Surface fusion of clay products can occur with prolonged exposure to temperatures above 1 000°C and thin walled blocks have been known to fail as a result of differential temperature stresses. Generally, however, ordinary clay products behave very well, having been manufactured at temperatures in excess of those normally encountered in building fires and special *refractory bricks* withstand extremely high temperatures.

Calcium silicate bricks These compare favourably with clay bricks.

Brickwork Failure in clay, calcium silicate or concrete bricks and mortar is often caused by expansion of steelwork.

Concretes Ordinary Portland cement disintegrates at 400–500°C but the performance of concrete depends very much upon the presence of reinforcement and upon the type of aggregate.

In general, concretes made from ordinary cement and stone aggregate begin to lose strength well within the range of ordinary building fire temperatures. (High alumina cement with special refractory aggregates is used where exceptional resistance to high temperatures is required – see page 157.) In reinforced concrete there is little residual strength above about 600°C. Steel and concrete have about the same coefficient of expansion but as steel is a good conductor of heat fire resistance depends largely upon the thermal insulation provided to the steel by the concrete *cover*. Aerated concrete and lightweight aggregate concretes are good in this respect, but with dense aggregates, for periods of fire resistance above 2 hours in reinforced beams and $1\frac{1}{2}$ hours in prestressed beams, supplementary reinforcement is required in the cover.

Aggregates are placed in two classes, in descending order of merit:

Class 1
Lightweight aggregates Concretes made with pumice, foamed blast-furnace slag, sintered pulverized fuel ash, expanded clay, well burned clinker and similar non-combustible aggregates have lower thermal movement than dense concretes and retain a higher proportion of their initial strengths in fires.

Dense aggregates Crushed brick, blast furnace slag and crushed limestone.[1]

Class 2
All crushed natural stones other than limestone including granite and gravel, consisting of flint and crushed quartzites, tend to break up and spall violently and are the least reliable aggregates in common use.

Gypsum products Plasters have the advantage that they are continuous and free from joints through which flames can pass, although differences in expansion tend to fracture the bond between plasters and backgrounds. Heat is used in expelling the water of crystallization from gypsum plasters but they remain non-combustible, even with inclusions of organic fibre or powder up to 5 per cent by weight, and gypsum plaster can resist fire for considerable periods if it is retained by a mechanical key such as expanded metal lathing. Lightweight aggregates for plaster, in particular expanded vermiculite, give superior adhesion, insulation and fire resistance.

Plasterboard is classed as *combustible* due to the paper liner, but because the liner is thin the board has a *Class 0* grading for *spread of flame*. The BS 476 Part 6 *Fire propagation index* is 9–11. In appropriate thicknesses and suitably fixed, plasterboard provides a substantial degree of fire resistance – see also *MBS:Finishes*, chapter 2.

---

[1]The freedom from spalling of limestones, which is an advantage in reinforced concrete, does not apply to block walls.

Metals Those used in building, are *non-combustible*, but they lose strength, and aluminium, lead and zinc melt within the range of building fire temperatures. Expansion can be troublesome and high thermal conductivity causes the temperature of surfaces remote from a source of heat to approach the temperature on the reverse and cause fires to spread.

Steel The behaviour of mild steel is interesting. Up to 250°C it increases in strength; but it returns to its normal value at 400°C followed by rapid loss of strength, so that at 550°C referred to as the *critical temperature*, the yield strength is reduced to the working strength level and it begins to fail.

Fire Research Technical Paper no: 15 by MOT and Fire Offices' Committee, Joint Fire Research Organization showed that in a building fire unprotected steel members on which flames did not impinge rose to only 300°C.

High-strength alloy steels follow a similar pattern but cold-worked high-strength steels show a markedly greater reduction in strength at high temperatures and the *critical temperature* for prestressed concrete tendons is 400°C.

Thermal expansion of 11×10^{-6}/deg C means that a 10 m member expands more than 50 mm when heated from room temperature to 550°C. Steel sheet cladding fixed in accordance with CP143:Part 10:1973 resists the passage of fire well.

Unprotected solid steel columns more than 150 mm diameter can have half an hour *fire resistance* due to their high heat capacity, but generally, structural steelwork must be considered to have no fire resistance – light steel frameworks often collapse within twelve minutes of the commencement of a fire. Generally, structural steelwork must be protected with fire-resisting encasements.

In appropriate thicknesses and correctly fixed these include: dense and lightweight concretes, gypsum plasterboards (see *MBS Finishes*, chapter 2) and other non-combustible boards, intumescent coatings – see below and even timber – see page 39.

However, in certain multistorey car parks, steelwork which is outside and at least 460 mm away from external walls where windows occur, does not normally need protective encasement. Further reductions in the protection required are likely to be made as a result of investigations which are proceeding in respect of fire loadings of buildings and the performance of steel in fires.

The JFRO is able to advise as to any protection necessary in particular cases.

Wrought iron behaves similarly to steel.

Cast iron Cast iron columns often survive in building fires after steel columns have collapsed but if water jets are applied to cast iron sections brittle fracture is likely. Thermal expansion 10.8 $\times 10^{-6}$/deg C is about the same as steel and it is recommended that cast iron should be similarly protected.

Aluminium Structural alloys lose strength rapidly when heated, the critical temperature being about half that for structural steel. Thermal expansion is about twice that for steel, thermal conductivity is higher and heat capacity is lower. The melting point, about 650°C, is low and the *Building Regulations 1985* require a higher standard of protection for aluminium structures than for steel. Aluminium sheet cladding does not survive for long in fires.

Glass Although glass is non-combustible, it transmits heat readily and may shatter unpredictably at an early stage in a fire. Double glazing is no better than single glazing and toughened glass is not fire resistant (see page 258). In suitable forms, wired glass, electro-copper glazing and hollow glass blocks are fire resisting – see pages 257, 265 and 266. When glass is tested for fire resistance the insulation requirement is sometimes waived, it being assumed that combustible materials are unlikely to be stored near it. Wired glass in roofs has a P60 BS 476:Part 3:1976 *External fire exposure roof test* rating.

CP 153 Part 4:1972 deals with *Fire hazards associated with glazing in buildings.*

Glass fibre, slagwool and rockwool products Resin bonded glass fibres are *combustible* and have a fire propagation index of about 7 and the fibres melt at about 600°C whereas slagwool is fairly stable up to about 900°C.

Asbestos products Pure asbestos is practically indestructible by fire, but the resistance of products to fire depends largely on the proportion of asbestos fibre which they contain and small proportions of organic materials in some products makes them *combustible*. The health risk is referred to on page 238.

Asbestos cement products to BS 690 Part 3:1973 containing only 12 to 15 per cent by weight of asbestos fibre tend to shatter when they are heated, sometimes explosively, depending on their moisture content, and make no significant contribution to fire resistance in typical structures.

Asbestos insulating boards and asbestos wallboards (BS 3536:1974) containing at least 50 per cent by weight of asbestos fibre, retard the passage of fire and were used for protecting steelwork and for improving the fire resistance of structures and doors.

Vermiculite products Exfoliated vermiculite used as an aggregate in plasters and renderings having binders of gypsum plaster or cement/lime, bond well, even to smooth surfaces, and adhere well in fires. Sprayed vermiculite plasters and renderings can be used to protect reinforced and prestressed concrete, asbestos cement sheets and steelwork.

Bituminous products Bitumen melts and flows readily when it is heated and is then easily ignited but as the bitumen content reduces, eg in asbestos based felts, products are less easily ignited and burn less readily. Mastic asphalt is difficult to ignite and burns only when it is melted after long exposure to heat. The external performance of bituminous felt and mastic asphalt roof coverings is improved by a finish of stone chippings, screed, tiles or slabs.

Paints Generally paint films are *combustible* and may increase the rate of *spread of flame* over surfaces (BS 476 tests).

However, being thin, paints can make only a very small contribution to the *fire load* and on *non-combustible* substrates such as steel, most paints are in Class 1 *very low* flame spread (BS 476) and being less than 0.8 mm thick they fall in

Class 0 of the Building Regulations 1976, see page 37.

Performance of ordinary paints varies considerably according to the type of binder, the pigment content and on the characteristics of the background on which they are applied. On *non-combustible* substrates resistance to flame increases in the following approximate order:

Cellulose nitrate – readily ignited, extremely flammable.

Polyester and alkyd resins – not difficult to ignite, usually flammable.

Oil paints

Epoxide resins – difficult to ignite, burn sluggishly.

Washable distempers

Non-washable distempers

Vinyl resins

Urea and melamine formaldehydes – difficult to ignite.

Phenol formaldehyde – difficult to ignite, burn slowly or self extinguishing.

Chlorinated rubber – used in flame retardant paints.

Chlorinated paraffin wax and *antimony oxide* – used in flame retardant paints.

When applied to *combustible* substrates certain paints can reduce the spread of flame, and delay but never prevent, their ultimate ignition. One form evolves smothering gases. Such coatings 0.7–1.5 mm and 1.2 mm thick can protect steelwork from the effects of fire for 0.5 and 1.0 hours respectively. A new formulatioin is durable externally.

Intumescent paints expand when heated forming an insulating porous char. This action is invaluable as a coating on honeycombs to form fire and air stops in ducts.

(i)   Appearance

The appearance of buildings derives from their structural forms, relations with adjoining buildings, the spaces enclosed and the textures, patterns and colours on surfaces, all as modified by artificial and natural light. On the other hand, 'good' appearance is subjective. Some prefer the comfort of obviously safe structures, clearly stated patterns, durable materials and historical associations. Others find aesthetic satisfaction in 'cliff hanger' structures, visual confusion and the excitement of impermanence. In the choice of materials, however, economic considerations generally favour investment in materials having low maintenance requirements – see page 16.

Relevant references are:

BRE Digest 149: *The co-ordination of building colours.*

BS 5252: 1976 *Framework for colour coordination for building purposes* and the related colour matching fan

BS 4800: 1972 *Paint colours for building purposes*

BS 5378 *Safety signs and colours*, Parts 1 and 2: 1980.

Forests cover more than one-third of the land surface of the world, and eight per cent of Britain. Detailed consideration of the effects of trees on the landscape and on buildings is given in BS 5837:1980 *Code of practice for trees in relation to construction*.

This chapter deals with felled timber for use in buildings.

The hundreds of available timber species vary widely in their properties and appearance. In addition, within any one species there is often a wide variation between trees growing in difference climatic conditions and on different soils, and between parts of trees. This variability presents problems in economic conversion and utilization, but in recent years knowledge of the properties of timber has increased and improved techniques have been developed for laminating, jointing and framing, seasoning and for protection against fungi, insects and fire. In consequence, timber continues to satisfy requirements for performance and cost in a wide range of uses as a structural material, components and sometimes as a decorative finish.

Information concerning the properties and uses of timbers is provided by the Building Research Establishment, the Princes Risborough Laboratory, Buckinghamshire, the Timber Research and Development Association (TRADA), Hughenden Valley, High Wycombe, Buckinghamshire and the *British Woodworking Federation*, 82 New Cavendish Street, London, W1M 8AD.

References include:

Timber-Properties and Uses, W P K Findlay, Crosby Lockwood Staples
Timber – its structure and properties, H E Desch, revised by J M Dinwoodie, Macmillan
A handbook of softwoods PRL (HMSO)
A handbook of hardwoods PRL (HMSO)
BS 6100, Part 4, section 4.1 *Characteristics and properties of timber and wood-based products*.

BS 565:1972 *Glossary of terms applicable to timbers, plywood and joinery*.
Timber selection by properties – the species for the job Parts 1 and 2, HMSO

Other references are given at the relevant points in this chapter.

Selection of timbers

Selection of timber depends upon many factors, in particular:
Strength, moisture movements and dimensional stability. Availability of species, sizes and sections. (This varies widely from place to place and time to time)
For external uses: natural durability and ease of preservation
Appearance

Information to aid selection of timbers is contained in the above references and table 32, which should be read in conjunction with tables 14 and 15 and figure 5.

Hardwoods and softwoods

These are botanical terms and do not always relate to hardness. Not all hardwoods are hard – balsa is very soft.

Hardwoods (angiosperms) are from *broadleaved* trees most of which are *deciduous*, although holly certain oaks and the majority of tropical trees are *evergreens*. Hardwoods include the densest (see figure 5 page 52, strongest (see table 15, page 54 and most durable timbers (see tables 32 and 28). Some hardwoods contain resins and/or oils which interfere with the hardening of paints and many such as teak and makoré include materials, eg silica, which make working difficult.

The cheaper hardwoods approximate in cost to the more costly softwoods. Cost also varies with species, quality, availability and dimensions. Narrow and short stock is less costly while extra long lengths, and in some species wide boards, are more costly.

Softwoods (gymnosperms) Not all softwoods are soft, some softwoods are very hard, eg yew, strong and durable while certain hardwoods have greater moisture movement than any softwoods – see table 16, page 57. Softwoods are all, for practical purposes, derived from *coniferous* trees which are mainly evergreens and grow chiefly in the Northern Temperate zone. Softwoods comprise about 75 per cent of the timber used in the UK although the number of genera is much smaller than the hardwoods. Examples with their Latin *genera*[2] and *species*, and BS 589 names (in bold type) include:

'fir', also called British Columbian pine, is neither a fir nor a pine and similarly Parana 'pine', Scots 'fir' and Western red 'cedar' do not belong to the genera their names suggest. *Picea abies* and *Abies alba* are both called 'whitewood', while *Pinus sylvestris* and *Sequoia* are both sometimes called 'redwood'. There is no 'deal' timber as such, in fact a 'deal' was a 9 in. × 3 in. section, but *Picea abies* is called 'white deal' and *Pinus sylvestris* is called both 'red' and 'yellow deal'.

British Standards 881 and 589: 1974 *Nomenclature of commercial timbers including sources of supply* give Latin, 'standard' and common names for hardwoods and softwoods respectively. Specification by British Standard names is recom-

Pinus spp **pines**	**Redwood** Red or yellow deal. Red pine 'Red' or 'yellow' Norway 'fir'	imported	*Pinus sylvestris*
	Scots pine Scots fir	home-grown	
	American pitch pine Longleaf pitch pine Southern yellow pine		*Pinus palustris* and *Pinus elliottii*
	Yellow pine **White pine** Quebec pine Weymouth pine		*Pinus strobus*
Abies spp **firs**	**Whitewood** **Silver fir**		*Abies alba*
Picea spp **spruces**	**Whitewood**[1] White deal Norway spruce	imported from the Continent	*Picea abies*
	European spruce Common spruce	home-grown	
	Douglas 'fir' British Columbian 'pine'		*Pseudotsuga taxifolia*
	Western red 'cedar'		*Thuja plicata*
	Parana 'pine'		*Araucaria angustifolia*
	Western hemlock		*Tsuga hetrophylla*
	Larch, European, etc		*Larix spp*
	Cypress		*Cupressus spp*
	Yew		*Taxus baccata*
	Sequoia Californian redwood		*Sequoia sempervirens*

[1] Often qualified by geographical origin   [2] Colloquially *genera* are referred to as '*species*'

Nomenclature of timbers

The examples of softwoods given above show that the naming of timbers can be confusing. Douglas

mended, and if it is desired to be doubly sure that the required timber is obtained, the Latin name should also be given. In addition, because the quality of timbers varies with locality of growth, and with the methods of conversion and selection employed by shippers, it is desirable to state their origin as well.

Anatomy of timber

The tree is a complex plant which uses salts from the soil and carbon dioxide from the air to manufacture food materials by the action of sunlight on chlorophyll in the leaves, and to do this the tree sometimes extends 90 m above the ground, thereby providing an excellent structural material.

The trunk (or bole) and branches grow outwards around a leading shoot by adding new rings of timber. Usually one ring is added each year, but as this is not always the case it is better to refer to the rings as *growth rings* rather than annual rings. The more rapid the growth, the wider are the growth rings, and when trees of a particular species are compared, the wider the growth rings, the less dense and strong is the timber. These rings consist of minute tubular or fibrous *cells* tightly cemented together and each ring has two parts, the *early wood* or *springwood* and the *late wood* or *summerwood* which grows more slowly and is often denser, darker and narrower than the early wood (see figure 2). Timbers in which the early wood contains larger pores than the late wood are known as *ring porous* and those in which the pores are equal in size in both zones are called *diffuse porous*.

As most trees mature, for each new ring which is added forming a band of *sapwood*, reserve materials such as starch are extracted from an inner ring (or they are changed into more durable substances) and a *heartwood* core is formed. The band of sapwood varies widely in width from 25 to 152 mm, or more in some tropical hardwoods. Mechanically, there is no significant difference between sapwood and heartwood but in most species sapwood is lighter in colour and because it contains sugars, starch and water it is more attractive to fungi and certain insects. *Sap-stain* is sometimes a means of recognising sapwood. The outer surface of the tree is protected by bark, which in the case of the Plane tree is renewed periodically. The bark of cork oak trees in Spain and North Africa is used for thermal insulation

and for floor and wall coverings. See page 23.

The term *grain* refers to the general direction or arrangement of the fibres and other wood elements (or cells) but it is sometimes used to describe structural or ornamental features of the timber. Radially disposed *parenchyma* storage cells are sometimes seen as narrow *rays* on end grain surfaces. They are more evident on radially cut surfaces and particularly so in true oaks as decorative *silver grain*. As no cut surface is truly radial, the rays appear to be discontinuous – see figure 9, page 63.

Natural defects

Natural defects may be described as features which develop in the living tree, or soon after it is felled, which may detract from the usefulness of the timber.

Other defects which occur at later stages are:
Conversion defects – page 65
Seasoning defects – page 61.
Deterioration in use – page 70.

For painted joinery it is often economical to remove defects such as dead knots and resin pockets and to replace them with plugs or patches. Building up members by lamination also makes for more economical use of timber which contains defects.

Brittleheart (soft-pith, soft, spongy, or punky-heart)

This defect, found in the centre of many tropical trees, breaks with a brittle fracture. It can be detected by raising the grain with the point of a knife and should be avoided where strength is of importance. BS 1186 disallows soft-pith in surfaces which are to receive final decoration.

Sapwood

Today most softwoods which are used in building come from trees which are immature and consequently have a high proportion of sapwood. Paint has proved to be an inadequate protection against the growth of fungi on sapwood in timber which is used externally and it must be treated with preservative – see *Preservation* page 78.

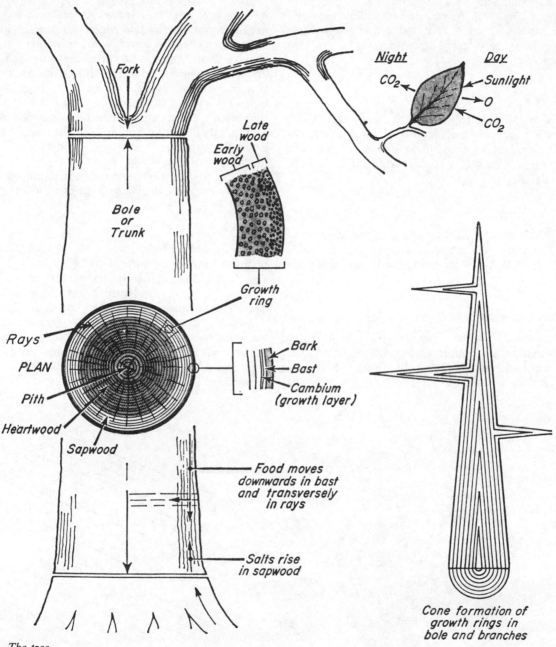

2   The tree

Sapwood which is discoloured by fungi which live on the cell contents of freshly felled timber but which cannot survive in seasoned timber – see page 46 (eg *blueing* of softwood), is as strong as ordinary timber but takes up moisture more readily and in damp conditions increases the risk of attack by destructive fungi. Internally, *sapstain* is usually considered to spoil the appearance of unpainted joinery.

Sapwood is also more attractive to insects than

heartwood and BS 1186: 1986, Part 1 *Specification of Timber in Joinery* does not allow sapwood in hardwoods other than beech and birch unless they have been treated against Lyctus beetles – see page 76.

Wide growth rings

These indicate rapid growth resulting in thin-walled fibres or a smaller proportion of the denser latewood, with consequent loss of density and strength – see pages 46, 53 and 69, and BSs 1186 and 5268.

Spiral grain

CP 112: 1971 lays down a rule for the rejection of spiral grain, which makes structural timber liable to distortion in seasoning and unsuitable for squaring.

Reaction wood

The wood, which is denser and stronger than normal timber, counteracts gravity, wind or other forces which tend to bend the trunk or branches. Reaction wood has the effect of throwing the heart off-centre, although this may be corrected by later growth. In hardwoods reaction wood usually occurs on the side which is in tension and is known as *tension wood*, while in softwoods it usually forms on the compression side and is known as *compression wood*. Unlike normal timber, movement along the grain is high and especially where it is concentrated on one side of a thin piece it often causes distortion and splitting, particularly during seasoning. Sawn surfaces have a woolly appearance and lead to difficulties in machining and finishing.

Upsets are fibres which have been damaged by shock or crushing during growth or in felling.

Fissures

These defects include *checks, splits, shakes* and resin pockets – see figures 3 and 4.

Checks are small surface splits usually caused by too rapid drying – see figure 7, page 61.

Splits extend from face to face – see figure 3.

Shakes   Ring shakes are due to stresses in the standing tree, in felling or in seasoning. Heart shakes indicate the presence of incipient decay in trees which have reached or passed maturity – see figure 3.

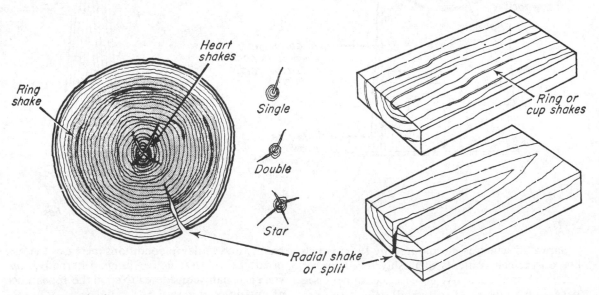

3   *Shakes and splits*

BS 1186 does not allow splits or ring shakes in any joinery timber. It also limits the size of other shakes and of checks in joinery 'ordered as selected for staining', and in 'surfaces not intended to receive final decoration' respectively.

CP 112 and BS 4978: 1973 (and to 1979) limit the size of fissures in structural timber – see table 13, page 53.

Resin pockets These fissures in softwoods which contain resin are not permitted by BS 1186, and CP 112 does not allow 'substantial exudations of pitch or resin from the faces of laminating grades'.

Rind galls are surface wounds which become enclosed in growth.

Burrs are swellings comprising highly contorted grain resulting from many undeveloped buds which form over a wound. In trees such as oak and walnut they are often exploited as decorative veneers.

Curl is a decorative effect revealed by skilful conversion of a crotch where fibres of a branch lock with those of a trunk.

Knots

A knot is the part of a branch which became enclosed in a growing tree – see figure 4.

Where the fibres of a branch are completely continuous with those of the tree a *live knot* results and where the fibres are continuous with those of the tree to the extent of at least three-quarters of its cross sectional perimeter the knot is *intergrown*.

A *dead knot* has fibres intergrown with the surrounding wood to the extent of less than a quarter of the cross sectional perimeter and if it is more or less surrounded with bark or resin it is an *encased knot*. Dead knots are sometimes *tight* but they are often *loose*. Knots attacked by fungus are

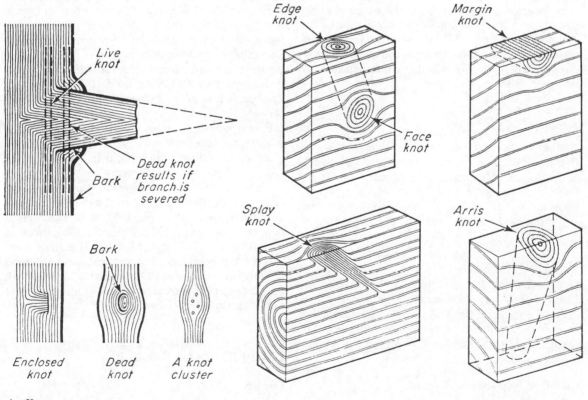

4 Knots

termed *unsound* or *decayed*. An *enclosed knot* does not appear on the surface of timber. *Pin knots* are 6 mm or less in diameter. A *knot cluster* consists of two or more knots arounds which the wood fibres are deflected.

BS 1186 does not allow exposed, decayed or dead knots in joinery, and because they are often considered to be unsightly, it does not allow any knots in surfaces which are ordered as 'selected for staining'. The standard allows only sound tight knots not more than 19 mm in diameter in the exposed surfaces of hardwood sills and limits the size of sound and tight knots and clusters in other surfaces.

Knots are hard, and cause uneven wear in flooring and are difficult to work. In softwoods they often contain resin which must be sealed before the wood is painted – see *MBS: Finishes*, chapter 6.

Knots, particularly edge and arris knots, reduce strength mainly in tension, but not in resistance to shear and splitting. CP 112 *stress grading rules* measure only the visible surface area of knots but those in BS 4978: 1973 limit the ratio of knot area to total area (KAR) at any cross section, the proximity of knots to the edges and their 'longitudinal separation'. CP 112 also contains rules for laminated timber.

Fungal decay

Timber which is seriously affected must be rejected, but *dote*, an early stage of decay which shows as whitish streaks or patches (*pocket rot*), may be acceptable for timber which is to be painted or out of sight, provided there is no softening of the fibres *(punk)* – see page 72.

Insect damage

Occasional exit holes of pinhole borers (see *Insects* page 76) are not generally regarded as defects in structural timber and they are acceptable in joinery which is to be painted or which is out of sight (BS 1186).

Chemical composition of wood

The walls of timber fibres are composed of cellulose and hemicelluloses and these are bonded together essentially by lignin.

Softwoods generally contain more lignin than hardwoods but otherwise the proportions of the constituents do not vary much from species to species.

Average proportions in dry wood are:
45 to 60 per cent cellulose
10 to 25 per cent hemicelluloses
20 to 35 per cent lignin.

Minor constituents, which vary considerably in nature and amount from one species to another, are responsible for some of the characteristic features of different species. Thus, resins, tannins, alkaloids, turpentine and rosin colour timber. Starch in sapwood is attractive to fungi but tannins and other phenolic componds in hardwoods, and oil infiltrates in Western red cedar, are toxic to insects and fungi.

Most woods are slightly acidic and produce acetic acid if stored in damp conditions. Timbers such as oak and Western red cedar contain tannin and thuyaplicins which corrode metals. Tannin in wood in contact with iron or iron componds, particularly in damp conditions, causes dark stains. Gums and resins adversely affect working properties and ability to take glue and surface finishes, while silica in some hardwoods blunts tools.

Some woods contain colourless components which may become pink when acted upon by acids such as those which occur in synthetic resin glues used for adhering veneers.

Tannins and sugars in wood can inhibit the setting of Portland cement in the manufacture of wood-wool slabs (see page 40 or in lightweight concrete (see page 184). Phenolic substances in certain woods such as iroko and Brazilian rosewood have been found to prevent the hardening of polyester varnishes. Oily timbers such as teak may have to be degreased before gluing or before applying surface finishes.

PROPERTIES OF TIMBERS

Density

Table 32 classifies some timbers in respect of

density, durability, resistance to impregnation and working properties.

The weight of wood tissue is about $1506 \, kg/m^3$ for all species, but the densities of timbers vary widely. As shown in figure 5 most seasoned timbers fall within the range $385–835 \, kg/3m^3$ and are light in weight compared with stones (about $2082–3204 \, kg/m^3$, common metals ($2640–11373 \, kg/m^3$) and plastics ($900–1400 \, kg/m^3$).

Within the range of moisture contents 5 to 25 per cent, the weight of timbers varies approximately 0.5 per cent for every 1 per cent variation in moisture content.

Thermal insulation

Timber is a good insulator. Conductivity (k) is $0.144 \, W/m$ deg C and transmittance (U) for 102 mm thickness is about $1.19 \, W/m^2$ deg C for timber weighing $481 \, kg/m^3$ with 20 per cent moisture content and for timber weighing $561 \, kg/m^3$ with 12 per cent moisture content.

Thermal movement

The coefficient is $30–60 \times 10^6$ deg C across the fibres and about one tenth as much parallel to the fibres. Expansion joints are not normally required even in large structures.

Behaviour in fire

See pages 34, 39 and 70.

Chemical resistance

A high cellulose and lignin content and a low hemicellulose content, low permeability, straight grain and small moisture movement contribute to good chemical resistance, criteria which are best met by softwoods such as sequoia, pitch pine, Douglas fir and southern cypress, and a few hardwoods including teak, iroko, purpleheart and greenheart.

Compared with metals, wood has good resistance to alkalis and weak acids. Sources of alkalis include casein and phenol formaldehyde glues while sources of acids include surplus hardeners in synthetic resin glues, hydrochloric acid arising from chlorine fumes in the damp conditions obtaining in swimming baths, and sulphurous acid from flue gases.

Ammonia turns oak brown, hence *fumed oak*, an effect which may be accidentally induced by the decomposition of animal glue.

Inorganic salts from soils or damp masonry, probably play a very minor role in the decomposition of wood.

Effect of exposure to light Wood darkens when exposed to light and suffers *photodegradation*, mainly due to visible wavelengths in sunlight. Consequently, clear finishes do not perform well on wood and there is a need for effective photostabilisers – see MBS *Finishes*, chapter 6.

Strength

Timber has a high strength:weight ratio both in tension and compression, and is elastic. It is able to sustain greater loads for a short while than it can over long periods so that in deriving working stress values from test results the rate of straining must be taken into account.

Generally, strength increases with density, particularly within a species and the NHBC specification stipulates minimum densities for each structural timber. Strength reduces as moisture content rises. Most strength properties of timber containing more than 28–30 per cent moisture (the *fibre saturation point*) are only about two thirds of those in timber at 12 per cent moisture content and one-third of corresponding properties in oven-dried timber. Table 14 shows the lower stresses allowed by CP 112 for timbers containing more than 18 per cent moisture.

A 1°C temperature rise reduces strength by about 0.3 per cent. BS 5628 refers to loss of strength in fire – see page 39.

As a natural product, timber is subject to considerable variations in strength. Properties vary widely between species, between trees of any one species, in different parts of a tree and in different directions. Thus, along the grain tensile strength may be as much as thirty times that across the grain, and two to three times the compressive strength.

Moreover, strength is reduced by the particular defects contained in each piece of timber. Historically, very crude methods of predicting the strength of timber specimens necessitated factors

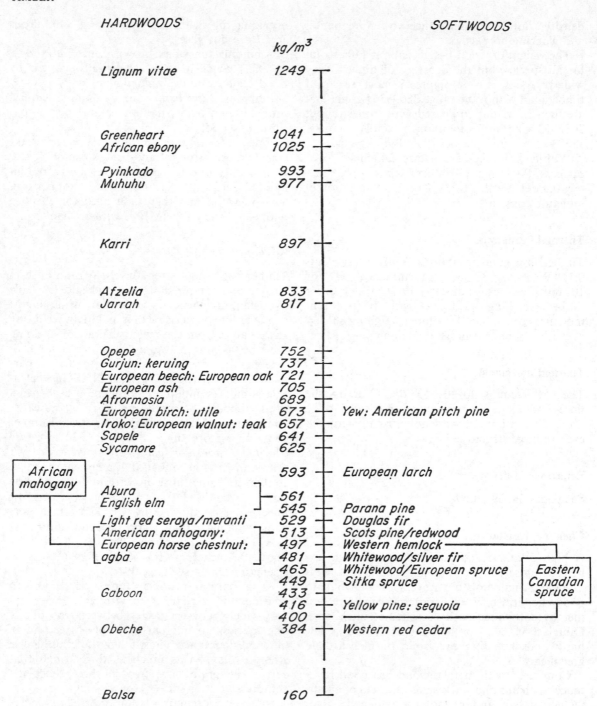

HARDWOODS   SOFTWOODS

kg/m³

Lignum vitae   1249

Greenheart   1041
African ebony   1025

Pyinkado   993
Muhuhu   977

Karri   897

Afzelia   833
Jarrah   817

Opepe   752
Gurjun: keruing   737
European beech: European oak   721
European ash   705
Afrormosia   689
European birch: utile   673   Yew: American pitch pine
Iroko: European walnut: teak   657
Sapele   641
Sycamore   625

African   593   European larch
mahogany
Abura   561
English elm   545   Parana pine
Light red seraya/meranti   529   Douglas fir
American mahogany: 513   Scots pine/redwood
European horse chestnut: 497   Western hemlock
agba   481   Whitewood/silver fir
                                       465   Whitewood/European spruce   Eastern
                                       449   Sitka spruce   Canadian
Gaboon   433   spruce
                                       416   Yellow pine: sequoia
                                       400
Obeche   384   Western red cedar

Balsa   160

(Densities vary as much as 300 per cent largely due to differences in proportions of early wood and late wood)

5   *Average densities of common timbers at 15 per cent moisture content*

of safety which were often very generous, and therefore wasteful when applied to structural members.

References

BRE Digest 287 *Structural timber* is a guide to specification. It deals with strength, moisture content and preservation and it refers to more detailed sources of information.

The Digest states that where applicable BS 5291:1984 *Finger joints in structural softwoods* must be complied with.

CP 112 is *The structural use of timber* and BS 5268 *CP for the structural use of timber* will cover the major structural uses of timber. Parts include:

Part 1 *Limited state design, Materials and workmanship* – for later publication

Part 2: 1984 *CP for permissible stress design, materials and workmanship* will replace CP 112, Part 2

Sections of this Part deal with timber, glues, laminated timber, tempered hardboard and joints.

Part 3 will supersede CP 112

Part 5 will supersede CP 98, *Preservation*

Part 6 *CP for timber frame wall design* and

Part 7 *Calculation basis for span tables* will be published later.

4 Stress grading

Today, BS 4978: 1973 and 1976; *Timber grades for structural use*, CP 112: 1971, *The structural use of timber* and BS 5756: 1980 provide rules for stress grading, ie for assessing the strength of individual pieces of timber so that they can be used to optimise their potential as engineering material. The economies in timber sizes which can result from the use of a higher stress grade for structural members is evident in the *Building Regulations 1985*, Approved Document A 'deemed to satisfy' sizes.

There are two methods of stress grading:

1 Visual stress grading

This method predicts the effects on strength properties of permitted defects by measuring their sizes and positions.

CP 112 contains rules for measuring the visible surface area of knots, and rules for some other defects are summarized in table 13.

Sapwood, including blue-stained sapwood, scattered pinholes and small occasional wormholes are allowed but not active infestation, fungal decay, brittleheart, serious spiral grain and other abnormal defects which reduce strength.

In CP 112 *clear pieces* – ie those free from all visible defects – of names species, are allocated

Grade	Minimum number of growth rings per 25 mm[1]	Maximum slope of grain[1]	Maximum amount of wane expressed as a fraction of width of surface in which it occurs	Maximum size of fissures and resin pockets expressed as a fraction of the thickness of members[2]
75	8	1 in 14	0.1	0.3
65	6	1 in 11	0.1	0.4
50	4	1 in 8	0.2	0.5
40	4	1 in 6	0.2	0.6

[1]Subject to rules for measurement
[2]Deeper fissures are permitted outside the middle half of the depth of the end cross section and at a distance from the end equal to three times the depth of the piece for compression members

Table 13   CP 112:1971 Rules for visible defects in structural timber

Softwood species Group	Name and origin		Density[2]	Grade	Bending and tension parallel to grain N/mm²	Compression parallel to grain N/mm	Compression perpendicular to grain N/mm²	Shear parallel to grain N/mm²	Modulus of elasticity Mean N/mm²	Modulus of elasticity Minimum N/mm²
S1	Douglas fir, Pine pine	imported	590, 720	Basic	13.8 (17.2)	9.7 (13.1)	1.72 (2.48)	1.38 (1.52)	9 000 (9 700)	4 500 (4 800)
				75	10.3 (12.1)	7.2 (9.3)	1.52 (2.21)	1.03 (1.14)		
	Douglas fir, Larch	home-grown	560, 560	65	9.0 (10.3)	6.2 (7.6)	1.52 (2.21)	0.90 (0.97)		
				50	6.9 (7.9)	4.8 (5.5)	1.31 (1.93)	0.69 (0.76)		
				40	5.5 (6.2)	3.8 (4.5)	1.31 (1.93)	0.55 (0.62)		
S2	Western hemlock – unmixed, – commercial; Parana pine; Redwood; Whitewood[1]	imported	540, 530, 560, 540, 510	Basic	11.0 (13.8)	8.3 (11.0)	1.38 (2.07)	1.38 (1.52)	6 900 (8 300)	4 100 (4 500)
				75	8.3 (9.7)	6.2 (7.9)	1.17 (1.72)	1.03 (1.14)		
	Canadian spruce; Scots pine	–home-grown	450, 540	65	6.9 (7.9)	5.2 (6.6)	1.17 (1.72)	0.90 (0.97)		
				50	5.5 (6.2)	4.1 (4.8)	1.03 (1.52)	0.69 (0.76)		
				40	4.5 (5.2)	3.1 (3.8)	1.03 (1.52)	0.55 (0.62)		
S3	European spruce; Sitka spruce	home-grown	380, 400	Basic	7.6 (10.3)	5.5 (8.3)	1.03 (1.52)	1.10 (1.24)	5 900 (6 900)	3 100 (3 800)
				75	5.5 (6.6)	4.1 (5.2)	0.90 (1.31)	0.83 (0.89)		
	Western red cedar	–imported	380	65	4.8 (5.5)	3.4 (4.1)	0.90 (1.31)	0.69 (0.76)		
				50	3.8 (4.5)	2.8 (3.1)	0.76 (1.10)	0.55 (0.62)		
				40	3.1 (3.4)	2.1 (2.4)	0.76 (1.10)	0.41 (0.45)		
	Greenheart		1060	Basic[2]	37.9 (41.4)	27.6 (30.3)	6.21 (9.31)	4.83 (5.52)	17 200 (18 600)	12 400 (13 400)
	Opepe		780		25.5 (31.0)	22.1 (24.8)	5.52 (8.27)	3.45 (4.14)	12 400 (13 800)	7 580 (9 300)
	Karri		930		22.1 (26.2)	16.5 (22.1)	4.83 (7.24)	2.48 (2.76)	13 800 (15 500)	8 270 (9 660)
	Afromosia		720		22.1 (26.2)	15.9 (22.1)	4.14 (6.21)	2.62 (2.76)	10 300 (12 100)	6 890 (7 930)
	Teak		720		22.1 (26.2)	16.5 (22.1)	4.14 (6.21)	2.34 (2.62)	11 000 (12 400)	6 890 (7 930)
	Iroko	imported	690		20.7 (23.4)	15.2 (19.3)	4.14 (6.21)	2.34 (2.62)	8 960 (10 300)	5 860 (6 890)
	Jarrah		910		19.3 (23.4)	15.9 (20.7)	4.14 (6.21)	2.34 (2.62)	10 300 (12 100)	6 890 (7 930)
	Sapele		690		19.3 (23.4)	15.9 (20.7)	4.14 (6.21)	2.34 (2.62)	9 660 (11 000)	6 210 (6 890)
	Gurjun/Keruing		720		17.2 (22.8)	13.8 (19.3)	3.10 (4.49)	2.34 (2.62)	12 400 (13 800)	8 270 (9 300)
	Abura		590		13.8 (16.5)	10.3 (13.8)	2.34 (3.45)	2.07 (2.41)	8 300 (9 300)	4 500 (4 830)
	African mahogany		590		12.4 (15.2)	9.66 (13.1)	2.07 (3.10)	1.72 (1.93)	7 930 (8 620)	4 140 (4 490)
	Red meranti/ red seraya	imported	540		12.4 (15.2)	9.66 (13.1)	1.79 (2.62)	1.52 (1.72)	7 580 (8 270)	4 140 (4 490)
	European ash		720		17.2 (22.8)	9.66 (15.2)	3.10 (4.49)	2.76 (3.10)	10 000 (11 400)	6 550 (7 240)
	European beech	home-grown	720		17.2 (22.8)	9.66 (15.2)	3.10 (4.49)	2.76 (3.10)	10 000 (11 400)	6 550 (7 240)
	European oak		720		15.9 (20.7)	9.66 (15.2)	3.10 (4.49)	2.48 (3.10)	8 620 (9 660)	4 490 (5 170)

[1]Picea abies and Abies alba   [2]The 75, 65, 40 and grades are omitted from this table.

Table 14   Stresses and moduli of elasticity for 'green' timbers having moisture content exceeding 18 per cent and for 'dry' timbers having moisture content not exceeding 18 per cent, given in parentheses From CP 12:1971

basic stresses. In practice, timber contains defects and these, are taken into account in the reduced stresses given for grades: 75, 65, 50 and 40.

For convenience in design, softwoods having similar properties are grouped, but hardwoods vary so widely that grouping is not practicable – see table 14.

Stresses are given for *green* timber, ie having more than 18 per cent moisture content, and for *dry* timber.

An amendment to CP 112 gives stresses for Canadian timbers graded under the *Structural joists and plans* section of the 1970 NLGA rules, grouped as: Douglas fir – larch, Hem – fir, and Spruce – pine – fir, (2, 4 and 12 species respectively).

BS 4978 employs different rules for assessing the effects of defects on the strength of a specimen. There are stipulations for fissures, wane, resin pockets, distortion and worm holes. The effects of knots are assessed by *knot area ratios* (KAR) which take into account their geometry within the section being considered.

The BS 4978 visual grades are:
(GS) General structural grade,
(SS) Special structural grade,
and three grades for timber for laminating.

Members of the TRADA visual grading scheme mark timber with the TRADA-MARK.

BS 5756: 1980 is entitled: *Tropical hardwoods graded for structural use*. It states that BS 4978 is not entirely suitable for grading tropical hardwoods, and provides rules for identifying a single visual stress grade *Hardwood structural* (HS) grade in supplies of timber where their exact end uses are not known. The BS applies to all types of structural members, whether in bending, compression or tension.

Part 2 of BS 5756 will give grade stresses and moduli of elasticity for individual timbers, and groups of timbers.

2 Machine stress grading

BS 4978 states conditions for approval and control of machines which measure modulus of elasticity and mark pieces at 1500 mm intervals, and at their ends, with an overall stress grading.

Machine grading is more rapid, and more accurate than visual grading, which is labour intensive and unable to assess the effects on strength of internal defects. Machine gradings are designated: MGS, MSS, M75 and M50 which are equivalent to the GS, SS, 75 and 50 visual gradings.

A machine stress grading system is being developed for tropical hardwoods dealt with by BS 5756.

At present, timber is available in GS, SS and M75 grades in preferred sizes:
38 and 50 × 75, 100 and 125
38, 50 and 68 × 150, 175, 200 and 225 mm.

Working stresses are found by applying *modification factors* to the table stresses, to allow for different conditions of loading and service.

Effects of variations in moisture content

Drying timber from the *green* to the normal seasoned condition reduces its density by 50 per cent or more with consequent shrinkage and increases in strength properties, thermal insulation, resistance to decay, and suitability for impregnation, painting and gluing.

Unless it is hermetically sealed on all sides, timber acquires a high moisture content when part of it is in contact with water or a damp material, but less obviously, being hygroscopic, it takes in or gives off moisture vapour until it reaches *equilibrium* with the humidity of the surrounding atmosphere. Temperature is relatively less important, at average relative humidities the moisture content of timber decreases only a few per cent between 16 and 38°C.

Equilibrium moisture contents for a typical species at various relative humidities at 16°C are shown in figure 8, and FPRL leaflet no. 47 (revised March 1967).

The movement of timbers shows *equilibrium moisture contents* ranging from 9 per cent for African blackwood, to 14.5 per cent for South American cedar, where the relative humidity is 60 per cent and the temperature is 25°C.

Moisture content is expressed as the weight of water in timber as a percentage of the weight of the dry wood. Thus:

MC per cent =
$$\frac{\text{weight of specimen} - \text{dry weight of specimen}}{\text{dry weight of specimen}} \times 100$$

To determine a representative moisture content of a member, where possible a sample should be taken at least 325 mm from the end. It is carefully weighed, to obtain the wet' weight, and to obtain the 'dry' weight it is repeatedly dried and re-weighed until the weight remains constant. If carefully performed as described in BS 4978, the test is very accurate. More conveniently, but less accurately, a moisture meter can be used to measure the electrical resistance between two points in a piece of timber and its moisture content can be read off from a scale calibrated for the species being tested. Surface moisture may bear little relationship to the moisture content of the interior of a large section of timber and a meter equipped with points which penetrate to the core, may be necessary.

Movements

As timber dries from the green condition, shrinkage starts when the cell walls begin to dry (at *fibre saturation point*), a process which can be reversed by re-wetting the timber.

Variations in size in response to variations in moisture content, which in turn, result from changes in atmospheric humidity, or from direct wetting are known as *moisture movements* – not to be confused with movements of moisture.

The extent of movements, and the ratios of movements in different directions, vary between species – see table 15.

Typically, shrinkage from *fibre saturation point* (27 per cent moisture content), to *oven dry* (0 per cent), is: for tangentially sawn members 10 per cent, and for radially sawn members 5 per cent. As, however, timber is rarely cut either truly tangentially or truly radially, the actual movement across the grain is usually an intermediate value. Longitudinal movement is negligible, about 0.1 per cent.

Thus, movements for a 1 per cent change in moisture content are: 0.37, 0.185 and 0.0037 per cent, respectively.

As an example, in an average heated room where timber varies in moisture content from 14 per cent in the autumn to 10 per cent at the end of the winter, tangentially sawn timber shrinks 1.48 per cent or 14.8 mm per metre, and radially sawn timber about half as much.

The fact that movement across the grain is at least 50 times greater than movement in its length often becomes apparent where wide rails shrink in their width leaving the ends of stiles projecting.

Incidentally, thermal expansion is relatively small, and causes timber to move in the opposite direction to drying movements.

The PR Laboratory uses the sum of the percentage radial and tangential movements resulting from a change in the relative humidity of air from 90 to 60 per cent, to group timbers into three broad classes:

Class 1　*Small movement* – less than 3.0 per cent
Class 2　*Medium movement* – between 3.0 and 4.5 per cent
Class 3　*Large movement* – more than 4.5 per cent.

Table 15 shows the radial and tangential movements and the *combined movement classes* (T + R) of some common timbers. (It will be seen that hardwoods are found in all classes but there are no softwoods in the *large movement* class.)

Other timbers which have small movement values include: abura, gedu nohor, lauan, limba, meranti, missanda and muhuhu.

Stress setting

Where the movement of timber is restrained, either by compression or tension, it becomes set in sizes which are permanent for the particular combination of atmospheric humidity and temperature prevailing at the time of *stress setting*. Thus, unless the *permanent set* is relieved by steam treatment, subsequent movement of timber starts from the *stress set* size, and with any change in atmospheric humidity and temperature the final size is smaller for compression set timber, and larger for tension set timber, than it would otherwise have been. H E Desch illustrates the effect of compression setting by reference to the premature installation of seasoned timber in a 'wet' building. Thus where flooring strips, which have been dried to suit a centrally heated building are installed before the building has dried out they take up moisture; but having been cramped in position and nailed they cannot expand fully and become set in widths narrower than they would normally

Movement of timber with relative humidity change from 90% down to 60% at 25°C +

Movement class	TIMBER	Tangential (T) mm per m*	Radial (R) mm per m*	T+R Per cent*	T−R Per cent*
CLASS I 'Small movement.' T+R less than 3·0%	Muninga	6	5	1·1	·1
	Western red cedar-imported	9	4·5	1·35	·45
	Afzelia SW	10	5	1·5	·5
	Iroko	10	5	1·5	·5
	Idigbo	10	6	1·6	·4
	Indian rosewood	10	7	1·7	·3
	Agba	13	6	1·9	·7
	Teak	12	7	1·9	·5
	Afrormosia	13	7	2·0	·6
	Obeche	12·5	8	2·05	·45
	European spruce (whitewood) SW	15	7	2·2	·8
	'African walnut' SW	13	9	2·2	·4
	Mahogany–Central American	13	10	2·3	·3
	Mahogany–African (Khaya ivorensis)	15	9	2·4	·6
	'Cedar–South American'	15	10	2·5	·5
	Balsa	20	6	2·6	1·4
	Yellow pine–home grown SW	18	9	2·7	·9
	Rhodesian 'teak'	16	10	2·6	·6
	Douglas fir SW	15	12	2·7	·3
	Western red cedar –home grown SW	19	8	2·7	1·1
	Western hemlock SW	19	9	2·8	1·0
	Avodiré	18	10	2·8	·8
	Guarea	16	12	2·8	·4
	Makoré	18	11	2·9	·7
CLASS II 'Medium movement.' T+R 3·0–4·5%	Utile	16	14	3·0	·2
	Ash (European)	18	13	3·1	·5
	Sapele	18	13	3·1	·5
	Mahogany–African (Khaya grandifoliola)	18	13	3·1	·5
	Scots pine (redwood) SW	22	10	3·2	1·2
	Serbian spruce SW	23	13	3·6	1·0
	Pyinkado	21	15	3·6	·6
	European walnut	20	16	3·6	·4
	English elm	24	15	3·9	·9
	Keruing	25	15	4·0	1·0
	English oak	25	15	4·0	1·0
	Caribbean pitch pine SW	26	14	4·0	1·2
	Black poplar	28	12	4·0	1·6
	Sycamore	28	14	4·2	1·4
	Parana pine SW	25	17	4·2	·8
	Dahoma	28	15	4·3	1·3
	Jarrah	26	18	4·4	·8
CLASS III 'Large movement.' T+R more than 4·5%	Ramin	31	15	4·6	1·6
	Turkey oak	33	13	4·6	2·0
	East African olive	29	17	4·6	1·2
	Black wattle	35	12	4·7	2·3
	Yellow Canadian birch	25	22	4·7	·3
	Beech	32	17	4·9	1·5
	Japanese ash	35	15	5·0	2·0
	Gurjun	33	20	5·3	1·3

* 1 mm per m. = 0.1 per cent linear movement   *The more favourable tangential-radial values are shown in heavy type

+ The atmosphere relative humidity change approximates to the conditions to which air-seasoned wood is subjected when it is 'manufactured' and used in a house

Table 15   Moisture movements of common timbers

SW Softwoods   Tangential (T) ▬▬▬   Radial (R) ▭▭▭

assume. When the building dries, the strips shrink as from the set size and their final width is less than it would have been if compression setting had not occurred.

Distortions

A small change of shape is often more objectionable than relatively large movements in all directions. Distortion can result from the application of external forces but we are concerned here with distortion which occurs because timber does not shrink equally in all directions when it dries – aggravated sometimes by defects such as knots or reaction wood.

WC Stevens[1] has shown that *cupping* of plain sawn boards and *diamonding* of squared timber (see figure 6) are mainly due to the excess of tangential shrinkage (in the direction of the growth rings) over radial shrinkage, and not upon the ratio of the two as has often been supposed.

Table 15 shows the difference between the tangential and radial movements in some common timbers for a 90 to 60 per cent change in the relative humidity of the atmosphere and these figures can be taken as an index of the tendency to cupping of clear specimens. It will be seen that although the tangential and radial movements of yellow Canadian birch are large, the difference between them is small so the tendency to cupping with changes in moisture content is likely to be small.

Other distortions may be due to differential shrinkage or expansion in pieces of timber containing distorted or curved grain.

Spring and *bow* occur in the same direction as the curvature of grain in relation to the edges or faces of straight pieces and *twist* may result where spiral growth occurred in the tree. Twist may also occur in a plain sawn board which is not cut parallel to the heart of a tree – see figure 7.

UTILIZATION OF TIMBER

Minimization of changes in moisture content

Apart from obvious precautions such as the avoidance of unnecessary wetting of timber,

[1]'Distortion of timber' by W. C. Stevens MA, AMIMechE, *Wood* June 1946

changes in moisture content are minimized by (1) impregnation, (2) surface coatings and (3) seasoning.

1 *Impregnation* of timber with a solution of resin is at present limited to proprietary flooring and small objects such as table knife handles.

2 *Surface coatings* Effectiveness in reducing the rate of intake or loss of water or water vapour varies widely as shown in table 18 but all external surfaces of general joinery should be at least primed with a 'pink' lead primer – see BS 5082:1974 and BS 5358:1976. Sealing is particularly important on surfaces which will be in contact with construction before it has dried out and two full coats of aluminium based primer should be applied to all surfaces which will be in contact with external walls.

3 *Seasoning* is the controlled reduction of moisture content to a level appropriate to the end use, the advantages of which have been listed under *The effects of variations in moisture content*, page 55.

Moisture contents for end uses

In traditional buildings a certain amount of shrinkage can be tolerated in carcassing timbers and these are not normally fully seasoned to the moisture content they will assume in service. Movement of good class flooring, joinery and furniture, however, can be serious. Ideally timber for such purposes should not be installed until the building is 'dry' and its moisture content should be the mean of the values it will have in use. If this is done movement will be limited to that occasioned by variations in the humidity of the air which surrounds it.

BRS Digest 99 (first series) *Light cladding* Part II June 1957 stated that the moisture content of timber in a well heated building is likely to be 15 per cent in the late autumn and 8 per cent in midwinter and that an average seasonal movement of 13.3 mm per metre should be expected. Externally, moisture contents of 15 per cent in the summer (or 10 per cent in thin boards in very hot weather) and 22 per cent in the winter should be expected and an allowance of 17 mm per metre

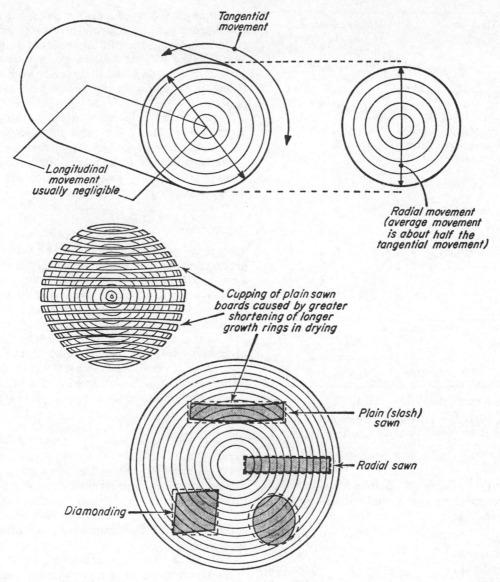

6   *Movements in drying*

should be made for the large moisture content change.

Figure 8 recommends moisture contents for various uses with the approximate equivalent relative humidities of air at 16°C based on information contained in CP 112 *Structural use of*

timber; PRL leaflet no. 9 (revised 1963) *The moisture content of timber in use* and on BS 1186: 1971: Part 1 (amended) *Quality of timber in joinery*.

The equilibrium moisture contents of different timbers vary at 60 per cent relative humidity from

Treatment	Moisture content after 12 days in saturated atmosphere (95–100 per cent relative humidity) per cent
Untreated	28·4
2 coat surface applications:	
Nitrocellulose wood finish	23·9
Long oil phenolic/tung oil varnish	21·0
Short oil alkyd resin varnish	20·6
Chlorinated rubber modified alkyd resin medium oil varnish	16·3
Epoxy resin varnish	12·0
Hard heavy bituminous paint	8·2
Aluminium leaf wood primer	3·7
Normal 3 coat oil paintwork	8·0 approx.
Wax or other impregnants	18·2–24·9

Research report C/RR/2 from TDA

Table 16   Moisture absorption of beech sapwood specimens after treatment with surface coatings and impregnants.

9.5 to 15.0 per cent. Timbers with equilibrium moisture contents significantly lower than the average include:

Afzelia	9.5 per cent
Western red cedar (imported)	9.5 per cent
Loliondo	10 per cent
Muninga	10 per cent
Teak	10 per cent

For purposes such as interior joinery and hardwood flooring these timbers should be seasoned to slightly lower moisture contents than those generally recommended.

Methods of seasoning

Sawn softwood is imported either 'green' or 'shipping dry' ie surface dry only.

If drying is too rapid the outer parts, in particular unprotected ends, shrink before the interior, surface checking and splitting results and ring and heart shakes may extend. In extreme cases the surface *case hardens* and the interior *honeycombs* or the timber *collapses* and becomes useless – see figure 7.

Some timber species are more difficult to season satisfactorily than others and those containing defects such as *reaction wood* are liable to distort.

Timber must be stacked, supported and sometimes restrained, so as to minimize distortion during seasoning.

Air seasoning

Timber is protected from rain and from the ground and stacked so that air can circulate freely around all surfaces and so that the risks of *degrade* and of attack by fungi and insects are minimized. In this country a moisture content of 17 to 23 per cent is usually attained with little risk of the process being too rapid except at the ends of timbers, and these can be protected. In favourable summer conditions, thin softwoods can be air-seasoned in weeks but in less favourable conditions some hardwoods require a year or more.

Kiln seasoning

Figure 8 shows that artificial means of seasoning must be used to achieve the moisture contents needed for joinery and furniture in modern buildings.

Timber can be kiln seasoned from the 'green' condition, but kiln seasoning may follow air seasoning.

Adherence to a precise schedule of humidity and temperature enables any moisture content to be achieved without significant *degrade*. *The timber drying manual* BRE Report 74. HMSO gives kiln schedules for various species. 25 mm hardwoods can be seasoned in days to months according to the species. BRE Report 74 also describes *Radio frequency drying* and other methods.

'Water seasoning'

This terms is a misnomer for the process by which logs are kept under water to preserve them from

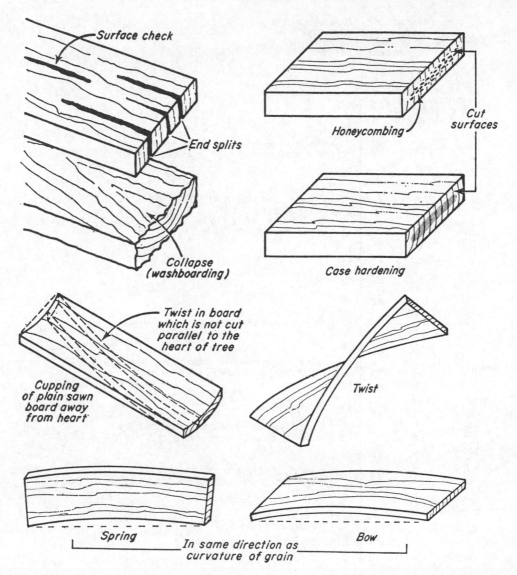

7   *Seasoning defects*

attack by insects and fungi, and ring porous hardwoods are sometimes immersed in running water to wash out the sap which is attractive to Lyctus beetles.

Care of seasoned timber

It should be clearly understood that seasoning is a reversible process and its cost is wasted if timber gets wet on the site. Close piling and covering with tarpaulins delays the absorption of atmospheric moisture, particularly in the interior of a pile. If timber having 10 to 12 per cent moisture content, which is appropriate for a normally heated building, is installed before a building is dry it can well acquire a moisture content approaching 20 per cent. Hence, before delivery and installation of high class joinery, buildings

61

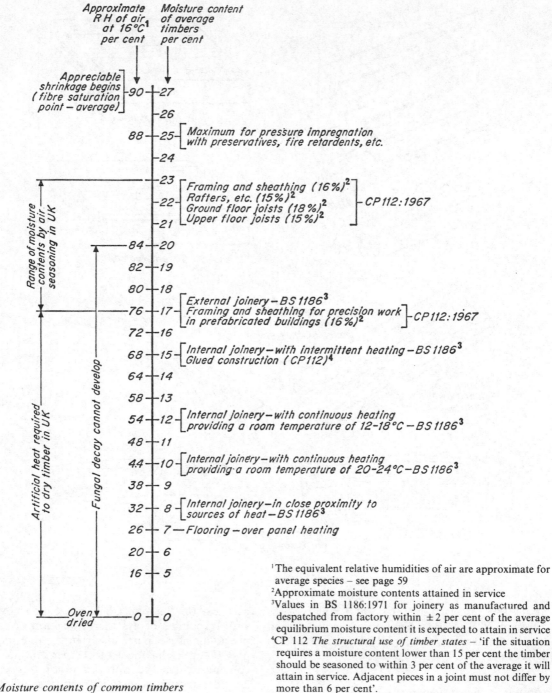

8   *Moisture contents of common timbers*

[1] The equivalent relative humidities of air are approximate for average species – see page 59

[2] Approximate moisture contents attained in service

[3] Values in BS 1186:1971 for joinery as manufactured and despatched from factory within ± 2 per cent of the average equilibrium moisture content it is expected to attain in service

[4] CP 112 *The structural use of timber states* – 'if the situation requires a moisture content lower than 15 per cent the timber should be seasoned to within 3 per cent of the average it will attain in service. Adjacent pieces in a joint must not differ by more than 6 per cent'.

should be glazed, if necessary dried by temporary heating, and tests on central heating should be completed.

Minimization of the effects of movements in service

Timber which has been seasoned will nevertheless respond to variations in atmospheric humidity. The advantages in the selection of a timber with relatively small moisture movement and to a lesser degree a small difference between radial and tangential movements, have already been considered. The juxtaposition of different species having widely varying movements is best avoided and the high cost of radially converted boards may be justified in some cases.

Generally, timber should not be restrained, and the drawing board with battens held by screws in slots, and tongued and grooved 'V' jointed boards are good examples of the application of this principle. 'Solid' timber table tops should always be fixed so that they are free to move. Tenons should be narrow so they do not loosen if they shrink, and similar traditional precautions should

be adopted. Tangentially sawn members such as skirtings and cover moulds should be fixed so that they will 'cup' towards the wall or other rigid surface.

Methods of conversion

The cutting of logs into sections before seasoning is known as *conversion*. Subsequent re-sawing and shaping is commonly called *manufacture*. Basic methods of cutting are *sawing*; *peeling* for producing plies for plywood; *slicing* of thin decorative veneers and *cleaving* or splitting, as used generally before the development of tempered steel saws, and now for palings.

Planed timber surfaces are said to be *wrought* (wrot).

Sawn timber

Figure 9 shows three ways in which the direction of the growth rings can relate to surfaces: ie, as *end grain*; more or less parallel to surfaces in *plain sawn* boards and at right angles to the surfaces of *radial sawn* boards. Where the growth rings meet

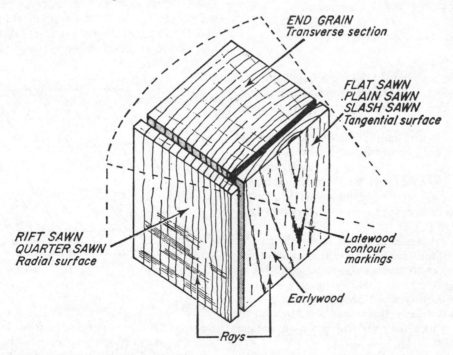

9   *Surfaces of cut timber*

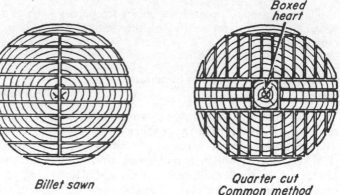

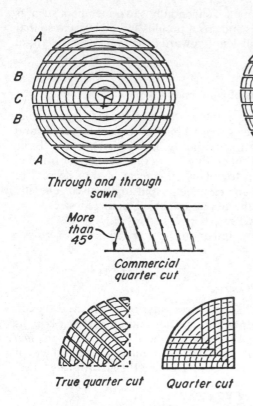

10　*Methods of sawing timber*

at an angle less than 45° on at least half of a surface, boards are known commercially as *rift sawn* in softwoods and *quarter sawn* in hardwoods.

Hardwoods are mainly converted by *through and through* cutting and figure 10 shows that this produces plain sawn boards (A) and others in which parts remote from the heart can be said to be either commercially quarter sawn (B) or in the case of the centre board (or *crown plank*) truly quarter sawn (C).

Radially sawn boards shrink less in their width (see figure 6) and are less liable to cup and twist; they are easier to season and wear more evenly than plain sawn boards. Unfortunately however, methods of cutting which produce a high proportion of quarter sawn timber are wasteful and the extra cost is justified only where the advantages are important.

Where the centre of the tree contains defects such as soft pith or heart shakes it is often cut out in one piece known as a *boxed heart*. An example is shown in figure 10.

Conversion defects

These include

Wane which refers to the bark or rounded timber below the bark which results from over-economical conversion of softwood logs. Some wane is acceptable for structural timber (BS 4978: 1973 and CP 112: 1971 – see page 53 – roof boarding and on the underside of floor boarding (BS 1297: 1970 (1980)).

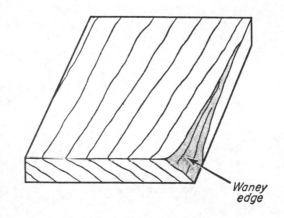

Sloping grain

Because the tree is a cone-like form some slope of grain is bound to show relative to the surfaces of all converted timber and the familiar 'contour' markings on flat sawn boards result. Pronounced slope of grain (diagram below) results from spiral growth or from conversion which is not parallel to the axis of the tree. BS 1186 limits the slope of grain with surfaces to 1 in 8 for hardwoods and 1 in 10 for softwoods. BS 4978: 1973 gives maxima of 1 in 6 GS grade and 1 in 10 for SS grade. CP 112 maxima are given in table 13, page 53.

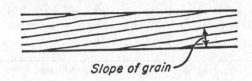

Slope of grain

Raised grain

Where plain sawn boards containing more than about 12 per cent moisture are machined, the knives force the latewood bands into the early-wood bands. Later the compressed early wood recovers and raises the latewood above the surface.

Available sizes of timber

Commercially available sizes should be used wherever possible.

Sizes of softwoods

BS 4471 *Dimensions for softwood* Part 1: 1978 *Basic sections* specifies basic ranges of metric sizes as at 20 per cent moisture content – see table 17. Compared with the imperial sizes (which continue to be imported from Canada), the metric sizes are up to about 5 per cent smaller. On the other hand, the tolerances stipulated by BS 4471 are much

Normal sources	Thickness[2]	Width[2]								
	mm	75	100	125	150	175	200	225	250	300
Europe	16	×	×	×	×					
	19	×	×	×	×					
	22	×	×	×	×					
	25	×	×	×	×	×	×	×	×	×
	32	×	×	×	×	×	×	×	×	×
	36[1]	×	×	×	×					
	38	×	×	×	×	×	×	×		
	40[1]	×	×	×	×	×	×	×		
	44	×	×	×	×	×	×	×	×	×
	50	×	×	×	×	×	×	×	×	×
	63		×	×	×	×	×	×		
	75		×	×	×	×	×	×	×	×
North and South America	44[3]	×	×	×	×	×	×	×	×	×
	100		×		×		×		×	×
	150				×		×			×
	200						×			
	250								×	
	300									×

[1]These thicknesses are unlikely to be available.
[2]The sizes given are for 20 per cent moisture content. For every 5 per cent additional moisture content up to 30 per cent sizes must be 1 per cent greater and for every 5 per cent moisture content less than 20 per cent sizes may be 1 per cent less.
[3]Canadian commercial hemlock (*Hem-fir*)

Table 17   BS 4471: Part 1: 1969 Basic cross-sectional sizes of sawn softwoods at 20 per cent moisture content

tighter so that the modulus of section of the metric sections is not less.

Lengths: Basic lengths are from 1.800 to 6.300 m in increments of 300 mm although pieces more than about 5 m long are scarce and costly.

Permissible deviations on sizes as originally produced are:

	Minus mm	Plus mm	
Thicknesses and widths: not more than 100 mm more than 100 mm	1 2	3 6	Minus on not more than 10 per cent of the pieces in any parcel
Lengths	none	un-limited	

Re-sawing allowance Where smaller sizes are produced by resawing from larger sizes a maximum reduction of 2 mm is allowed on each piece. This reduction is not additional to the reductions allowed in table 18. Such pieces are designated 'resawn ex larger'.

Finished sizes Table 18 shows, for certain products, the maximum permitted reductions of sawn sizes by processing two opposed faces, applicable where reductions are not otherwise given in other relevant British Standards.

Regularizing is the sawing and/or machining of softwoods so that all pieces in particular batches are of uniform widths and thicknesses.

Reductions of 3 mm for pieces up to 150 mm wide and 5 mm for greater widths are subject to a permissible deviation of ± 1 mm. Reductions are not additional to the 'permissible deviations on sizes as originally produced or to the 2 mm allowance for resawing'.

Small resawn sections

BS 4471: *Dimensions for softwood* Part 2: 1971 (1977) gives finished dimensions – see table 19. Permissible deviations on the sections as ori-

Finished thickness	Finished widths mm				
	22	30	36	44	48
6	×		×		
14			×		×
17	×		×		×
22	×	×	×	×	×
30			×	×	×
36			×		×
44					×
48					×

Table 19 BS 4471:Part 2:1971 *Finished widths and thicknesses of small resawn softwood sections*

[1]A machine process by which thicknesses and width are made uniform throughout the length of a rectangular section of timber.

Purpose	Sawn widths and thickness mm				
	15–22	*more than 22–35*	*more than 35–100*	*more than 100–150*	*more than 150*
Constructional timber-surfaced			3.0	5.0	
Floorings[1]					
Matching interlocking boards[1]		4.0		6.0	
Planed all round					
Trim	5.0		7.0		9.0
Joinery and cabinet work	7.0		9.0	11.0	13.0

[1]Reduction in width is overall the extreme size including any tongue or lap

Table 18 Reductions of sawn softwood sizes to finished sizes by processing (planing) two opposed faces

ginally produced are from minus 0.5 to plus 3 mm, not more than 10 per cent of pieces in any parcel having a minus deviation in cross section.

Sizes of hardwoods

Where a large quantity is required, special sizes can be obtained from home grown or imported logs, but most hardwood is imported in the sawn condition in random widths and lengths as given in table 20.

In certain species *strips* and *shorts* are cheaper than larger sizes.

Strips, in this context, are generally 25 × 51 to 140 mm wide in lengths of 8 m upwards. In some cases 32, 38 and 51 mm thicknesses are obtainable.

Shorts are lengths 0.9 to 1.7 m. Generally 25 × 75 mm and wider and 32, 38 and 51 mm thick are available in some species.

BS 5450: 1977 *Sizes of hardwoods and methods of measurement* gives *Basic cross sectional sizes of sawn hardwoods* having 15 per cent moisture content, although not all the sizes are likely to be available in any one species – see table 21.

Permissible deviations from basic widths and thicknesses of hardwood sections vary from minus 1 to plus 12 mm.

Basic lengths of hardwood sections are multiples of 100 mm, but not less than 1 m. No minus deviations are allowed on lengths.

Rules are given for measuring random widths, and waney edged boards.

Reductions from basic sawn sizes for processing two opposed faces of hardwoods vary with the cross sectional size, process and product as shown in table 18.

For moisture contents above or below 15 per cent, because hardwood species vary considerably in their response to variations in moisture content

Source	Imported condition	Species	Widths[1] mm	Lengths[2] m
West Africa	logs	various	559 min dia	3.7–9.1
	sawn		152 min min average 229	1.8 min min average 2.7
Malaya	sawn	keruing	152 min min average 178	1.8 min min average 3.7
		ramin meranti	152 min min average 229	1.8 min min average 2.7
Japan	sawn	oak elm maple	152 min	1.8 min
North America	sawn	oak	152–356	1.8–4.9
		Canadian yellow birch	152 min	2.4 min
		rock maple	102 min	1.8 min

[1] Widths are measured to the nearest 10 mm
Lengths are measured to the nearest 100 mm

Table 20   Sizes of hardwoods (conventional)

Thickness	Width (mm)									
mm	50	63	75	100	125	150 and 175	200	225	250	300
19			×	×	×	×				
25	×	×	×	×	×	×	×	×	×	×
32			×	×	×	×	×	×	×	×
38			×	×	×	×	×	×	×	×
50				×	×	×	×	×	×	×
63						×	×	×	×	×
75						×	×	×	×	×
100						×	×	×	×	×

Table 21   Basic sizes of sawn hardwoods at 15 per cent moisture content

End use or product	Reduction from basic size to finished size[1]				
	For basic sawn sizes of width or thickness (mm):				
	15 to 25	26 to 50	51 to 100	101 to 150	151 to 300
	mm	mm	mm	mm	mm
Constructional timber, surfaced	3	3	3	5	6
Floorings, matchings and interlocked boarding, planed all round	5	6	7	7	7
Trim	6	7	8	9	10
Joinery and cabinet work	7	9	10	12	14

[1]Finished sizes after processing are allowed ±0.5 deviation

Table 22   Reductions of sawn hardwood sizes to finished sizes by processing (planing) two opposed faces

(see table 15), radial and tangential sizes should be calculated.

Grading of timber

Timber being a very variable material, freedom from undesirable characteristics, strength properties and appearance must be assessed for parcels of a particular species, and ideally for individual pieces.

Grading methods can be described as being either *quality grading* or *stress grading*. The latter systems are described on page 53.

Quality grading

These methods can be further divided into *traditional* and *BS 1186* systems.

Traditional quality grading

These systems of grading depend upon the appearance of pieces rather than upon their strength. The traditional systems for marketing ordinary sawn timber are complicated, but either a *defect* or a *cutting system* of grading is used.

Defects grading systems

Most softwoods are graded by exporters' *defects systems* which specify maximum sizes or degrees of defects for each grade in relation to the width and thickness of a board. Table 23 shows the approximate relationships between the grades of the main systems.

The 'cuttings' grading systems

Most hardwoods are graded by a *cutting system* in which each grade must provide a minimum fraction of the total plank area as *acceptable cuttings* measured in rectangular units of 308 × 308 mm. Also the minimum number and size of *cuttings* is specified – depending on the size of a plank. High grades require few and large cuttings which add up to a high proportion of the total plank area. There are three sets of hardwood grading rules applicable in the UK:

1 **BS 4047: 1966 (1980)** *Grading rules for sawn home grown hardwood*. This standard describes:
   Grades *1–4* under the *cutting system* and Grades *A–C* under the *defects system*.
2 The National Hardwood Lumber Association (NHLA) Rules.
   These rules are widely used in North America. The main grades are: *First, Second, Select* and

nos 1 and 2 Common. (Grades *3A* and *3B Common* and *Below grade* are not exported).

West African sawn hardwood graes are based on the higher NHLA grades and designated: *FAS (first* and *second), No. 1 common and select FAQ (fair average quality)* for logs.

3 The Asia-Pacific Regional Grading Rules for sawn hardwoods – other than teak.

These are four basic grades: *Prime, Select, Sandard* and *Serviceable,* the latter for local consumption only. Borer infested wood is graded under the defects systems.

In practice the rules are used as a general guide only and business is transacted on the basis of mutual understanding between the vendor and purchaser.

BS 1186 Grading for joinery timber

BS 1186 *Quality of timber and workmanship* comprises:

Part 1: 1986 *Specification for timber,* and Part 2: *Quality of workmanship.*

In 1952, BS 1186 recommended only, that the traditional grades shown in table 24 were suitable for joinery. Part 1 of the current standard, however, avoids all reference to commercial gradings but gives rules for:

Class 1 – high quality for specialized joinery.

Classes 2 and *3* – for general purpose joinery using species such as European Redwood and Whitewood or hardwood (not including sections up to 15 mm^2).

Class CSH — for 'clear' grades of softwoods and hardwoods.

Sections up to 15 mm^2 must be class 1 or CSH. The full specification should be consulted.

Rules include:

1 Sizes and distribution of knots relative to sizes of finished pieces of timber for various surfaces. Unsound, dead and loose knots are allowed only in concealed or semi-concealed surfaces.

2 Checks and shakes may not exceed

System		Grades and uses			
Norway, Sweden, Finland, Poland and Eastern Canada The rules are rather general and are interpreted differently by the various mills. Shipping marks indicate the quality and origin	—	*I* sometimes *II* marketed *III* as unsorted *IV* (U/S)	often marketed as unsorted (U/S)	*V* the grade generally used for carassing	*VI* *Utskott* *or* *wracks*
USSR	—	*I, II, III* usually marketed as *Unsorted (U/S)*		*IV*	*V* *wracks*
Brazil (Parana pine)	*no. 1 no. 2*	—		—	—
British Columbia and Pacific Coast of North America *(R list)* Separate grading systems are employed for specific end products, eg door stock and flooring	*no. 1 clear* *no. 2 clear* *no. 3 clear*	usually sold as *no. 2 clear and better*	*Select merchantable*[1] No. 1 merchantable[1]	*No. 2 merchantable*[1] sometimes used for carcassing in this country	*No. 3 common*

[1]Sometimes sold as a grade with a percentage of the grade below

Table 23   Approximate equivalent softwood grades

Redwood or whitewood	Swedish and Finnish	Better shipments of *Unsorted*
	Polish	Best shipments of *Unsorted*
	Russian	*Unsorted* White sea or Kara sea
Whitewood	Norwegian	Best shipments of *Unsorted* planed boards
Douglas fir and		*Clear and door stock;*
Western hemlock	North American	*Select merchantable*
Sitka (silver spruce) and		*Clears* and *Select merchantable*
Western red cedar		
Western white pine		*1st, 2nd, 3rd;*
		Log run (clears in, culls out)
Parana pine	South American	*Prime* (80 per cent *first*, 20 per cent *second* export grade)

Table 24   Traditional softwood grades for joinery

300×0.5 mm or more than one quarter of the depth of the piece.

3 Not less than six growth rings per 25 mm for external uses, and four internally, measured over 75 mm and 25 mm from the heart.

4 Slope of grain not more than 1 in 10 in softwoods and 1 in 8 in hardwoods.

5 Pitch pockets must be cut out and filled.

6 Boxed heart allowed in softwoods only, if there is no shake on exposed surfaces.

7 Sapwood, including discoloured sapwood is allowed in certain cases, but not in hardwood surfaces exposed to the weather.

8 All timber must be free from decay, and insect damage, except pinhole borer holes if filled in.

9 Laminating, finger jointing and edge jointing, which increase effective sizes and stability may be allowed if they are not too conspicuous.

10 Moisture movement characteristics of species are important. BS limits for moisture content are shown in figure 8, page 62.

11 The character of grain must be the same on all surfaces and matched as far as possible.

12 Plugs and inserts must not be wider than 12 mm.

DETERIORATION OF TIMBER

In favourable conditions timber remains in good condition indefinitely, but causes and effects of deterioration are summarized in table 25.

The causes of deterioration considered below are: weathering, fire, fungi, marine borers, insects.

Weathering

'Weathering' is an imprecise term used to describe the various effects resulting from the exposure of timber externally, the most obvious of which is loss of colour. All timbers become uniformly grey as a result of action of ultra-violet light and flowing water and are roughened by rain and wind-borne abrasives. Other effects may include colour changes caused by algae, moulds and chemical fumes, and cracks resulting from constant wetting and drying.

Fire

The behaviour of timber in fire is considered on page 39. Timber is combustible and evolves smoke and toxic gases. Most timbers have *Class 3 (medium) Spread of flame* (BS 476). Western red cedar, obeche, poplar and willow are the only commonly used woods weighing less than 400 kg/m^3 and therefore in *Class 4 (Rapid spread of flame)* – (BS 476). Surfaces char in fire at 0.5–0.7 mm per minute. On the other hand, unburnt timber retains its strength and there is negligible expansion in length, so that typical timber mem-

Cause		Effect
Fire		charring
Mechanical:	excessive loads	fracture
	abrasion	loss of surface
	erosion by rain, dust and sand	loss of surface
Water:	flowing water	leaching of soluble colour erosion of surface
	wetting and drying	expansion and contraction causing mechanical failure in forms of cracks
Sunlight		fading of colour embrittlement of surface
Chemicals – see page 51		discoloration
		complete disintegration
Bacteria		surperficial discoloration
Fungi – see page 72 (a) in moist conditions: moulds		superficial discoloration
'dry rot' *(merulius lacrymans)*		
(b) in wet conditions microscopic rots		complete disintegration (in advanced stages)
visible 'wet rots', eg, cellar fungus		
Marine borers, eg Teredo		tunnels
	Limnoria	tunnels (external)
Insects – see page 76 Termites (not UK)		irregular honeycombing or wide channels
	Beetles	tunnels and exit holes

The right-hand brace from "abrasion / loss of surface" down through "Chemicals – complete disintegration" is labelled 'weathering'. A second brace from "Bacteria" through "Fungi – moulds" is labelled 'weathering'.

Table 25   *Causes and effects of deterioration*

bers survive longer in fires than equivalent unprotected steel members.

When standard steel and laminated timber roof beams were subjected to a standard time/temperature curve test the steel beam sagged within minutes and collapsed in 30 minutes while the timber beam had deformed only 57.2 mm and at mid-span 75 per cent of the original section remained undamaged.

BS 5268: *Code of practice for the structural use of*

Species	Charring in 30 minutes[1] (mm)	Charring in 60 minutes[1] (mm)
All structural species listed in table 1, CP 112: Part 2: 1971 except those below:	20	40
Western red cedar	25	50
Oak, utile, keruing (gerjun), teak, greenheart, jarrah	15	30

(linear interpolation or extrapolation is permissible for periods between 15 and 90 minutes).
[1]The rates of charring for solid timber also apply to laminated timber members joined with: resorcinol, phenol, phenol-resorcinol, urea and urea-melamine formaldehyde adhesives.

Table 26   Rates of charring of timber

timber: Part 4: 1978 *Fire resistance of timber structures* enables the effective sizes of residual sections after charring in fire, to be estimated.

As timber beneath the charred layer does not lose significant strength, with known rates of charring, initial sizes of structural members can be calculated which would continue to support design loads after various periods of exposure to fire. Metal fastenings must be within the residual section and be protected, eg with timber of appropriate thickness, securely fixed.

Fungi

A reference is: *Decay of timber and its prevention*, HMSO. Fungi are simple plants without leaves or flowers, which consume 'ready-made' organic matter and therefore require no chlorophyll or sunlight. The spores (or seeds) are microscopically small but become visible in a mass as an extremely fine powder. They are produced in immense numbers and dispersed through air or sometimes by insects or animals. The hyphae which permeate wood are also minute and these are sometimes visible as masses of mycelium, although there are often no visible growths. Water is necessary for fungal growth but it is not true as has sometimes been stated, that repeated wetting and drying favour growth, in fact dry rot cannot live in such conditions.

Table 27 lists the four essentials for the growth of fungi and means of denying them, the more generally practicable of which are underlined.

In eradicating wood-destroying fungi it must be appreciated that they can survive for long periods at very low temperatures and in dry conditions. For example dry rot can survive for about 12 months and wet rot up to seven and a half years in dry timber at 20°C

Requirements for growth	Means of denial
Suitable 'food', ie timber. In all species growth is more vigorous in sapwood than in heartwood.'	Choice of heartwood only of 'durable' timber – see page 73. *Preservation of timber* – see page 78
Suitable moisture content of timber At least 20 per cent moisture content is required for colonization although fungi cannot live in saturated conditions	*Maintaining a low moisture content* Ponding, ie submersion in water
Suitable temperature Optimum temperature is about 23°C. Growth is about twice as rapid at 21°C as at 10°C	Very high temperature (Few fungi can grow at a temperature above 38°C) Very low temperature (growth stops at about 0°C)
Oxygen	Seal timber eg, in metal. Ponding, ie submersion in water

Table 27   Requirements for growth of fungi

Durability of timber

The term *durability* as applied to timber is generally synonymous with resistance to fungal attack.

The resistance of heartwood varies as much as 30:1 between the species, but because the sapwood of all species has poor resistance, classifications usually relate to heartwood only. It is important to realise that by no means all hardwoods are durable, so that merely to specify *hardwood* without further qualification can result in *perishable* timbers such as birch, beech or ramin being supplied.

In order to classify timbers in respect of durability the Princes Risborough Research Laboratory places $50.8 \times 50.8 \times 609.6$ mm heartwood specimens in soil leaving 228.6 mm above the ground and records their lives. (See FPRL Record no. 30.) Some classifications are given in table 28 and in table 32 pages 85–95. They can be used to forecast the performance of timbers of the same species and size where they are used in contact with the ground in this country, and their relative durabilities where they are used in less severe conditions.

The natural durabilities of timbers can be raised by *preservation*, generally more readily with the less naturally durable species see page 79.

BRED 296 is *Timbers: their natural durability and resistance to preservation treatment.*

Types of fungi

The more important fungi can be classified as either:
(a) Non-destructive fungi or (b) Destructive fungi.

(a) Non-destructive fungi

Moulds There are many species of moulds which indicate that surfaces are damp but do not damage timber. The black or greenish powder can usually be brushed off when it is dry. A mould of the penicillium type which stains oak pale yellow (hence *golden oak*), can grow at 43°C and may therefore present problems in seasoning.

Staining fungi These fungi feed mainly on the starch and sugar in sapwood cells but not on the cell walls, so that timber is not appreciably weakened. However, affected wood takes up moisture most readily and paint is sometimes pushed off external joinery. In softwoods *sap-stain* is usually grey-blue but other colours include blue, black, green, purple and pink, and more rarely brown.

(b) Destructive fungi or 'rots'

Each fungus has characteristic features and effects, but in general, wood is often discoloured and shows softening (*brashness*) and serious loss of strength, even where the fungus is at the incipient stage of growth. Shrinkage may be apparent as waviness in painted surfaces and later is cracking. Decayed timber loses weight and may be attacked by species of insects which would not bore into sound wood. It is very absorptive and ignites more readily.

It is important to distinguish between *true dry rot* and the *wet rots. Dry Rot in Buildings. Recognition, prevention and cure* PR Laboratory leaflet No. 6 (amended 1968) is a useful reference (the leaflet also deals with wet rots).

Dry rot This vigorous strand former is represented in this country only by *Serpula lacrymans* (formerly called *Merulius*) – see BRED 299 *Dry rot: its recognition and control.* Dry rot usually spreads more extensively, causes more damage and is more difficult to eradicate than the wet rots. The optimum moisture content for the growth of both dry rot and wet rots is 30 to 40 per cent, but unlike wet rots dry rot cannot colonize very damp wood. Dry rot will eventually die if the moisture content of timber falls below 20 per cent. It is generally found in timber which has been moistened, often by contact with damp brickwork, and in very stable conditions, but very rarely out of doors. It is killed at 40°C sustained for 15 minutes and is inactive at temperatures below 20°C.

When growing actively, in damp conditions cotton-wool-like masses develop. In drier conditions a grey skin with occasional patches of yellow or lilac forms.

Colonization of timber is by minute *hyphae* and these can spread over surfaces and through minute cracks in brickwork, plaster and other materials which provide no nourishment. The *strands* which are grey and vary from thread to pencil thickness convey nutrient materials which

Classification	Hardwoods	Softwoods
Very durable more than 25 years	Afrormosia Afzelia Guarea Greenheart Iroko Jarrah Makoré Muninga Opepe Purpleheart Pyinkado Teak	
Durable 15–25 years	Agba Chestnut, sweet Idigbo Mahogany, American Meranti, red Muhimbi Oak, European Utile	Western red cedar Yew Sequoia (Californian redwood)
Moderately durable 10–15 years	African walnut Avodiré Ayan Dahoma Gurjun, Indian Keruing, Malayan Mahogany, African Oak, Turkey Walnut, European	Douglas fir Larch, European and Japanese Pine, Caribbean pitch Pine, maritime Sequoia (home grown) Western red cedar (home grown)
Non-durable 5–10 years	Elm, Dutch Elm, English Elm, Wych Oak, American red Obeche 'Silver beech' Seraya, white Sterculia, yellow	Douglas fir (home grown) Fir, balsam Fir, grand (home grown) Fir, silver (home grown) Hemlock, western Pine, Scots (home grown)[1] Redwood (imported)[1] Spruce, Eastern Canadian Spruce, European Spruce, Sitka (home grown) Whitewood
Perishable less than 5 years	Abura Alder, common Ash, European Balsa Beech, European Birch, European Birch, yellow Horse-chestnut, European Lime, European Plane, European Ramin Sycamore Willow, crack Willow, white	

[1]Pinus sylvestris. (All sapwood is either non durable or perishable)
From Forest Products Research Record no. 30 *The Natural Durability of Timber 1959 (revised)*

Table 28   Durability classifications of heartwoods

support the hyphae in colonizing suitably moist timber. Occasionally the strands enable the hyphae to colonize nearby timber which would otherwise be dry but which is prevented from drying out by the atmosphere, perhaps by gloss paint. Here, the necessary moisture content is partly conveyed by the strands and partly synthesized by the fungus itself.

The *fruit bodies*, marked by wide shallow pores, carry minute rusty-red spores. Decay usually goes on out of sight but the *fruit body* tends to emerge into rooms in order to distribute its spores.

The presence of fungal decay may be detected by waviness in surfaces or by the softness of wood when it is prodded with a bradawl, but claims to identify dry rot by odour are of doubtful veracity.

Decayed wood is brown and divided by deep cracks both along and across the grain and the texture is dry and powdery.

Eradication of dry rot In order of importance the operations are:

1 Eliminate all sources of moisture which are supporting the rot.
2 Dry out the timber and building thoroughly. If this is done rapidly and dry conditions continue, growth will cease but normally rapid drying cannot be obtained and it is necessary to trace and destroy all growths, which would otherwise continue to grow.
3 In addition, visibly infected wood and about 300 to 500 mm of adjoining wood should be removed and burnt at once. With expert advice it may be possible to retain infected timber by boring holes at an angle to the grain and repeatedly filling them with preservative solution.
4 Non-combustible surfaces in the vicinity of the attack should be sterilized with a blowlamp, or better a brazing lamp, until they are too hot to touch. They should then be treated with a solution of 50 g sodium pentachlorophenate or sodium orthophenylphenate per litre of water. Any remaining sound timber should receive two or three coats of a preservative and new timber should be preserved, preferably by vacuum pressure methods.

Wet rots Like dry rot, wet rots are most active in wood with a moisture content of 30 to 40 per cent and they can continue to develop in timber with a moisture content down to about 21 per cent. The most common wet rots found in buildings are:

Cellar fungus (Coniophora cerebella) Decayed wood often resembles wood attacked by dry rot but it is darker and where surface cracks occur across the grain they are usually shallower. Often, however, decay is internal without any visible growths and the fruiting body, which is a thin skin with small rounded lumps or pimples olive green in colour darkening with time to dull olive brown, is rarely found in buildings. The strands, which are never thicker than thin twine, usually darken to dark brown or almost black but they sometimes fail to darken where they occur under impervious floor coverings. *Paxillus panuoides* is similar to cellar fungus but it has paler coloured strands; yellow or violet rather fibrous superficial growth and a small yellowish stalkless mushroom-like fruit body with gills on the underside. The wood is discoloured reddish-brown.

The rots so far described are *brown rots* which consume only the cellulose.

White rots break down both cellulose and lignin. The wood darkens at first but later becomes much lighter than normal, and 'lint-like'. When *white rot* occurs in patches it is known as *pocket rot*. Growth which is usually external by *Phellinus megalaporous* (or *cryptarum*) has often caused serious damage to oak in old buildings.

Soft rot is caused by micro-fungi which grow slowly from the surfaces in wet conditions, on cooling towers, bases of oak posts and on marine structures. It is not always easy to distinguish superficial soft rot from *weathering*.

Eradication of wet rots Defects giving rise to excess moisture must be corrected and the building must be dried. If the damp area can be dried rapidly it may only be necessary to replace seriously weakened timber and to treat lightly decayed timber with a combined fungicidal and insecticidal preparation. If rapid drying is not possible it is not necessary to sterilise brickwork or other masonry, but otherwise procedure should be as for dry rot.

Marine borers and wood destroying insects

An excellent reference is *Insect and marine borer damage to timbers and woodwork*, J D Bletchley, HMSO.

Marine borers

Marine borers such as *gribbles* a crustacean) and *teredo* (a mollusc) do not live in fresh water but they are extremely destructive, particularly in warm salt waters. No timber is immune from attack but resistant species include ekki, greenheart, opepe, pyinkado, teak and totara (a softwood).

Wood-destroying insects

Termites (incorrectly called 'white ants' although like them they have a highly complex social organization) are extremely destructive in the tropics and a serious problem in many of the warmer temperate countries. Fortunately termites do not live in the UK.

Beetles In the UK beetles are the chief pest, and they, together with wood wasps which sometimes emerge in buildings are discussed below.

Unlike wood-destroying fungi, beetles rarely cause structural failure. They are not so dependent upon moisture as fungi, although dampness favours attack. Few timbers are immune but normally only sapwood is attacked.

The stages in the life of a typical wood-destroying beetle are:

1. Eggs are laid in cracks or crevices, or in the case of the *powder post beetle* in pores, but not on polished or painted surfaces.
2. Eggs developed into *larvae* which are curved fleshy white grubs (*wood worms*) and enter the wood leaving holes too small to be seen with the unaided eye. They tunnel mainly in sapwood for one or more years leaving excreted *frass* behind them.
3. When fully grown the larva forms a chamber close to the surface of the wood, in which it becomes a *pupa*.
4. The pupa becomes a *beetle* and in the spring or summer 'eats its way through the thin outer skin of timber (and sometimes emerges through paint, linoleum and even lead sheet) leaving an *exit* or *flight hole* which is usually the first indication of beetle activity.
5. The beetle may fly. Mating takes place within a few days. Beetles can be identified by:

    (a) the beetle or larva
    (b) frass
    (c) flight holes.

In the UK the principal wood-destroying beetles in buildings are: (a) the common furniture beetle, (b) death watch beetle, (c) power post beetle and (d) the house longhorn beetle.

(a) *The common furniture beetle (Anobium punctatum)* This beetle attacks softwoods and hardwoods and is responsible for most of the damage occasioned by beetles to structural timbers and joinery in buildings in this country. It was often found in plywood bonded with animal and vegetable glues, particularly birch and alder (modern plywood is immune), and also in wickerwork. The eggs are minute and being laid in batches of only two or four, in cracks or crevices, they cannot be detected. The larva remains in damp wood for up to three years, and longer in dry wood, leaving frass which has a sandy feel. The beetle is dusty brown or blackish brown in colour and 3 to 5 mm long. It leaves an exit hole approximately 2 mm in diameter and might be seen in the early summer flying, or crawling on surfaces.

(b) *Death watch beetle (Xestobium rufovillosum)* This beetle is mainly confined to large sections of hardwoods such as oak that have been softened by decay (in the south but rarely in Scotland). The larvae, which are similar to those of the furniture beetle but larger when fully grown, leave a frass containing bun-shaped pellets which are visible to the naked eye. The beetle which is brown to chocolate brown and mottled in colour about 8.5 mm long, leaves a larger hole than the furniture beetle. It attracts a mate by rapid succession of tapping sounds. The life cycle is three years, and much longer in sound timber.

(c) *Powder post beetle (Lyctus brunneus)* This beetle attacks the sapwood of most hardwoods which have a high starch content and large pores in which the female can deposit her eggs. Starch content reduces with age and beech, birch and other fine-pored timbers are immune.

The bore dust is like flour to touch. The beetle which is reddish brown is about 4 to 5 mm long and leaves a flight hole about 1.6 mm in diameter.

(d) *House longhorn beetle (Hylotrupes bajulus)* The grub, which is up to 30 mm long, causes serious damage in the sapwood of seasoned softwoods, particularly in roof timbers. It is sometimes audible and bulging of timber surfaces often indicates the presence of tunnels near the surface. Frass takes the form of wood fragments and small cylinders of wood particles. The house longhorn beetle is black, 10 to 20 mm long, and leaves large oval shaped exit holes – usually far apart. The major diameter is 6.0 to 9.5 mm. The life cycle is 3 to 11 years, 6 years is a fair average. On the continent the beetle is established as a very serious pest, but in this country it is mainly confined to the northern parts of Surrey where the Building Regulations 1985 require roof timbers to be treated with certain specified preservatives. Because the beetle could become the most serious wood-destroying pest here, any infestation should be reported to the Princes Risborough Laboratory of the BRE.

Ptilinus pectinicornis This insect is less important than those so far described but it sometimes found with furniture beetles attacking hardwoods in outbuildings and in fine grained hardwood flooring such as maple. The beetle has interesting antennae, comb-like in the male and saw-like in the female, and the latter is unusual in that it bores tunnels in which to lay its eggs.

Ambrosia beetles (pinhole and shot-hole borers) These insects of the *Scolytidae* and *Platypodidae* families damage tropical trees soon after felling but do not persist in seasoned timber and damage is never of structural significance. The female beetle bores straight tunnels usually at right angles to the grain and in so doing introduces fungal spores. The larva does not tunnel (there is no frass) but feeds on the fungal growth which sometimes stains fair-sized areas of wood beyond the tunnels. Exit holes and tunnels are 0.5 to 3 mm in diameter. The smaller holes can usually be filled so they are not even seen, even in polished timber.

Bark borers (Ernobius mollis) This beetle attacks the bark and the outer sapwood of softwoods. Its tunnels may be seen in waney edges of converted timber and sometimes beetles emerge through finished surfaces.

Weevils *(Pentarthrum and Euophryum spp)*

Boring weevils are usually found attacking very damp and decayed wood. They are small, brownish and have a snout-like projection of the head. The damage resembles that of the common furniture beetle but tunnels are bored by both weevil and larva, they are smaller and the frass is darker and finer, containing pellets which are more nearly circular in shape. Flight holes are irregular and about 0.8 mm across.

Wood wasps

Female wasps of the *Sirex spp*, including the black and gold *Sirex Gigas* which is up to 44.5 mm long, lay their eggs deeply in the wood beneath the bark of softwoods. They sometimes emerge from recently installed timber, but they do not breed in buildings.

Eradication of wood-destroying insects

If it is not certain that an insect infestation is already extinct, treatment should be undertaken as soon as possible and it should be repeated until no more exit holes or frass appear. Widespread infestation should be dealt with by specialists. Seriously attacked timbers should be removed and burnt and replaced with timber which has been treated to prevent attack. Timbers which are to remain in position should be liberally brushed or sprayed with a liquid insecticide and it should be fed generously into open joints, splits and shakes in order to kill eggs, newly hatched larvae and sometimes emerging beetles.

Surface applications also prevent further eggs being laid. Insecticide should be injected into exit holes to kill eggs which may have been laid there, and larvae in the tunnels.

A *woodworm fluid* should contain one of the powerful chlorinated hydrocarbon insecticides such as dieldrin or gamma BHC in a penetrating organic solvent. Proprietary formulations also contain preservatives against fungal decay. Home made preparations are not recommended.

It is important to remember that insecticides may be injurious to humans and animals, and should not be tasted, inhaled or allowed to remain on the skin. Also initially, they present a fire hazard.

PRESERVATION OF TIMBER

This section deals with the treatment of timber with toxic chemicals to protect it from attack by both fungi and insects, rather than with specific treatments for the eradication of established fungal or insect attacks.

Information and advice is available from the Princes Risborough Laboratory of the BRE and from the British Wood Preserving Association, 62 Oxford Street, London W1.

Important references include:

BS 4261:1985 *Glossary of terms relating to timber preservation.*

BS 4079:1966 *Plywood for marine craft* is treated against attack by marine borers.

BS 1282:1975 *Classification of wood preservatives and their methods of application.*

BS 5268: *Code of practice for the structural use of timber* Part 5:1977 *Preservative treatments for constructional timber.* This BS assesses the risks of fungal decay or insect attack during the service life of each of the main structural elements in buildings.

BS 5589:1978 is a *Code of practice for preservation of timber* which includes recommendations for external woodwork, in buildings, fencing and timber in contact with sea or fresh water.

BS 5666: *Methods of analysis of wood preservatives and treated timber* (many parts). It emphasises that all preservatives are potentially toxic to man and the environment, and that when treated timber is mechanically worked the dust is dangerous.

In new buildings protection of wood against insects is usually less justified than against fungi, except where abnormal risks obtain, as in the tropics against termites and in the part of South East England where the Building Regulations require softwood roof timbers to be protected against the House Longhorn beetle.

Success in preservation depends on the timber species, the size and condition of the specimen, the effectiveness of the preservative and the resulting depth of penetration and the amount which is retained.

Where timber may be damp in service, as in the case of external joinery, even if it is painted, either an inherently durable timber or a less durable, but preservative-treated, timber must be used. Incidentally the latter often has a lower cost.

In some cases the cost advantages of preservation are obvious, for example non-durable timber telegraph posts which have been impregnated with preservative often remain sound for more than sixty years, whereas untreated poles would decay after a few years. Good construction in buildings should not normally put even *perishable* timbers at risk, but with the smaller sizes and increased sapwood content of structural timbers today, a cost of preserving them of less than one per cent of the total cost of building a typical house is justified. BS 5250:1975 *Code of basic design data for the design of buildings : the control of condensation* recommends that all constructional timber in roofs should be pressure impregnated with preservatives. BS 5268:Part 5:1977 identifies four categories of need for preservation of structural timbers, which this author interprets in table 29.

The code also gives guidance on the preservatives and methods of treatment appropriate to various durabilities and conditions of use.

Under the Building Regulations 1976 the heartwood of the Western red cedar and Sequoia (Californian redwood) and of many hardwoods could be used in their natural state for external boarding but Douglas fir, Western hemlock, European and Japanese larch, Redwood (European) or Scots pine, Sitka spruce and Whitewood (or European spruce), Abura and Elm must be subject to prescribed preservative treatments additional to any protection which they may be given by paint.

Conditions of use	Preservation
A Negligible risk, even for timbers with low inherent resistance to biological degradation or where the consequences of attack are either acceptable or make the cost of preservation generally unfavourable	Unnecessary
B Low risk or where remedial action is simple	An 'insurance' against the cost of repairs
C Where experience has shown that there is an unacceptable risk of decay, whether due to the nature of the design or the standard of workmanship and maintenance. Where there is a substantial risk of decay or insect attack which, if it occurs, would be difficult and expensive to remedy	Necessary
D Where timber cannot be protected by design from a continually hazardous environment. Where there is a high risk of decay or insect attack in structures the collapse of which would constitute a serious danger to persons or property	Essential

Table 29   Need for preservation of structural timbers

PRL Technical note 24, August 1967 stated that Baltic redwood sapwood is liable to decay whenever it remains wet for an appreciable time, and windows, most of which are made in this wood, sometimes decay within 5 or 6 years. As it is generally neither practicable to use only heartwood, nor realistic to rely on good standards of building practice and of maintenance, the PRL strongly recommended that all *non-durable* softwoods and hardwoods used for painted external joinery should receive preservative treatment. As the conditions here are not so severe as for timber in contact with the ground or for timber which is in the open and unpainted, 'it is believed that the lighter type of treatment such as immersion will give sufficient protection, although a higher factor

of safety will be obtained from the heavier impregnation treatments'.

BRE Digest 73 deals with *Prevention of decay in external joinery*.

Absorption of preservatives

Permeability of species varies widely and the Princes Risborough Laboratory places them in four classes some examples of which are given in tables 31 and 32. It will be seen that resistances of heartwood to preservation processes are not directly related to natural durabilities – see BRED 296.

Preservatives penetrate most readily into the end grain of timber and penetration is usually greater into quarter-sawn than flat-sawn surfaces. Sapwood is nearly always easier to treat than heartwood. For example the sapwood of oak is *permeable* but heartwood is extremely resistant to treatment.

It is impossible to force in any more liquid into wet timber so it is essential that moisture content is reduced to about 28 per cent before preservation treatment is attempted. Timber, particularly for surface treatment, must be surface dry and clean.

Treatment of plywood

Effective preservative treatment of plywood is possible if the wood is sufficiently permeable and provided the glue can withstand the temperature and moisture conditions during the process. BS 5268 includes recommendations for the preservative treatment of structural plywood, although the technology being less advanced, they are less detailed than those for timber – see also page 98.

Methods of preservation

References include:
BS 5268 *CP for the structural use of timber:* Part 5:1977 deals with *Preservative treaments for constructional timber*
Specifications of the British Wood Preserving Association (BWPA).
BRE Digest 201 describes *Wood preservatives: application methods.*
A BRE report is: *Methods of applying wood preservatives* by D F Purslow, HMSO.

Conditions	Minimum durability classification of untreated timber	Preservative treatment for timber of any durability classification
Permanently 'dry' eg internal joinery	Sapwood or heartwood of any timber	None required against fungi
Intermittent dampness: eg timber exposed to weather[1] or condensation 1 where it can redry rapidly, eg unpainted weather boarding	Heartwood only of moderately durable timbers[2]	Brush or spray repeated every few years, dipping subsequently brush or spray treated, or steeping
2 where rapid drying is not possible	Heartwood only of durable or very durable timbers[2]	Steeping or pressure impregnation
Permanent dampness: eg timber in contact with the ground or where dampness may arise by accident and the cost of repair would be high	Heartwood only of very durable timbers[2]	Pressure impregnation

[1] The only untreated timber permitted by the Building Regulations 1976 for external wall boarding (whether painted or not) was heartwood of certain moderately durable (or better) hardwoods, Western red cedar and sequoia (Californian redwood)

[2] These timbers may be difficult to obtain and more costly than timber treated with preservative

Table 30   Practical application of durability data

Processes vary from superficial treatments of limited protective value, to pressure impregnation. As, however, complete penetration is rarely obtained even by pressure impregnation all work in cutting should be performed before treatment or where this is not practicable the exposed surfaces, especially end grain, should be dealt with as a separate operation.

Processes are now discussed in ascending order of effectiveness:

1 Brush and spray

The liquid should be flooded over surfaces so they absorb as much as possible. Externally, the treatment must be repeated every two or three years.

2 Deluging, dipping and steeping

Organic solvent type preservatives are usually employed but any preservative can be used. Preheating of suitable types assist penetration. Dipping should be for at least 10 seconds for very small sections; a 3-minute dip is often specified. Large sections should be dipped for 10 minutes or more. The Building Regulations 1976 required complete immersion for at least 10 minutes in certain specified preservatives as protection against house longhorn beetles in certain areas.

Deluging on a production line is usually equivalent to a short dip, Steeping of permeable timbers for several days may give quite deep penetration and protection sufficient even for timber which will be in contact with the ground.

Resistant timbers however show only surface penetration even after several weeks.

3 The hot-and-cold open tank method

This simple method gives protection to sapwood

	Heartwood	Sapwood
Permeable (P) May be completely pressure-treated and heavily impregnated by the open tank method	Beech (UK) **P** Birch (European) **P**	Scots pine (UK) Yellow pine (Canada) Redwood (Russia, Finland) Beech (UK) Oak (European)
Moderately resistant (MR) Fairly easily pressure impregnated to give 6 to 19 mm lateral penetration in approx 2 to 3 hours	Scots pine (UK) **ND** Yellow pine (Canada) **ND** Redwood (Russia, Finland) **ND** Elm (UK) **ND**	Douglas fir (UK) Larch Sitka spruce
Resistant (R) Difficult to pressure impregnate and require lengthy treatment. Often maximum penetration possible is 3 to 6 mm. Incising of surfaces may increase depth of treatment	Douglas fir (UK) **ND** Western red cedar (UK) **ND** Spruce (UK) **ND**	Western red cedar (UK)
Extremely resistant (ER) absorb only a small amount after prolonged pressure treatment. Virtually impenetrable laterally	Afrormosia (Guana) **VD** Mahogany (Uganda, Nigeria) **MD** Oak (European) **D** Teak (Burma) **VD**	

[1] Durability classification of heartwood:

P	perishable	**D**	durable
ND	non-durable	**VD**	very durable
MD	moderately durable		

Table 31   Absorption of preservatives. Information from PR Laboratory

and permeable heartwood comparable to that given by pressure impregnation but is usually confined to the treatment of fence posts. The timber is submerged in a tank of suitable preservative which is then heated to between 80 and 90°C and kept at that temperature for several hours. Absorption by the timber must then be allowed to take place as the liquid cools.

4 Pressure impregnation

This process provides the deepest penetration and is essential for long life of most timbers in direct contact with the ground, in sea water and in similar environments. The timber is sealed in a pressure vessel, air is removed under vacuum; preservative is forced in under strict control and a second vacuum stage removes excess liquid. More than two hundred specialized plants are available in this country. The process is described in BS 913:1973 *Wood preservation by pressure creosoting* which specifies minimum pressure periods and minimum net retentions. Whenever possible all sapwood must be penetrated, but species such as beech and birch, which are both perishable and permeable, must be fully penetrated. Before treatment, resistant timbers, including Douglas fir, for uses such as bridge decking, more than 76 mm thick, must be *incised* to a depth of 20 mm, the knife cuts to be in the direction of the grain and 25 mm apart, staggered and in rows at 60 mm centres.

BS 4072:1974 *Wood preservation by means of water-borne copper/chrome/arsenic compositions* includes descriptions of composition of preservatives and methods of application by vacuum/pressure or pressure impregnation.

5 The diffusion process

This is the only means by which certain water-borne preservatives can be introduced into green timber. After spraying or dipping in a concentrated boron salt solution the timber is close piled for several weeks during which the preservative diffuses into the timber. However, the salts are water soluble and the process is unsuitable for timber which will be exposed externally.

6 Plug inserts

Preformed plugs can be effective inserted in drilled holes, eg on the bottom rails of sashes.

7 Injection

Rentokil Ltd inject preservative into pre-drilled holes in existing joinery – in particular window cills.

Holes are drilled at intervals, suitably formulated fungicide in injected through non-return valves and the holes are then capped or plugged.

Preservatives

In addition to penetrating well, and protecting timber from attack by fungi and insects over long periods, preservatives may be required to be non-toxic to plants and animals, odourless, free from detrimental effects on adhesives, paints and polishes and generally they must not 'bleed' or be washed out by rain. Water repellents in some preservatives reduce the movement of timber and its tendency to crack and allow water to enter leading to decay of the unprotected interior.

Care should be taken in using preservatives, eg goggles should be worn when spraying them.

BS 1282:1975 *Classification of wood preservatives and their methods of application* lists the following types:

TO Tar-oil types

TO 1 Coal tar creosote BS 144:1973, for pressure impregnation

TO 2 Coal tar creosote BS 3051:1972 for brush application.

Good quality creosote is a very effective and inexpensive preservative which is resistant to leaching and particularly suitable for use externally. Tar-oil types are not readily flammable and after a few months weathering they probably do not increase the fire hazard. They are not generally corrosive to metals. On the other hand they are difficult to paint over satisfactorily, and inclined to 'bleed' and to discolour adjacent porous materials. Its odour makes it unsuitable internally, especially near food, and the preservative injures some forms of plant life.

OS Organic solvent types

These consist of the following preservative substances in organic solvents.

OS 1 chlorinated naphthalenes and other chlorinated hydrocarbons

OS 2 (a) copper naphthenate (green in colour) BS 5056:1974 describes preservatives for application by double vacuum or for use on small absorbent timbers for a wide range of environments and hazards.
(b) zinc naphthenate

OS 3 pentachlorophenol and its derivatives or mixtures of organic solvent types.

BS 5707 is entitled: *Solutions of wood preservatives in organic solvents*. The parts are:

Part 1: 1979 *Solutions for general purpose applications including timber that is to be painted.*

Part 2: 1979 *Pentachlorophenol solution for use on timber that is not to be painted.*

Part 3: 1980 *Methods of treatment.*

Tri-butyl-tin oxide *will doubtless be included in future revision of the specification.*

The organic solvents may be volatile or relatively non-volatile petroleum fractions. Specific insecticides and water repellant materials can be incorporated. Organic solvent preservatives do not corrode metals, they are suitable both internally and externally and penetration is superior to that of creosote; preserved timber can usually be painted when the volatile solvents have evaporated and the wood is then no more flammable than untreated wood. Some are colourless, although pigments may be added to indicate the extent of their penetration or for decorative purposes. Some have an odour which taints food. Although they are costly, organic solvent preservatives are now being widely used for vacuum impregnation of mass produced external joinery.

WB   Water-borne types

These consist of the following preservative substances dissolved in water:

WB 1   copper-chrome

WB 2   copper-chrome-arsenic (the most commonly used type) BS 4072:1974 describes wood preservation with water-borne preservatives of this type.

WB 3   fluor-chrome-arsenate-dinitrophenol

WB 4   others, eg copper sulphate, sodium fluoride, sodium pentachlorophenate, zinc chloride, organic mercurial derivatives.

Water-borne preservatives have the advantages that they can be over-painted, do not 'bleed' when they are dry, are odourless and non-combustible, but after treatment normal seasoning is necessary. Those containing two salts become insoluble in water. Apart from some of the WB 4 class they are not corrosive. Water-borne preservatives are widely used for pressure impregnation and deep penetration is obtained in permeable timbers.

Properties of some common timbers

Table 32 which follows gives properties of, and suggests uses for, some common timbers.

Colour (1)	BS name	Latin name	Hardwood (H) or Softwood (S) / density (2)	Origin	Natural durability (3) / Resistance to impregnation with preservatives (4)	Texture
BROWN light *continued*	Agba	*Gossweileroden-dron balsamiferum*	H/m	W Africa	D/R	Medium
	Beech	*Fagus spp*	H/h	Europe Japan	P/P	Fine
	Cedar–English	*Cedrus spp*	S/m	England	D/R‡	Fine
	Chesnut–sweet	*Castanea sativa*	H/m	Europe	D/Ex	Medium
	Elm	*Ulmus spp*	H/m	Europe, USA Japan	N/M (Europe)	Medium
	Loliondo	*Olea welwitshii*	H/vh	E Africa	M/Ex	Medium/fine
	Oak	*Quercus spp Q robu and Q Sessliflora (UK)*	H/h	Europe America Japan	D/Ex† (Europe/ America)	Coarse
	Olive–East African	*Olea hochstetteri*	H/vh	E. Africa	M/–	Fine
	Tasmanian 'oak'	*Eucalyptus spp obliqua*	H/h	SE Australia	M/R	Coarse/medium
	White seraya	*Panashorea spp*	H/m	Sabah	N/R or Ex	Medium
	Western hemlock	*Tsuga heterophylla*	S/m	Canada USA	N/R	Medium/fine
	Yellow pine	*Pinus strobus*	S/l	E Canada	N/M†	Fire

Table 32   Properties of some common timbers classified by approximate colours

rking properties	Uses (6)			Price range (7)	Remarks (1)
ood	Ext J F Int J* P	Flg n/P	V	Medium	Appearance resembles light mahogany; has wide sap-wood band; exudes gum and resinous odour some-times present Very stable
ood exceptionally good steam bending properties	Int J F	Flg h/P l/I	V	Medium – imported Low – home grown	Becomes deep reddish brown when steamed. Relatively hard Shows medullary rays. Large movement Attracts common furniture beetle Availability very good
ood	Int J F		V	Medium	Scented timber mainly used for moth-proof furniture linings; sometimes extremely knotty
ood good bending properties	Int J* F	Flg	V	Medium	Colour and texture resembles oak and it also corrodes and is stained by iron; but it is easier to work and has no silver grain; good grades in short supply
edium	Int J F	Flg	V	Low/medium	High resistance to splitting
ood/medium	F	Flg h/P l/I		Medium/high	Similar to East African olive (see below); highly resistant to abrasion
edium/difficult good bending properties	Ext J F Int J P	Flg h/P	V	Medium/high	Corrodes and is stained by iron; quarter sawn timber shows characteristic silver grain; sapwood sometimes highly susceptible to lyctus attack; Japanese oak is paler in colour and milder in texture
ifficult		Flg h/I	V	Medium/high	Flooring equal to rock maple in resistance to heavy traffic
edium	F	Flg l/I	V	Medium/high	Resembles English oak, but is stronger in bending, tougher and stiffer
edium	Int J F	Flg l/I		Low	
ood	Ext J F Int J* Str (S2)	Flg l/P		Low	Widely used in UK for carcassing; moderately soft; excellent for staining or polishing but if incorrectly seasoned inclined to distort and to show grain
ery good	Int J patterns			High	Excellent quality; sometimes pinkish; very stable, soft even texture; suitable for carving and pattern making, but inferior grades mainly imported

continued . . .

Colour (1)	BS name	Latin name	Hardwood (H) or Softwood (S) / density (2)	Origin	Natural durability (3) / Resistance to impregnation with preservatives (4)	Texture
WHITE	Ash	Fraxinus excelsior	H/h	Europe	P/M	Coarse
	Birch	Betula spp	H/h	Europe	P/P	Fine
	Maple rock,	Acer saccharum	H/h	Canada USA	N/R	Fine
	Obeche/ Wawa	Triplochiton scleroxylon	H/l	W Africa	N/R*	Medium
	Sycamore	Acer pseudo-platanus	H/m	Europe	P/P	Fine
	Whitewood/ European spruce	Picea abies	S/l	Europe	N/R	Fine
yellowish	Afara/Limba	Terminalia superba	H/m	W Africa	N/M	Medium
	Avodiré	Turraeanthus africanus	H/m	W Africa	N/Ex	Medium
	Ramin/ Melawis	Gonystylus spp	H/h	Sarawak Malaya	N/P	Medium
	Western white pine	Pinus monticola	S/l	Canada USA	N/M	Fine
BROWN	Abura	Mitragyna ciliata	H/m	W Africa	N/M*	Medium/fine
light	Afrormosia	Pericopsis elata	H/h	W Africa	V/Ex	Medium/fine

Table 32   Properties of some common timbers classified by approximate colours

Working properties (5)	Uses (6)			Price range (7)	Remarks (1)
...ood very good bend- ing properties	Int J F		V	Medium	Extremely tough and flexible; used for parallel bars. In short supply
...ood good bending properties	Int J F P	Flg l/P	V	Low	Small logs only, therefore mainly used for plywood
...edium good bending properties	Int J	Flg h/P h/I		Medium	Good resistance to abrasion; may show colour markings. Turns well
...ood	Int J F P			Low	Rather too soft for joinery subject to wear; not suitable in damp conditions; sapwood sometimes highly susceptible to blue stain, pinhole borer and lyctus attack
...ood excellent turning	Int J P	Flg	V	Medium	Some has attractive figure
...ood	Int J Str (S2)	Flg		Low	Has natural lustre; liable to contain resin pockets and hard dead knots
...ood	Int J	Flg l/P	V	Medium	Sapwood sometimes highly susceptible to blue stain, pinhole borers and lyctus attack. Very stable
...edium	Int J F		V	Medium	Quarter sawn wood often mottled
...edium	Int J F	Flg n/P		Medium	Pale colour; subject to blue stain
...ood	Ext J Int J			High	Stable and suitable for good class joinery but exclusion of large knots involves high wastage; sometimes pinkish
...edium/difficult	Int J F	Flg l/P		Low	Plain appearance. Variable in colour with a tendency to be pinkish
...edium	Ext J F Int J* Str	Flg n/P	V	Medium/high	Similar to teak but is not oily; tends to darken on exposure and to corrode and be stained by iron Sapwood clearly distinguished Small movement Availability reasonable

continued ...

Colour (1)	BS name	Latin name	Hardwood (H) or Softwood (S) / density (2)	Origin	Natural durability (3) / Resistance to impregnation with preservatives (4)	Texture
BROWN light *continued*	Teak	*Tectona grandis*	H/h	Burma and Thailand	V/Ex	Coarse/medium
yellowish	African 'walnut'	*Lovoa trichilioides*	H/m	W Africa	M/Ex	Medium
	Muhuhu	*Brachylaena hutchinsii*	H/vh	E Africa	V/R	Fine
reddish	Afzelia/Doussié	*Afzelia africana and spp*	H/vh	W and E Africa	V/Ex	Coarse/medium
	'Cedar' – Central and South American	*Cedrela spp*	H/m	Central and South America	D/Ex	Coarse/medium
	Cherry	*Prunus avium*	H/m	Europe and Japan	M/M	Fine
	Danta	*Nesogordonia papaverifera*	H/h	W Africa	D/R	Medium
	Gedu nohor/ Edinam	*Entandrophragma angolense*	H/m	W Africa	M/Ex	Medium
	Guarea	*Guarea spp*	H/m	W Africa	D/Ex	Medium
	Agitong Gurjun/ Keruing Yang	*Dipterocarpus spp*	H/h	India, Malaya, Thailand and Philippines	M/R	Medium
	Kapur (Borneo camphorwood	*Dryobalanops spp*	H/h	Malaya Sabah	V/Ex	Medium
	Mahogany – African	*Khaya spp*	H/m	W Africa	M/Ex	Medium
	Mahogany – American	*Swietenia macrophylla*	H/m	Central and South America	D/Ex	Medium/fine

Table 32   Properties of some common timbers classified by approximate colours

Working properties (5)	Uses (6)			Price range (7)	Remarks (1)

Medium
  hard on cutting
  edges
	Ext J F
	Int J*
	P
		Flg
		n/P
			V	High/very high	Price varies considerably with size; contains oil; has stained masonry. Golden brown, darkens on exposure. Very stable

Medium	Ext J F Int J	Flg l/P	V	Medium	Irregular dark streaks tending to brown, but colour fades. A most attractive timber.
Difficult		Flg h/I		Medium	Sapwood lighter colour; usually in short lengths. Limited supply
Medium/difficult	Ext J Int J*	Flg n/P		Medium	Similar performance to teak but rather heavier and harder; not easy to glue; exudes yellow dye when wet
Good	Int J F			Medium/high	Similar appearance and working properties to a softish Honduras mahogany; limited supplies
Medium	Int J F		V	Medium	Chiefly used for high class work
Good		Flg h/P, l/I		Medium	Sometimes dark brown Hard wearing
Medium	Int J F	Flg l/P	V	Medium	Slightly inferior to sapele and utile Lower price
Good	Ext J F Int J* P	Flg n/P	V	Medium	Of same type but less stable than mahogany; gum exudation may present difficulty in painting
Medium	Ext J Int J*	Flg h/P		Low	Exudes gum; difficult to paint
Medium	Ext J Int J	Flg		Low	Has pin holes Stains in contact with iron
Medium very poor bending properties	Ext J F Int J P	Flg l/P	V	Medium	Wide differences between types in different areas; some woolly and have interlocking grain Small movement. Availability—good
Very good	Ext J F Int J patterns		V	Very high	Becomes golden on exposure to sunlight; very stable; ideal for pattern making. Availability good

continued . . .

Colour (1)	BS name	Latin name	Hardwood (H) or Softwood (S) / density (2)	Origin	Natural durability (3) / Resistance to impregnation with preservatives (4)	Texture
BROWN reddish continued	Plane	*Platanus acerifoia*	H/m	Europe	P/–	Fine
	Rhodesian 'teak'	*Baikiaea plurijuga*	H/vh	Rhodesia and Zambia	V/Ex	Fine
	Sapele	*Entandrophragma cylindricum*	H/m	W Africa	M/R	Medium/fine
	Sequoia (Californian redwood)	*Sequoia sempervirens*	S/l	California	D/M	Fine/coarse
	Utile	*Entandrophragma utile*	H/h	W Africa	D/Ex	Medium
pinkish	Douglas fir British Columbian pine	*Pseudotsuga taxifolia*	S/m	Canada and USA	M/R	Medium/fine
	Gaboon	*Aucoumea klaineana*	H/l	Equatorial Africa	N/M	Medium
	Niangon	*Tarrietia utilis*	H/h	W Africa	MD/R	Coarser than African mahogany
	Pitch pine	*Pinus palustris P. caribaea and spp*	S/h	USA and Central America	M/M (P. caribaea) R (P. palustris)	Fine
	Scots pine (home grown) Redwood (imported)	*Pinus sylvestris*	S/m	Europe	N/M	Fine

Table 32   Properties of some common timbers classified by approximate colours

Working properties (5)	Uses (6)			Price range (7)	Remarks (1)
Good	Int J		V	Medium	Quarter sawn timber is known as 'lacewood'
Difficult		Flg		Medium	Sometimes marked with irregular black lines or flecks; sapwood pale in colour; stains in contact with iron
Good	Ext J F Int J P	Flg n/P	V	Medium	An alternative to mahogany; has attractive stripe figure where quarter sawn
Good	Ext J F Int J P	Flg n/P			Straight grained: non resinous Appearance resembles Western Red Cedar with contrasting early wood and latewood. Near white narrow sapwood. Quality varies
Good	Ext J F Int J P	Flg n/P		Medium	Similar to sapele but straighter grained and more stable
Good	Ext J Int J Str (S1) P	Flg 1/P	V	Medium	Clear grade; free from knots; gives excellent finish but inclined to show grain; corrodes iron; flooring should be rift sawn
Medium	P			Low	Tones to light pinkish brown on exposure; low strength; fine sheen when sanded; rarely imported in the solid
Medium	Ext J F Int J			Medium	Similar to African mahogany; sapwood is paler than heartwood; has a greasy feel
Good	Ext J Int J* Str (S1)	Flg		Low/medium	Resinous; does not readily accept paint and inclined to show grain; in short supply
Good	Ext J Int J Str (S2) P	Flg 1/P		Low	Nomenclature see page 45—main timber used in UK for joinery and cladding. Earlywood and latewood distinct colours and sapwood lighter. Remarkably wide distribution of growth areas causes variations in texture, density and number and size of knots. Home-grown Scots pine slightly superior mechanical properties but sapwood is wider

continued . . .

Colour (1)	BS name	Latin name	Hardwood (H) or Softwood (S) / density (2)	Origin	Natural durability (3) / Resistance to impregnation with preservatives (4)	Texture
BROWN dark	American walnut (black walnut)	*Juglans nigra*	H/h	E USA and E Canada	D/Ex†	Rather coarse
	Ekki	*Lophira alata*	H/exh	W Africa	V/Ex	Coarse
	Iroko/ Mvule	*Chlorophora excelsa*	H/h	W Africa and E Africa	V/Ex	Medium
	Muninga	*Pterocarpus angolensis*	H/m	E Africa	V/R	Medium
	Panga panga	*Millettia stuhlmannii*	H/vh	E Africa	V/Ex	Rather coarse
RED	Jarrah	*Eucalyptus marginata*	H/vh	S W Australia	D/Ex	Coarse/medium
	Makoré	*Tieghemella heckelii*	H/m	W Africa	V/Ex	Fine
	Meranti/Seraya	*Shorea spp*	H/m	Malaya and Sabah	M–D/R or Ex	Medium
Blood red to dark brown	Padauk	*Pterocarpus spp*	H/vh	W Africa Burma	V/Ex	Coarse/medium
			H/h	Andamans	V/M	
YELLOW	Idigbo	*Terminalia ivorensis*	H/m	W Africa	D/Ex	Medium
ORANGE	Opepe	*Nauclea diderrichii*	H/h	W Africa	V/M	Medium
VARIE- GATED COLOURS	Australian blackwood	*Acacia melanoxylon*	H/h	Australia	D/Ex	Medium

Table 32   Properties of some common timbers classified by approximate colours

Working properties (5)	Uses (6)			Price range (7)	Remarks (1)
Good	Int J F		V	High	Sapwood light colour
Difficult	Heavy construction	Flg		Medium	Extremely heavy
Medium/difficult due to calcareous deposit	Ext J Int J P	Flg n/P	V	Medium	Initially lightish yellow; although not as strong as teak it is often used as a cheap substitute for that timber. Small movements Availability good
Good	Ext J	Flg n/P	V		
Difficult		Flg		Medium	Alternate bands of dark and light colour
Difficult	Ext J	Flg n/P	V	Medium	Pale sapwood, fire resistant Used in marine works Limited supply
Medium/difficult dust sometimes irritant	Ext J F Int J* P	Flg n/P	V	Medium	Corrodes iron Requires tipped tools for cutting
Good	Ext J Int J	Flg l/P		Low/medium	Good general purpose joinery wood; apt to contain pin holes. Darker, heavier, harder, stronger and more durable material also available Availability good
Difficult	Ext J F		V	High	
Medium	Ext J Int J*	Flg n/P	V	Medium	Appearance resembles oak and is used in lieu, rather variable density and hardness; slight corrosive effect on metals stained by iron. Stains yellow in contact with water
Medium	Ext J Int J	Flg n/P		Medium	Tendency to irregular grain not therefore suitable for small sections; darkens on exposure to rich orange-brown. Limited supply
Medium	Int J		V	High	Gold or red to dark brown with dark markings

continued . . .

Colour (1)	BS name	Latin name	Hardwood (H) or Softwood (S) / density (2)	Origin	Natural durability (3) / Resistance to impregnation with preservatives (4)	Texture
VARIE-GATED COLOURS *continued*	Mansonia	*Mansonia altissima*	H/m	W Africa	V/Ex	Fine
	Rosewood	*Dalbergia spp*	H/vh	India and Honduras	V/–	Medium
	Walnut – European	*Juglans regia*	H/h	Europe	M/R	Medium
	Western red 'cedar'	*Thuja plicata*	S/l	Canada and USA	D/R	Coarse/medium
	Parana 'pine'	*Araucaria angustifolia*	S/m	Brazil	N/M	Fine

(1) The *colours* given are approximations for newly cut heartwoods. They may vary from tree to tree and there may be wide variations within one tree. Colours change with exposure to light and all timbers become grey after exposure to the weather.

(2) Densities for seasons timbers

	kg/m³			kg/m³
/l light	320–480	/vh very heavy		800–1 040
/m medium	480–640	/exh exception-		
/h heavy	640–800	ally heavy		over 1 040

(3) Durabilities relate to *heartwood* – see page 74: (sapwood of all species is non-durable or perishable)

V very durable   N non-durable
D durable   P perishable
M moderately durable

(4) Resistance to impregnation with preservatives – see page 78:

P permeable   ‡sapwood permeable to
M moderately resistant   resistant
R resistant   †sapwood permeable
Ex extremely resistant   * often contain large amounts of permeable sapwood

Sources: The Timber Research and Development Association, The British Woodworking Federation BS 1186:1971 *Quality of timber and workmanship in joinery, Handbook of Softwoods* and *Hardwoods* and *Timber selection by properties – the species for the job* Volume 1: *Windows, doors, cladding and flooring,* Volume 2 *Furniture.*

Table 32   *Properties of some common timbers classified by approximate colours*

Working properties (5)	Uses (6)				Price range (7)	Remarks (1)
Good dust very irritant	Ext J Int J	Flg 1/P	V		Medium	Purple when fresh fading to light fawn on exposure
Difficult	Int J	F		V	Very high	*Indian rosewood:* dull brown, purple, black. *American (Rio) rosewood:* handsome showy markings, bronze-black
Medium	Int J P	F		V	High	Considerable variation in colour; greyish brown with darker markings; English walnut is the most beautiful; mainly used for veneers; solid timber therefore in short supply.
Good	Ext J Int J* cladding				Low/medium	Light pink to chocolate brown; extremely stable and resistant to decay; may be too soft for joinery; corrodes and is stained by iron; strength relatively low
Good	Int J P	F 1/P	Flg		Low	Brown to bright red and dark streaks; hard; gives smooth finish; nearly all free from knots\nDifficult to dry and tends to warp if not correctly seasoned.

(5) Working properties include: nailing, planing, sawing and gluing.

(6) The *uses* listed are for suitable qualities of the respective species

Legend:
Str — structural
Ext J — external joinery – sapwood must be treated with preservative or excluded; heartwood may require to be treated with preservative and/or painted, varnished etc – see page 78.
Int J — internal joinery
Int J* — internal joinery not including draining boards
F — furniture
Flg — flooring –
1/P – n/P – h/P: light/normal/heavy pedestrian traffic
1/I – h/I: light/heavy industrial traffic—see *MBS: Finishes*
V — veneers
P — plywood

3 Boards, slabs and panels CI/SfB R

Preformed boards, sheets, ceiling tiles, slabs and panels continue to reduce 'wet' processes on the building site, and the variety of products increases.

A detailed reference is *The Building Board Directory*, Benn Publications.

Products include:
 Fibre reinforced composites – see chapter 10
 Synthetic resin bonded laminates – see chapter 13
 Cellular plastics boards – see chapter 13
 Plasterboards.

BS 1230 *Gypsum plasterboard*, Part 1:1985 describes boards suitable for wall and ceiling linings for either direct surface decoration or gypsum plaster. The BS now includes moisture resistant and water repellent types and a board with improved fire properties – see *MBS Finishes*, chapter 2.

This chapter deals with:
 1 Plywood
 2 Blockboards and laminboards – page 99
 3 Densified laminated wood
 4 Particle boards – page 100
 5 Fibre building boards – page 105
 6 Vermiculite and perlite boards – page 107
 7 GRC boards
 8 GRG boards
 9 Wood-wool slabs
 10 Compressed straw slabs – page 109
 11 Pancls – page 109

These products vary widely in their properties. For example, strength ranges from weak materials for thermal insulation or for sound absorption only, to boards and slabs which are suitable for flooring and roof decking supported on joists. Performances in fire also vary widely with materials and methods of fixing. *Surface spread of flame* ratings are relevant to use as linings. – see page 37

It is becoming increasingly common to fabricate composite boards and slabs an example of which could comprise a straw slab core which provides thermal insulation, with enamelled aluminium alloy sheets bonded to each side providing ready-decorated weather-resistant surfaces. Such a stressed skin composite would have a high strength:weight ratio. Perforations in 'hard skins' can admit sound to be absorbed by a spongy core, although with some loss of sound insulation which necessitates mass, eg as provided by lead-cored plywood, together with attention to detailing.

Table 33 gives the moduli of rupture of the common boards and slabs.

BS 4606:1970 gives *Recommendations for the coordination of dimensions in building – coordinating sizes for rigid flat sheets used in buildings*. This BS relates to BS 4330:1968, *Recommendations for the coordination of dimensions in building*.

	N/mm^2
Wood-wool slab	0.34– 1.72
Fibre insulating board	1.03– 3.45
Plasterboard	1.38– 11.03
Chipboard	2.76– 8.96
Medium fibreboard	5.55– 17.24
Asbestos wallboard	10.34– 20.68
Plywood	10.34–137.89
Asbestos wood	17.24– 27.58
Asbestos cement sheet (BS 690)	17.24– 34.47
Standard hardboard	24.13– 55.16
Tempered hardboard	27.58– 55.16

Table 33 Moduli of rupture of boards and slabs (approximate ranges)

1 PLYWOOD

BS 565:1972 *Glossary of terms relating to timber and woodwork*, defines plywood as 'an assembled product made up of plies and adhesives, the chief characteristic being the crossed plies which distribute the longitudinal wood strength'. Plies are 'peeled' from a pre-boiled log by rotating it against a knife. The resulting surface grain pattern is often less visually interesting than that of either plain sawn or quarter sawn timber.

Normally, the direction of the grain of each ply runs at right angles to that of the plies on each side so that strength properties are more uniform in the length and width of board than in the case of 'solid' timber, and moisture movement across the board is only about one-tenth that in the width of plain sawn timber.

Plies are not always of equal thickness and the 'inner' ply, or plies, can be of lower strength than those on each side. The strength and moisture movement properties of the plies on each side of the centre must be 'balanced', from which it follows that there must always be an odd number of plies.

Plywood is difficult to split and can be nailed or screwed close to the edges. It offers high resistance to 'pull-through' of screw and nail heads.

Plywood is stronger and stiffer than solid timber. It is used for structural purposes such as, flooring, including stressed skin units, and I and box beams.

BS 5268 is a *CP for the structural use of timber*, Part 2, Section 4 deals with *Plywood*. It gives moderating factors for properties in wet conditions and for durations of loading. It states that plywoods bonded with exterior type adhesive are

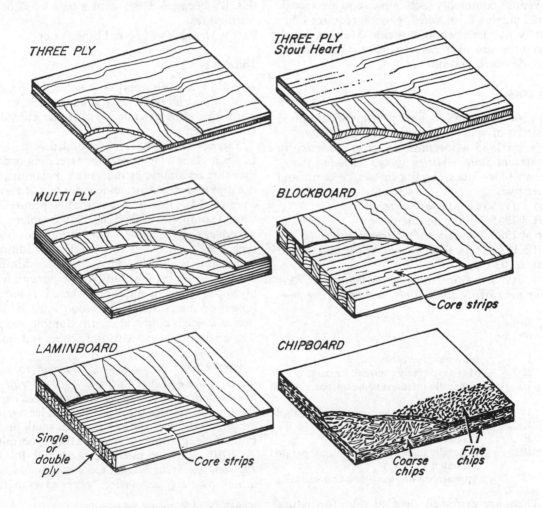

11 *Typical plywood, laminboard, blockboard and chipboard*

not necessarily suitable in damp conditions for long periods. Wood which is not inherently durable should be adequately treated against decay.

Fire properties are essentially those of the timber which is used but plywood flooring provides rather better fire resistance than t and g boarding of the same thickness. The *spread of flame* rating under BS 476 is *Class 4* where the surface veneer is less than 400 kg/m³, or is very thin. Most plywoods however achieve *Class 3* and suitable surface coatings and chemical impregnants bring plywood up to *Class 1*.

BS 5268: Part 4 deals with *Fire resistance of timber structures* including plywood.

Woods commonly used in plywood are gaboon and Douglas fir of which gaboon requires filling but is an easier wood to paint. Makoré, sapele and utile are also easy to paint. Other woods include beech and birch.

BS 6566

BS 6566 *Plywood* is a nine part specification the contents of which are summarised as follows:

Part 1:1985 a *specification for construction of panels and characteristics of plies, including marking*, includes rules for thicknesses of cores and outer plies.

Part 2:1985 is a *Glossary of terms*

Part 3:1985 *Testing and sampling*

Part 4:1985 *Dimensional tolerances*

Part 5:1985 *Moisture content* 6–14 per cent (tested at ±103°C)

Part 6:1985 is a *Specification for limits of defects for the classification of plywood by appearance*

Surface Grades	Exposures
I	May remain visible
II	May be directly overlaid, or painted
III	Generally intended to be unseen, painted or coated
IV	Appearance not the prime consideration
E	Intended to be seen
B	May remain visible
BB	Generally intended to be unseen, painted or coated
C	Appearance not the prime consideration

There are extremely detailed rules for natural defects and repairs and for joints.

Part 7:1985 gives four *Durability classes*:

G for general purposes

E high resistance to wood borers other than termites (but not to decay)

M moderate resistance to decay

H high resistance to decay

Classes M and H are not necessarily resistant to borers such as lyctus and termites

Preservative types and treatments are recommended

Part 8:1985 *Bond performance of adhesives for plywoods*

The requirements are similar to those in BS 1204 – see page 297 but the BR classification in this BS becomes CBR with a new *cyclical boil resistant test*

Part 9 *Strength* will be published later

Durability

Interior grade plywood is bonded with casein, soya, blood albumen and animal glues, or with synthetic resins extended with other substances (INT glues).

Exterior grade plywood is bonded with synthetic resin glues which, listed in ascending order of moisture resistance, include: urea, melamine, phenol and resorcinal formaldehydes (see *Adhesives*, chapter 14). For full external exposure glue should comply with BS 1203:1979 *Synthetic resin adhesives (phenolic and aminoplastic) for plywood* type *WBP* (water and boil proof). In addition to employing a durable glue, because durability depends upon the less durable component, where plywood will be in situations where it is liable to fungal or insect attack the wood must be heartwood which is either inherently durable (see page 73), or be protected with fungicides and insecticides.

Part 7:1985 *Resistance of plywood to fungal decay and wood borer attack*. 1088 *Marine plywood manufactured from selected untreated tropical hardwoods* requires the wood to be at least *moderately durable*[1] with the exception of small proportions of clean, sound sapwood which is permitted. A WBP[2] adhesive must be used and BS 4079 *Plywood made for marine use and treated against attack by fungi or marine borers* gives a range

[1] Durability of heartwood see page 74
[2] Adhesives see page 295.

of appropriate preservatives. BS 3842:1965 is entitled: *Treatment of plywood with preservatives*.

Types

In addition to the ordinary product, plywoods are made in the following forms:

With 'decorative' veneers on both faces or on one face with a non-decorative balancing veneer on the other side. Decorative veneers may be single sheet, side matched, book matched or random matched. *Plywood planks* simulate natural wood planking in various ways and some have pre-finished surfaces.

With etched and moulded surfaces provided by sand blasting or in some cases by a press which scorches parts of the surface.

Metal faced with galvanized steel, stainless steel, copper, aluminium and other metals. Where metal is applied to both faces and sealed at the edges plywood is protected from insects and fungi and has a very high strength:weight ratio.

Plastics faced boards are available with decorative surfaces and for concrete formwork.

Exterior grades are described in BS 1088 and 4079. BS 1088 – see *Durability*.

With tongued and grooved edges which provide perfect alignment at joints and dispense with the need for supporting noggings where the plywood is used for flooring, concrete formwork or wall and ceiling linings.

Lead cored for shielding against radio-active radiation or for sound insulation.

Sizes

Plywood is manufactured in many countries in a very wide range of sizes. Maximum sizes determined by the size of the press used are commonly 3084 × 1524 mm, occasionally 3099 × 1600 mm and much larger sizes are possible by scarfing joints.

Cross-grained plywood is made with the direction of the face grain across the width of the board, but in most plywood (known as *long-grain plywood*) the direction of the face grain is in the length of the board. In both cases the dimension first stated is that in the direction of the grain of the face plies. Thus dimensions are quoted as:

Long-grain boards – length × width
Cross-grain boards – width × length

Thicknesses range from 3 to 25 mm. (The thicknesses of USA and Canadian plywood are described in inches.) Manufacturers' products vary in the number of plies in a given thickness. Because the greater the number of plies, the stronger, the more dimensionally stable and more expensive is the product, in specifying plywood both the number of plies and thickness should be stated.

2 BLOCKBOARD AND LAMINBOARD

These boards are shown in figure 11. Relevant British Standards are:

BS 3444:1972 *Blockboard and laminboard*
BS 3583:1963 *Information about blockboard and laminboard*
BS 1186 *Timber for and workmanship in joinery*, Part 1:1986
BS 6566, Part 2:1985 defines these products.

Blockboard varies in thickness from 12 to 43 mm and 38 to 48 mm in 'blanks for flushdoors with a solid core consisting of blocks up to 25.4 mm wide, faced each side either with one thick veneer or two thinner veneers'. The grain in a single veneer runs at right angles to that of the core blocks so that a second veneer is necessary if the face grain is to run in the length of the boards. They are available in sizes up to 1.651 × 3.658 m.

Blockboard is useful for large and rigid panels to receive veneers or paint, particularly in shop fittings and furniture.

Laminboard is similar to blockboard but the core strips are narrow (up to 7.0 mm), and unlike some blockboards, they are continuously glued. Thus, laminboards are freer from surface distortions and are an excellent base for high quality veneers or for high gloss paint.

Battenboard, which has a core of strips more than 30 mm wide, is not allowed by BS 1186 for joinery.

3 DENSIFIED LAMINATED WOOD

These high quality products comprise wood veneers, usually with the grain running in the same direction, hot-pressed and bonded with synthetic resin adhesives. There are two main types, *Impreg* and *Compreg*. In the case of *Impreg* the veneers are impregnated with resin before pressing. Very high temperatures and pressures are employed and the process imparts high density combined with a high strength:weight ratio; high resistance to abrasion, and to the effects of temperature and moisture, and *Impreg* is particularly suitable for engineering applications such as gear wheels and pulleys.

Hydulignum is an example of *Compreg* developed and manufactured by Hordern-Richmond Ltd which uses resorcinol formaldehyde adhesive. It has been used for stair treads, handrails, counter tops and shopping bag rails in supermarkets.

4 PARTICLE BOARDS

Boards made from particles of wood, or other lignocellulosic fibrous materials, are of two main types: those bonded with synthetic resins, and others bonded with cement.

(a) Synthetic resin bound particle boards

References

BS 5669:1979 *Specification for wood chipboard and methods of test for particle board*
BRE Digest 239:1980 *The use of chipboard*
CPA Technical Manual:1980 Chipboard Promotion Association, Stocking Lane, Hughenden Valley, High Wycome, Bucks, HP14 4NU.

The main uses of chipboard are in furniture and built-in fitments, floors, partitions, fire doors, roof decking and formwork. The following notes give guidance on some of the types of chipboard available and their most important properties.

Chipboards are bonded with thermosetting synthetic resins which cure rapidly and irreversibly under heat.

Types

The main types of binder are:

Urea Formaldehyde (UF) Most frequently used, it causes no discolouration so boards are suitable for clear finishing, but UF bonded boards are only suitable in dry conditions.

Phenol Formaldehyde (PF) Used alone or in combination with UF resin (PF/UF) to give improved moisture resistance. Dark brown in colour.

Melamine Formaldehyde (MF) Used alone or in combination with UF resin (MF/UF) to give greatly improved moisture resistance. Although colourless, a colour dye is often introduced for identification.

Iso-cyanate Enables a high bond strength to be obtained with a smaller resin content.

There are several different ways of forming the boards:

Platen-pressed boards

Most chipboards are formed by pressing between pairs of heated platens, so that the particles lie in random fashion with their larger dimensions parallel to the surface of the board.

Types of platen-pressed board are:

Single layer – having consistent particle distribution and equal density throughout the thickness.

Three layer – having dense outer layers of fine particles for a smooth surface finish, with a core of coarser chips. Alternatively, the denser outer layers may contain long thin chips or flakes to increase bending strength.

Multi-layer – similar to three layer but with five or more layers or a continuous gradation of particle sizes (graduated-density boards).

Waferboards – made exclusively from large wood chips at least 32 mm in length and overlapping in adjacent layers.

Oriented structured boards (OSB) – layered to have the larger dimension of the chips pointing in the same direction within each layer and at right angles to the adjacent layer. As in plywood, strength is more uniform in the two directions.

Extruded boards

Extrusion of the chips between parallel heated platens allows for almost unlimited board length and thicker boards can be made, than with platen pressing. However, the process causes most of the particles to lie perpendicular to the surface and wood veneers or plastic laminates must be bonded to both faces, in order to add strength and rigidity. Thicker boards, or panels extruded with hollow cores are lighter and are suitable for demountable partitions.

Calendered boards

Another method is to form the boards around a heated cylinder and later to allow the board to lie flat. This *'Mende'* process is mainly for manufacturing thin chipboards (4–6 mm).

Specification

It is most important that the appropriate type of board, as described in BS 5669, is used:

Type I	Standard
Type II	Flooring
Type III	Improved moisure resistance
Type II/III	Combining the strength properties of Type II with the improved moisture resistance of Type III

The Standard gives dimension tolerances for: thickness, length and width, edge straightness and squareness.

It also gives mean quality levels and minimum acceptance levels for other properties which include: strength and stiffness, dimensional stability, edge screw holding and total extractable formaldehyde.

Additional requirements are:

Thermal conductivity must not exceed 0.14 W/m deg C.

Surface spread of flame must not be inferior to *Class 3* (BS 476: Part 7).

BRE IP3/85 *Wood chipboard, recommendations for use* by J M Dinwoodie should be consulted.

Sizes

BS 5669 does not give sizes, but recommends that the standard sizes of boards used in buildings designed in accordance with BS 4330:1968 should be those recommended in BS 4606:1970.

The most common sizes of boards for general uses are: 1220/1200 × 2440/2400 × 6–25 mm.

Floor grade boards are usually 600/610 × 2400/2440 × 18 or 22 mm. For boards with t and g edges the sizes are 'face' sizes.

A full range of available sizes and thicknesses is given in CPA data sheets:

CPA data sheet 1.01: *Standard boards (Type I)*

CPA data sheet 1.02: *Speciality boards*

CPA data sheet 1.03: *Flooring grade boards (Types II and II/III)*.

Appearance

Chipboards differ in surface fineness and porosity. This is often unimportant if boards are to be veneered. Standard boards require filling before being primed and painted. Boards are available with surfaces pre-filled and sanded, pre-primed or paper surfaced ready for gloss painting. Water based paints and emulsions swell the surface particles and produce a textured finish.

Boards are also available surfaced with decorative wood veneers, or plastic laminates, vinyl foils, phenol impregnated paper and textiles such as hessian – *see* CPA data sheet 2.03: *Finishing*

Properties

Density

Densities vary from less than 480 kg/m^3 to 800 kg/m^3. Strength and rigidity tend to increase with density but density is not a means of assessing quality.

Moisture content

As manufactured, the moisture content of boards is 7–13 per cent.

Moisture movement

Like timber, wood chipboards are hygroscopic but are dimensionally stable within small changes of moisture content.

Distortion may occur if:

1 boards are subject to deforming stresses, while wet,

2 the humidity differs on the two sides (eg a door to an airing cupboard),

3 the board is not 'balanced' (eg veneered on one side only),

4 flat surfaces, but not edges, are protected from absorbing moisture, with consequent swelling and protrusion at edges where boards are veneered.

New boards should be stored in dry conditions, on building sites before fixing they should be stacked loosely for as long as possible so they adjust dimensionally to the atmosphere in which they are to be used – *see* CPA data sheet 2.04: *Handling and storage.*

Durability

Resistance to moisture

Chipboard should not be used in conditions where it may remain wet for long periods (eg external cladding of buildings). Where humidity and temperature may fluctuate (eg bathrooms and kitchens) it is wise to specify a *moisture resistant grade*, – *see* CPA data sheet 2.05: *Moisture Resistant Boards*

Resistance to insects

Ordinary particle boards are not attacked by insects in temperate climates. Termite-resistant boards are available for use in the tropics.

Resistance to fungi

Chipboards are not likely to be damaged by wood-destroying fungi unless in permanently damp conditions above 20 per cent moisture content. Some moisture resistant grade boards are pre-treated with a fungicide. They are suitable for use in roofing, or where undetected moisture may be a hazard.

Resistance to fire

Like timber, chipboard is *combustible*.
Spread of flame is *Class 3* (BS 476:Part 7) for untreated chipboard. *Class 1* spread of flame can be achieved by impregnation with a flame retardant chemical or by a fire retardant coating.

Chipboard is used in fire doors and suitable constructions can resist one hour fire endurance (BS 476:Part 8).

The following data sheets deal with uses of chipboards in fire-resisting constructions.

CPA data sheet 3.04: *Wall linings and partitions*
CPA data sheet 3.05: *Ceiling linings*

Thermal movement

This is not normally significant and it does not affect the strength properties of wood chipboard.

Strength

Bending strength and stiffness properties depend on the type of board and its thickness. The range of mean quality levels for strength properties shown in BS 5669 (Table 1) indicate higher stress values for thinner boards and for *Type III*.

Type III (MF/UF) boards can fully recover their strength following a period of wetting, whereas *Types I and II* (UF) boards may show very little recovery – *see* BRE Digest 239 (figure 2).

Particle boards under heavy permanent loading may have a tendency to 'creep', a form of progressive deformation that is made worse by high or fluctuating moisture content. In dry conditions, where the board has not been over-stressed there should be no permanent deformation of the board after removing the load.

Machining

Working characteristics are more consistent than those of timber but cutting-tools are blunted more rapidly. High speed steel cutting edges are advisable for the denser particle boards. Planing tears surfaces and if it is necessary to reduce the thickness of boards they should be sanded.

Jointing

Because there is no end grain, particle boards can be glued at any angle with tongues or dowels for location purposes. Glues with a high water content should not be used.

Screws hold best in the faces of single layer high density boards, but holding properties are improved by using special double threaded screws. Where the load is considerable, wood plugs or nylon bushes should be glued in to receive screws.

Knock-down fittings have been specially developed to form right-angle joints – *See* CPA data sheet 2.02 : *Machining, Jointing and Fixing.*

Veneers

Bonding of veneers to chipboards is normally done with urea formaldehyde resin adhesive in a workshop and with a rubber-based contact adhesive on the site. Animal glues which contain water should not be used. Wherever a veneer is applied to a chipboard surface it is important that a balancing veneer is applied to the opposite surface – *see* CPA data sheet 2.02 : *Machining, Jointing and Fixing.*

Uses

The many uses for resin bonded wood chipboard in building today include:

Flooring

Chipboard is now the most commonly used material for flooring which is fixed to joists or battens. Large areas can be laid rapidly with relatively few joints and the cost 'as laid' is competitive with all other materials. Only boards complying with BS 5669 *flooring grade* should be used. *Type II* is suitable for dry site conditions: *Type II/III* should be specified where occasional wetting of the surface or high humidity is likely (eg kitchens and bathrooms). It should not be used in areas which are frequently wet (eg shower-rooms) unless the upper surface is 'tanked' with a continuous platics sheet material welded at the joints, and carried up walls.

BS 5669 specifies the following maximum spacing of joists for domestic floors:
18/19 mm thickness: 450 mm joist centres spacing
22 mm thickness: 610 joist centres spacing

Other combinations of board thickness and joist spacing may be used in more heavily loaded areas, such as offices, provided there is evidence of satisfactory performance. Laying instructions are given in Appendix C3.5 of BS 5669. If square-edged boards were to be used all edges must be supported by joists or battens and noggings. Boards with tongued and grooved edges are easier to lay. They require noggings only at the perimeter of the room and where corners of boards do not rest on a support. To avoid squeaking it is strongly recommended that profiled joints be glued and the boards should be firmly nailed or screwed to all supports.

Chipboard is used as a 'floating floor' on foamed polystyrene insulation which is laid directly on a concrete slab. All ground floor slabs must contain a damp-proof membrane – *see MBS Finishes* chapter 1.

Where boards are to be exposed as wearing surfaces at least one factory applied coat of polyurethane sealer is desirable as protection during building. Two factory costs provide an immediate finish, but a single factory coat may require two further site costs to give a satisfactory finish –

CPA data sheets: 3.01: *Wood chipboard as flooring on joists*
CPA data sheets: 3.02: *Wood chipboard as a floating floor*.

Furniture and fittings

BS 5669 Appendix C2 recommends higher values of tensile strength, surface soundness and screw-holding properties for panels surfaced with thin skins such as PVC, paper foil or direct painting.

Chipboards which are surfaced on both sides with stiff wood or plastic veneers at least 0.5 mm thick, and which are lipped with solid wood should comply with the basic quality levels of table 1 of BS 5669. Other requirements which may be specified are: initial deflection, and creep particularly for boards which will not have stiff surface veneers. Flatness and fineness of surfaces may need to be specially defined for boards which are to receive high quality thin finishes.

In 'wet' environments even *Type III* chipboards having improved moisture resistance, should be protected from moisture by both surface and edge sealing.

Shelving

BRE Digest 239 recommends that for industrial shelving *Type II* flooring grade boards should be used, or *Type II/III* where the relative humidity is likely to exceed 65 per cent. Board thickness should not be less than 18 mm and special boards are available in thicknesses 38 mm and 50 mm to support heavier loads. Spacing of supports is critical and may require to be as close as 300 mm. Design of heavily loaded industrial storage shelves should be undertaken by a professional designer.

Roofing

Flat roof decking BRE Digest 239 recommends that the choice of board must be restricted to MF/UF or PF moisture resistant types, BS 5669 Type II/III, preferably incorporating a fungicide. It is still necessary to provide weatherproof roof covering and to limit condensation below the deck.

Pitched roofs Chipboards, typically 12 mm

thick may be used for 'sarking' on pitched roofs. If UF bonded boards are used they should be protected from rain during construction. Prefelted boards and wax or bitumen coated boards fulfil this requirement. Improved loft insulation is likely to increase the moisture in the roofing boards and in such circumstances a moisture resistant chipboard incorporating a fungicide is recommended – *see* CPA data sheet 3.03: *Roofs*

Formwork for concrete

Tests carried out by the Cement and Concrete Association indicate that chipboards faced with phenolic film or melamine give a good surface finish to concrete and can be re-used several times. Boards are mainly 18 mm thick.

(b) Wood-cement particle boards

Wood particles have long been used as aggregate in some lightweight concrete blocks but boards comprising about 25 per cent by weight of wood chips bound together with cement, are a recent development.

These boards are denser than resin-bonded chipboards and sound reduction is greater, if the boards are suitable fixed. The material is *combustible* (BS 476) but spread of flame ratings are *Class 0* (Building Regulations 1976) and *Class 1* (BS 476). The boards have high resistance to insects, including termites.

Although they are heavier than resin-bound particle boards, and bending strength and impact resistance are slightly lower, BRE Digest 239 suggests that wood-cement particle boards have considerable potential as flat roofing decking. On the other hand, loss of strength on wetting and drying is at least as good as for MF/UF particle boards and their superior dimensional stability would reduce stresses transmitted to roofing felt. Also, wood-cement chipboards are very durable in wet conditions, the high alkalinity discouraging the growth of fungi.

BS 1142 Boards				Maximum mean percentage increase with 33–90% RH	
Type		Code	Normal Density kg/m³	Length and width	Thickness
Part 3: 1972	*Insulating* – Standard (softboards)	—	up to 350	0.37	
Part 2: 1971	*Medium boards* – low density	LM	350–560	0.30 0.40	5 5
	– high density	HM	560–800	0.25 0.30	7 10
Amd 1 1985	*Medium density fibreboards*	MDF	over 600	0.40 – up to 22 mm thickness 0.35 – over 22 mm thickness	6 5
1971	*Hardboards* – Standard	S	over 800	0.35	10
	– Tempered	T	over 960	0.30 0.35	10 15

Table 34   Types, densities and moisture movements related to relative humidities

The higher cost of wood-cement chipboards, compared with that of plasterboards, could be justified for uses such as linings to stud partitions in wet conditions, and where superior resistance to impact is required.

5 FIBRE BUILDING BOARDS

BS 1142: *Fibre building boards* describes products commonly known as 'fibreboards', which are made from wood or other vegetable fibres in most cases without any added binder. They range from soft boards having good thermal insulation, to 'hardboards' having properties comparable with plywood.

Information concerning 'fibreboards' is obtainable from the Fibre Building Board Development Organization Ltd. (FIDOR).

BS 1142, Part 3, 1972 is *Insulating boards* and Part 2:1971 (amd 1985), is *Medium board, medium density fibreboard (MDF) and hardboard.*

Manufacture

To make *Insulating board*, partially dried pulp is simply rolled to the required thickness and then reduced to the normal moisture content. Bonding depends mainly upon felting of the fibres and upon their inherent adhesive properties. In the case of *Standard hardboards* the cut sheets are hydraulically hot-pressed, and medium density fibreboards (MDF) and *tempered hardboards* also contain oils or resins.

Moisture movement

BS 1142 gives typical figures for moisture movements related to environmental relative humidities as shown in table 34.

As with timber, it is important to condition fibre building board to suit the humidity of the environment in which it will be used. The moisture content of the factory product is low, and it is necessary to increase the moisture content of the board, so that after fixing it will tighten on its fixings rather than expand and warp.

Sizes

At present, standard sizes are based on imperial sizes and a module of 305 mm:

Lengths and widths	Widths	Lengths
305	1 370	1 830
405	1 525	2 135
610	1 600	2 440
915	1 700	3 050
1 220		3 660

Boards are available to special order in *coordinated metric sizes* (mm):

$$
\left. \begin{array}{l} 600 \\ 900 \\ 1\,200 \end{array} \right\} \times \begin{array}{l} 1\,880 \\ 2\,400 \\ 2\,700 \\ 3\,000 \end{array}
$$

Thicknesses (mm) of standard sizes and coordinated metric sizes are given in table 35.

BS 1142 deals with MDF boards from less than 12–35 mm thick.

Types

Described in order of increasing density the main types are: insulating fibre building boards, wallboards and hardboards.

Insulating fibre building boards

Standard insulating board (softboard)

This is made from uncompressed wood, or sugar cane fibres, giving a low density (not more than 350 kg/m³) board with low thermal conductivity (k) not more than 0.06 W/m. The untreated board has a high rate of *surface spread of flame (Class 4 rapid*, BS 476) although this can be improved by impregnation, or by surface treatments (see table 12, page 38). Sound absorption of surfaces is good, but sound insulation is low, as may be expected of a low density product.

13 mm thick boards can be bent to a radius of 1 m without special preparation. It is used as a sheet wall and ceiling lining material (the surface can be plastered), permanent shuttering to concrete and an underlay to floor coverings.

Bitumen-bonded insulating boards are made in multiple layers bonded together with bitumen. They have lower water vapour diffusance than standard insulating boards but reduced water

Board	2·0 2·5 3·2 4·8	6·4 9·0 9·5	10·0	12·0 12·7	13·0	16·0 19·0	25·0
Insulating boards Wallboards		√	√		√	√	√
Medium boards		√	√	√		√	
Standard and tempered hardboards	√	√	√	√			

Table 35   Common thicknesses of fibre building boards

absorption. Certain forms of insulating boards heavily impregnated with bitumen are used to fill gaps between concrete slabs and to support sealing mastic which is poured into the remaining space.

Special forms of insulating fibreboards

Paper faced
Pulp faced   A layer of ground wood pulp gives a light coloured and smooth textured finish
Patterned with a course 'dimpled' pattern
Decorated with a primer, a finishing coat or both
Plastic faced
Wood veneered
Aluminium foil faced
Acoustic boards   These are low density boards perforated or grooved to give high sound absorption
Flame-retardant boards   Boards faced with vermiculite, painted with flame-retardant paint or impregnated with a flame-retardant chemical have a lower rate of spread of flame, in some case *Class 1* – see table 12, page 38.

Wallboards

Homogeneous fibre wallboard (or building board) is slightly denser than insulating boards (not more than 480.6 kg/m³) and thicknesses are normally 6.4 or 8.0 mm.

Production is small but uses include room linings and underlays for sheet floorings.

Medium boards and hardboards

Medium boards

Low density medium boards normally have density of 350–560 kg/m³, and high density boards, including *panelboard*, 560–800 kg/m³. Low density is suitable for 'pin-up boards'. As a wall or ceiling lining, or floor underlay, it is stronger and harder than insulating board, while providing better thermal insulation than standard hardboard. To condition medium boards (and insulating boards), they should be unpacked where they will be fixed, stored on edge for at least 24 hours, or preferably for two days, so as to allow air movement around the boards so that the moisture content of each sheet adjusts to the surrounding atmosphere.

Medium density fibreboard (MDF)

This product, described in the 1985 amendment to BS 1142, Part 2, is made by a dry process. The synthetic resin binder may give off irritating formaldehyde gas where large quantities of the board occur in poorly ventilated areas. *MDF* boards have densities over 600 kg/m³, and can be cut, machined and embossed to give a high quality finish. The BS gives a test for extractable formaldehyde, strengths and moisture movements for boards – see table 34.

Standard hardboard

This, generally made from wood fibres, has a

minimum density of 800 kg/m^3. In the ordinary product one side is smooth and the other usually has a mesh texture. Uses include wall and ceiling linings, flush door faces, floor underlays and light-duty flooring, and many uses in joinery where plywood of similar thickness might otherwise be used.

Special forms of standard hardboards include:

Perforated – pegboard linings for: suspending light objects, surfacing to acoustic materials and for ventilation

Enamelled – various finishes, normally stoved

Plastic paper laminate faced

PVC sheet faced

Wood veneered

Printed wood grains

Moulded and embossed in the press – patterns include tiles, 'reeds', flutes, leather grain etc.

Duo-faced – which is smooth on both sides

Pre-primed and sealed

Ivory faced – a layer of light coloured pulp is added to the surface

Flame-retardant – impregnated after manufacture

Bitumen impregnated.

Hardboard is despatched from the mill with a moisture content of between 5 and 8 per cent which is in equilibrium with an atmospheric relative humidity of about 65 per cent. It is recommended, particularly where boards will be used in damp surroundings, that 0.5 to 1.0 litres of water should be brushed on the backs before fixing. Subsequent drying out tightens the boards on their fixings.

Most hardboards can be curved without special preparation to a radius of about 300 mm. Radii of 200 mm can be obtained by cold/wet bending and 76 mm or less by hot/wet bending.

Tempered hardboard

This is generally made from wood fibres and has minimum density of 960 kg/m^3. It is made by impregnating newly pressed standard hardboard with oils or resins and then applying further heat treatment. As a result it has superior resistance to water absorption and has a lower equilibrium moisture content than timber, eg a 10 per cent moisture content in tempered hardboard relates to 18 per cent for timber in the same environment. The board is not recommended for permanent buildings in 'wet' conditions but it was the only fibre building board, acceptable for external wall claddings, by the Building Regulations 1976. Because starches and sugars have been removed, insects in UK will not normally attack tempered hardboard.

Tempered hardboard is stronger than standard hardboard and has structural potential. Permissible stresses and effective moduli for TE grade boards up to 8 mm thick can be obtained by applying modifying factors, including a significant allowance for creep, to the *dry grade* stresses given in BS 5268, Part 2, 1984, *The structural use of timber*.

The texture is as for *standard hardboards* but the surface is harder. The colour is usually dark brown.

Uses include floor wearing surfaces, working surfaces eg in workshops, I and box beams, linings for concrete formwork and exterior uses.

6 VERMICULITE AND PERLITE BOARDS

A fire clay board contains exfoliated vermiculite. It is therefore *non-combustible*, and primarily intended for fire-resisting construction. (The materials are also preformed as encasures for structural steelwork). The products may not be suitable for use at relative humidities exceeding 70 per cent.

Other similar boards contain perlite – see page 188.

7 GRC BOARDS

Boards made from glass fibre reinforced Portland cement – see page 236, have replaced asbestos wallboards and insulating boards and to some extent semi-compressed asbestos cement boards – see page 239.

8 GRG BOARDS

Boards made from glass fibre reinforced gypsum – see page 235 have good impact resistance and rigidity. They are *non-combustible* and fire resistant but not suitable in wet conditions.

Cores \ BS Types	HN, LN, HN/S, LN/S	H/I, H/S, LH, LH/S	H2
Plywood	Good interior, or better, grade	Hardwood with BR adhesive	
Blockboard	5 ply fully-glued Western red cedar BS 3583		
Laminboard	5 ply, good interior, or better grade, BS 3583		
Particleboard	Type 1 BS 5669 or boards with hardwood veneers		
		Cement-bonded wood chipboard	Moisture resistant chipboard Type III BS 5669
Fibreboard	Tempered or standard hardboard of homogeneous construction BS 1142, Part 2 or Medium density fibreboard	Tempered hardboard of homogeneous construction BS 1142:Part 2	
Asbestos cellulose board	Not less than 800 kg/m^3		
Non-combustible boards	BS 690:Part 2 or BS 3536:Part 2		
Non-asbestos boards		Certain mineral boards	

Table 36   Cores for BS 4965:1983 'Decorative laminated plastics sheet veneered boards and panels'

9 WOOD-WOOL SLABS

These are made of long wood shavings, coated with cement and compressed while leaving a high proportion of thermal insulating voids. Density is 400.5 to 480.6 kg/m^3 and thermal conductivity (k) 0.093 W/m deg C.

BS 1105:1978 describes *Wood-wool-cement slabs up to 125 mm thick*. Although *combustible*, they are not readily ignited. Spread of flame is *low*, Class 1, BS 476. As roof decking the 51 mm thick slab can span joists at 610 mm centres or between the metal edges of *channel reinforced slabs*. The open texture is a good base for plasters and renderings but the natural surface, which can be decorated by spraying with paint, provides good sound absorption.

Slabs are 610 mm wide in various lengths and in thicknesses from 24.5 to 101.6 mm.

The slabs do not rot. However, it is important to use the appropriate type of slab as designated by BS 1105:

There are three types:

Type A for non load-bearing uses, including permanent shuttering

Type B for Type A uses. Stronger than Type A and slabs not less than 50 mm thick are suitable for roof construction, if Safety precautions of construction (Working precautions) Regulation 66 are observed.

Type SB is stronger than Type A and has greater resistance to impact than Type B Slabs not less than 50 mm thick are suitable for roof construction and for the uses given for Types A and B.

BS 3809:1971 describes: *Wood-wool permanent formwork and infil units for reinforced concrete floors and roofs.*

10 COMPRESSED STRAW SLABS

BS 4046:1971 describes *Compressed straw building slabs.* These consist of straw compacted only by heat and pressure, surfaced with, and bound at the edges with paper.

Surfaces can be decorated direct, or receive gypsum board plaster. As partitions, they can be butt jointed with adhesive or be fixed in wood or metal sections.

Lightweight thermal insulating roof decking can span over joists at 600 mm spacings, but provision is essential for ventilation to comply with CP 114, Part 1, 1968. Slabs lose strength and decay in damp conditions.

Size The standard slab is 1200 mm wide × 2400 mm long × 58 mm thick.

11 PANELS

There is a wide range of standard and purpose-made panels, with solid or cellular cores, providing varying degrees of strength and resistance to humidity and wetting and to fire and sound transmission.

Timber selection by properties, Volume 2, *The species for the job*, HMSO, describes wood based panels.

For interior uses BS 4965:1983 *Decorative laminated plastics sheet veneered boards and panels* specifies materials and fabrication. The cured aminoplastic sheet laminates comply with BS 3794:Part 1 (see page 288), adhesives are specified and requirements for core materials are summarised in table 36.

The required properties for nine types are:

Heavy duty: H/1, H/2, H/S and HN

Light duty for situations with small risk of impact: LH, LN and LN/S

Resistant to sustained high humidity (80 per cent RH and higher) and to frequent wetting: H/1 and LH

Not resistant to ditto: HN and LN

Resistant to occasional high humidity and infrequent wetting: H/2

Class 1 Surface spread of flame (BS 476:Part 7): H/S, LH/S, HN/S and LN/S.

LH, LH/S, H/1 and H/S types are veneered both sides and HN, HN/S and LN/S types one or both sides.

The BS states that for use in humidities below 40 per cent RH all material should be conditioned.

4 Stones

This chapter deals with the properties of the various types of stones, known to the geologist as *rocks*, and their uses in building, as blocks, slabs, flags, setts, roofing slates and damp-proof courses.

Other uses for stone, mentioned in other chapters, include:

1 Aggregate for concrete, terrazzo, mortars, plasters, renderings, tarmacadam and mastic asphalt.
2 Granules for surfacing bituminous felts, etc
3 Powders for extending paint
4 Abrasives.
5 Rock wool for insulation.

Information

The Stone Federation and The Joint Committee on Building Stones, 82 New Cavendish Street, London W1M 8AD.

Useful *references* are:

Building Stones of England and Wales, Norman Davey, Bedford Square Press.
Stone in Building, J. Ashurst, F. Dimes, Architectural Press.
Stone for Building, Hugh O'Neill, Heinemann
A Future for stone, Hutton and Rustron, HMSO
Natural Stone Directory, Ealing Publications
BRED 269 *The selection of natural building stones*
BS 6100 *Glossary of building and civil engineering terms*, Part 5 Section 5.2:1984 defines *Types of stone, tools for stone, etc.*
BS 5628 *CP for the use of masonry, Part 1:1978 Structural use of unreinforced masonry*, supersedes CP111:1970
Part 2:1985 *Structural use of reinforced and prestressed masonry.*
Part 3:1985 *Materials and components design and workmanship* supersedes CP121 on block masonry.
BS 5390:1976(1984) *CP for stone masonry* supersedes CP121, 201 and CP121.202

CP298:1972 *Natural stone cladding* (*Non-load bearing*).

It should be noted that the term *masonry* is applied to natural, artificial, cast or reconstructed stonework and also to brickwork and blockwork.

Cost

In spite of considerable rationalisation and mechanisation of the industry, the cost of natural stone blocks and slabs is relatively high because great care must be taken not to damage stone in quarrying and because in stones such as slate there is a great deal of waste. Further, the cost of cutting, dressing and polishing is considerable, even with modern techniques, such as the flame cutting at 2 760°C of granite and slate. However, the *cost in use* of slabs of economic dimensions may be considered to be reasonable if account is taken of their prestigious appearance. This is generally durable and mellows with age, can be maintained, and recovered by cleaning, repolishing or reworking surfaces – see page 121.

Artificial stones

Casting in moulds is cost effective, particularly for mass production of sculptural forms, and reinforcement can be incorporated.

BS 6457:1984 is a *Specification for reconstructed stone masonry units.*

Crushed stones in white or coloured cement matrices can resemble dressed or rock faced natural stones in colour, textures and profiles, although weathering is usually less pleasing than that of natural stones.

Artificial marbles include traditional multicoloured gypsum plaster *Scagliola* and coloured plastics products for use internally.

Appearance

Natural stones usually retain their good appearance inside buildings. Externally if left to weather naturally, limestones improve in appearance and slate does sometimes, but granites and marbles seldom do.

The fact that no two blocks of stone are identical in appearance, even within one quarry, provides visual interest, but it means that even large samples can give only a general indication of colour, veining and texture.

The smoothness of materials determines the amount of light they reflect so that they darken when they are wet or polished and crystals, fossils, veins and other features become increasingly visible. For example, as quarried, Norwegian 'pearl' granites are almost indistinguishable from many featureless stones, but they become highly decorative when they are polished.

The principal descriptions of stone surfaces, in approximate order of increasing smoothness, are:

Rockfaced
Rough picked
Fair picked
Axed
Fine axed, dolly pointed; or flame textured (on granite and slate)
Split (riven)
Sawn, slightly ribbed or rippled
Sanded or shot sawn

Gritted
Eggshell, honed; or fine rubbed (on slate)
Polished.

The split or *riven* surfaces of quartzite, sandstone and slate vary in coarseness. *Knapped* flint surfaces are glassy.

Great care must be taken to protect surfaces during building operations, even dense stones can be readily stained.

Properties

Some properties of natural stones are summarized in table 37. It will be seen that natural stones are strong in compression. The thermal coefficient of expansion of limestone and marble is low, about that of bricks, but allowance for thermal movement must be made with sandstone, slate and granite.

Durability

Natural stones are generally extremely durable but deterioration may result from wrong choice of stone – see BRED 269 which classifies stone A–F and relates them to exposures. Other causes of deterioration include faulty design and workmanship, atmospheric pollution, soluble salt action,

	Density kg/m^3	Failing stress in compression N/mm^3	Thermal movement mm/m per $90°C$ per cent approx	Moisture movement mm/m for dry-wet change
Granites	2560–3200	105 335	0·93	none
Sandstones	2130–2750	27·5 195	1·0	approx. 0·7
Limestones	1950–2400	16·5 42·5	0·25 (porous limestone) 0·34 (dense limestone)	0·8 negligible
Slates	2800–3040	42·5 216	0·93	negligible
Marbles	2880		0·34	negligible
Quartzites	2630		0·90	none

Table 37   Properties of building stones

frost, solution, wetting and drying, rusting of ferrous metals and sometimes vegetation.

Atmospheric pollution Sulphur compounds, mainly sulphur dioxide, formed by burning coal and oil, produce sulphurous acid when they are dissolved in rain water. This reacts with carbonates in limestones, dolomites, calcareous sandstones and mortars and where they are not freely washed by rain a hard glassy skin tends to break away and sometimes blisters form. However, there is a very considerable difference in behaviour between calcareous sandstones which usually fail relatively quickly and Portland stone, on which the skin can remain intact for centuries – see table 10.

Soluble salt action Salts, often derived from soils, exert considerable expansive force when they crystallize and are the chief cause of progressive decay. A sodium sulphate crystallization test, although not reliable with burnt clay products, is useful for assessing the salt resistance of stones – see table 8, page 33.

Thermal movement Table 6, page 30 compares the movements of various materials. In recent years the differences between the movements of reinforced concrete and stone claddings, particularly dark claddings, fixed with 'tight' joints, has caused stresses which have caused fixings to fail. Compressible movement joints should be provided both vertically and horizontally at not more than 3 m intervals to accommodate expansion of the stone and shrinkage of the concrete.

Frost Damage, often rapid, may occur in stones which are frozen when they are very wet or where water is retained in cracks or in shells on horizontal surfaces but it should rarely affect normal walling in this country. It is not clear to what extent frost damage results either from the expansion of water or from a mechanism analogous to *frost heave* in soils. Assessments of likely behaviour of members of a few groups of stones can be deduced from their physical properties but there is no rapid method of establishing the frost resistance of other stones – see table 9. External exposure of samples in trays which hold water provides a long term test for frost resistance. Stones which survive for three years are probably resistant and those which are unaffected after seven years are almost certainly resistant.

Solution Limestone is slightly soluble in water and surfaces which are exposed to rain slowly dissolve away – see page 117.

Wetting and drying Contour scaling is characteristic of certain sandstones and sometimes occurs among the less durable limestones. It is mainly attributable to rainwater penetrating to a constant depth, drying out and in so doing bringing forward soluble matter in solution. A soft layer below a relatively hard outer skin results and eventually the latter breaks away following the contours of the surface.

Corrosion of metals The rusting and consequent expansion of ferrous metals can cause serious damage, and zinc salts released by galvanized steel have caused local decay.

Vegetation Lichens, mosses and virginia creeper do little harm but ivy has caused serious damage.

TYPES OF STONES

According to the manner of their geological formation all rocks, referred to in building as stones fall into one of three classes: igneous, sedimentary or metamorphic, each having recognisable physical characteristics.

The main stones used in building are discussed here in the following order:

Geological class	*Group*
Igneous	granites
Sedimentary	sandstones
	limestones
Metamorphic	slates
	marbles
	quartzites

Igneous stones

About 5 000 million years ago the earth was a ball

of gas and interstellar debris, and the interior is still white hot below about 200 km. Stones formed by the cooling of the molten magma, which include granites, syenites, dolerites, basalts and pumice, clearly, cannot contain fossils or shells.

Felspar is usually the chief ingredient in igneous stones, others are:

Quartz is an extremely hard and durable crystalline form of silica, usually transparent and colourless, but sometimes grey, pink, yellow or purple.

Mica is a silicate of aluminium with potassium (muscovite mica), which with the further inclusion of iron and magnesium (biotite mica) is soft and cleaves very easily into very thin flakes.

Hornblende. Essentially a silicate of calcium, magnesium and iron occurs as almost black six-sided prismatic crystals.

Augite is similar to hornblende but having eightsided crystals.

Iron pyrites is a sulphide of iron occurring in small specks or cubes of brassy colour and liable to decompose rapidly

Olivine, an iron magnesium silicate, is black, green or yellow in colour.

Asbestos is made up of various incombustible fibrous minerals related to either hornblende or olivine.

The structure of igneous stones is also dependent upon the rate at which they cooled. Thus:

Volcanic stones cooled rapidly and are very fine grained non-crystalline and glassy. In the main-volcanic stones are hard and difficult to work, unattractive in appearance and are used mainly for concrete aggregates and road metal. Basalt is such a stone. Pumice, which is porous, has been used as an aggregate for lightweight concrete.

Hypabyssal stones cooled slowly and consequently have a medium size crystalline structure. Porphyries are of this class. They take a good polish and some have been used for ornamental purposes but most are used as concrete aggregate or road metal.

Plutonic stones cooled very slowly at great depths below the earth's surface and a coarse crystalline structure results. Although the term 'granite' should strictly be limited to the acid rocks containing more than 66 per cent silica, the practice of calling similar igneous stones such as syenites granites, will be followed here.

Granites

Granites are a mosaic of crystals, sometimes several inches in length. The principal constituent is felspar, dull white to deep red in colour. Other ingredients are small grains of grey quartz and mica, which contribute 'sparkle'. Sometimes hornblende and, less frequently, augite, are present.

'Pearl' granites from Norway are highly decorative. Shap granite has unusually large red felspar crystals.

Granite has very high failing strength in compression, only a tiny fraction of which is normally employed in buildings. Because granites are extremely dense and hard, they are very costly to quarry, cut and surface, but they can be smoothed mechanically to a glass-like self polish. Being extremely resistant to knocks and abrasion, granite is eminently suitable for roadway kerbs, bridge cutwaters, jetties, spur stones to gateways, bollards, external steps to many important buildings, and heavy-duty paving as slabs or setts.

Partly because of its high density, between 2 460 to 3 200 kg/m^3 and its freedom from bedding planes or planes of cleavage, granite is very resistant to ordinary chemicals. Innumerable examples of the stone have retained a very high natural polish for a century or more in highly polluted industrial atmospheres which is a sure test of durability. No other stone can retain a natural polish for many years in such conditions and granite is used for external wall cladding where 'permanent' good appearance is required, eg, throughout the length of Victoria Street, London.

Most granites are virtually impermeable and serve as damp-proof courses provided the vertical joints between blocks are designed to prevent water movement.

On the debit side, staining of some granites which contain minute fissures has resulted from insufficient washing after acid treatment to remove iron derived from frame saws. Granite is less reliable in buidling fires than other stones, owing to the marked expansion of quartz and the differing expansion characteristics of the various constituents.

Some granites in common use, are listed under their predominant colours.

Colour of polished stone	Name	Country
Black	Andes	Brazil
	Bon accord	Sweden
Grey-black	Britts blue	South Africa
Grey	Rubislaw	Scotland
Light-medium grey	Creetown	Scotland
Light grey	Cornish	England
	Galloway	Ireland
Very light grey	Silver white	Norway
Green-black	Emerald pearl	Norway
Blue	Blue pearl	Norway
Red	Balmoral red	Finland
	Carnation red	Sweden
Pink	Peterhead	Scotland
	Shap	England

Table 38 Predominant colours of granite in common use

Sedimentary stones

Sedimentary stones are formed either from particles of older rocks which were broken down by the action of water, wind or ice, or from accumulations of organic origin. Sand and shingle are loose sediments, whereas the particles of sedimentary stones which are available as blocks and slabs have been cemented together by minerals originally carried in solution in water, and consolidated by super-imposed deposits.

Sediments were carried by water and sometimes by wind, and the tendency of particles to lay horizontally often produced *natural grain*. Changes in composition from time to time produced layers of differing character, including occasional *soft beds*. Sedimentary stones are therefore often *bedded* with *bedding planes* which may be only 0.2 mm apart. Bedding planes, or grain, although rarely visible to the naked eye in recently quarried stone, sometimes lack natural cohesion and are often potentially weak. To obtain the highest strength and durability in stone masonry the *natural bed* must be respected in fixing.

Where the natural bed is parallel to an exposed face stones often delaminate, and *face bedding*, as it is called, should not be allowed, unless the stone is known to be sufficiently resistant to this form of decay.

Stones are stronger when bedding planes are at right angles to the thrust, so that in normal columns and walling they should be disposed horizontally, but cornices should be *joint bedded* with their bedding planes parallel to the vertical joints.

Some stones known as *freestones* have widely separated bedding planes and layers of homogeneous material as much as 4 m deep, and these are invaluable for blocks which are exposed to the weather on all sides.

When first quarried, some sedimentary stones, in particular Bath stones, are more or less saturated with mineral matter in solution (*quarry sap*). To avoid frost damage stones should not be exposed to the weather before their water content has been reduced to a safe level.

The main types of sedimentary stones are shown in tables 39, 40 and 41.

Sandstones

Sandstones consist of fine or coarse particles of quartz often with particles of felspar or mica, bound together by a natural cement which latter is the chief factor in deciding strength, durability and colour. Classified according to the cement, sandstones are:

Siliceous sandstones are generally extremely acid-resistant and otherwise durable, but relatively difficult to work. Silica tends to give a light grey colour. Examples are Blue Pennant and Darley Dale.

Calcareous sandstones are cemented with calcite crystals of calcium carbonate, which in their pure form are white. The stone is more easily worked than siliceous sandstone but is less durable and should be used externally only in rural areas. An example is Reigate (Surrey) stone.

Dolomite or magnesian sandstones are bound with calcium carbonate and magnesium carbonate, both of which are white-buff. Examples are Red and White Mansfield.

Ferruginous sandstones contain oxides of iron, giving brown, red and yellow colours. As a class

Main constituents	Sedimentary stones			Metamorphic stones
	Unconsolidated	Bedded stones	Stones crystallised from solution	
	Gravel coarse aggregate for concrete	Pudding stones (conglomerates) too coarse for building work		
Silica	Sand fine aggregate for concrete, an ingredient in calcium silicate bricks Silt (fine sand) –too fine for building work	Sandstones wall cladding, paving	Flints undressed or knapped wall facings, and an ingredient in flint-lime (calcium silicate) bricks and vitreous ceramics	Quartzite (metaquartzite) wall cladding, paving, flooring
	Diatomaceous earth lightweight clay insulating bricks and blocks			
Silica and alumina (minute particles)	Clay clay bricks and other ceramic products an ingredients in Portland cement	Shale bricks and other ceramic products		Slates wall cladding, paving and floorings, roofing, dpcs
Calcium carbonate	Calcareous sediments, an ingredient in calcium silicate bricks, limes and Portland cement	Limestones (organically or oölitically formed) wall cladding, paving, concrete aggregate	Limestones (lagoon or cave formed) wall cladding, paving	Marbles wall cladding, paving and floorings, aggregate for terrazzo
Calcium sulphate			Gypsum alabaster, onyx, marble, gypsum plasters	

Table 39 Constituents, forms and uses of sedimentary and metamorphic stones

they weather well. Examples are Red Runcorn, Woolton, and Hollington (Coventry Cathedral).

The best sandstones are extremly durable but sandstones become dirty more readily and weather less attractively than limestones. Sandstones in industrial areas may decay rapidly if they receive washings from limestones. This was soon evident in the paving around the post-war Paternoster development near St Paul's Cathedral, London.

The bedding planes in sandstones are very closely spaced and they are often visible. Some can be split into flagstones and a layer of mica is often seen on the riven surface. *Face-bedding* often leads to delamination. After forty years or more, some sandstones show *contour scaling* in which the lines of spalling are independent of the direction of bedding.

Hard and durable fine grained sandstones which are particularly suitable for paving are called *York stones*. An example is Silex from Halifax, Yorkshire. They are durable, but may decay in towns where they receive washings from limestones – see above and page 117.

Gritstones contain angular particles. An example is Darley Dale, Derbyshire.

Apart from paving, steps and thresholds the use of sandstones is largely confined to the quarrying areas.

Examples of sandstones are listed under their colours in table 40.

Limestones

Limestones consist mainly of calcium carbonate in the form of calcite. They were formed by the deposit of solids mainly in lakes or seas, or by their deposit from solution:

1 *Oölitically formed stones* Most of the limestones used today, including the Portland and Bath stones, are oölites or roe-stones, so called because the structure resembles fish roe. In a typical oölite small calcium carbonate oöliths are formed by the deposit of concentric layers of calcite around fragments of shell or sand and these are cemented together by calcite in which fossils and sand grains may occur.

2 *Organically formed stones* These are accumulations of shells and other remains of animals and plants. Chalk and the much harder Hadene stone from Derbyshire are of this type.

3 *Crystallization from solution* This happens in pipes carrying hot water containing minerals and where dripping water forms stalactites and stalagmites. The most important limestone of this type is Travertine, often called a marble, which was formed by the evaporation of water around geysers. In this stone irregular voids were formed by gases.

Colours	Texture	Name of stone	Location of quarry	Density kg/m^3	Failing stress in compression N/mm^2
Yellowish-white	Fine	White Mansfield (Dolomitic)	Nottingham	2240	49·6
Cream	Close grain	Darney	Edinburgh	2400	89·6
Buff and white	Fine, close grain	Darley Dale (grit)	Derbyshire	2400	70·3
Buff to pink	Medium	Birchover	Derbyshire	2560	48·2
Light brown to cream	Medium/fine	Springhill Quarries	Yorkshire	2755	
Light brown	Fine even grain	Silex		2560	92·4
			Halifax		
Brown/grey	Fine grain	Elland Edge Flag rock	Yorkshire	2995	89·6
Golden brown	Fine even grain	Crosland Hill (millstone grit)	Huddersfield Yorkshire	2640	68·9
Brown and blue	Close grain	Mouslow (gritstone)	Derbyshire	2755	193
Blue-grey	Fine close grain	Blue Pennant	Pontypridd Glamorganshire	2690	172
Red/brown	Even grain			2130	27·6
White and red mottled	Fine/slightly Coarse	Hollington	Uttoxeter Staffordshire	2210	32·4
Red	Fine and even	Red St Bees	Cumberland	2145	53·1

From *Stone for Building* by Hugh O'Neill, Heinemann

Table 40 *Properties of some sandstones*

'Pure' limestone, such as chalk, is white or off-white, but other ingredients often colour the stone cream, yellow, brown, red grey and almost black. Coarsely crystalline calcite may impart a slight lustre.

Many limestones contain a proportion of magnesium carbonate and are called *Magnesium Limestones*. Where the content of magnesium carbonate is 45.7 per cent or more by weight the stone is known as *Dolomitic Limestone*. Other common ingredients include carbonaceous matter and iron oxides.

Properties of some English limestones are listed in table 41.

Limestones vary widely in hardness. Those of the Pennine Chain are extemely hard and suitable for concrete aggregates, road metal and paving slabs. Generally, however, limestones are less hard and easier to work than sandstones. Many are very suitable for carving and some are soft enough to be sawn by hand.

Limestones are soluble in water containing carbon dioxide so that they are self-cleansing where they are freely washed by rain and the more durable fossils and veins of calcite give visual character to some limestones as they weather. At the same time soot collects in protected parts giving the familiar black and white effcct which is considered by some to be an attractive feature of limestone buildings.

Limestones are attacked by acids to an extent which is related to their density. Sulphurous matter in soot reacts with calcium carbonate forming a glassy skin of calcium sulphate or with magnesium carbonate forming a skin of magnesium sulphate. These skins act as a binder for further soot and dirt and cause rapid decay of the softer limestones. Sulphur gases sometimes cause blisters to form in semi-sheltered parts and cavernous decay of magnesium limestones.

Rain washings from limestones, particularly magnesium limestones, cause decay of sandstones and of the less durable clay bricks.

Limestones are now being employed in a thickness of 25.0 mm as permanent shuttering on precast concrete wallslabs. The installed cost of concrete slabs faced with limestone is often very competitive with that of concrete slabs using the same stone as aggregate.

Varieties of limestones Limestone in this country is mainly confined to a belt from the Wash to Dorset, and most limestones used in buildings are obtained from the South West of England. The more important examples are:

Portland stone which is creamy white. It has excellent resistance to weathering, even in polluted atmospheres. It occurs in four principal beds the first of which to be exposed is the *cap* with a crushing strength in the region of 34.5 N/mm² now being crushed and used with white cement as aggregate for concrete. Below this is the *roach* bed 455 mm to 915 mm high. It has a fairly

Type	Weathering	Resistance to pollution	Density kg/m³	Porosity per cent	Absorption per cent	Failing stress in compression N/mm²
Portland: Dorset						
Roach	Excellent	Excellent	2090	21·4	4·3	38·7
Whitbed	Excellent	Excellent	2340	12·0	3·7	42·8
Basebed	Excellent	Excellent	2210	16·9	5·7	40·9
Bath Stones:						
Box Ground	Excellent	Excellent	2070	23·2	6·8	11·7
Monk's Park	Good	Good	2240	17·5	7·76	24·3
Clipsham: Rutland	Excellent[1]	Excellent[1]	2310	14·0	4·7	31·7
Doulting: Somerset	Excellent	Good	2400	17·0	8·6	21·3
Guiting: Gloucestershire	Excellent[1]	Good[1]	1970	20·0	9·7	16·3

[1]These remarks relate to the better quality material
Information from Stone Firms Limited

Table 41   *Properties of some limestones*

pronounced shell formation ranging from coarse to fine texture and when worked it gives a cellular surface. It is not very suitable for working fine arrises but for ashlar it provides textural interest and weathers extremely well. It was used for facing St Paul's Cathedral Choir School and *The Economist* building in London.

Whitbed which is 600 mm to 1 000 mm high and is close grained, fairly even in texture and contains some shells both throughout its mass and in thin layers.

Portland stones are very strong and Whitbed is rather stronger than the other Portland stones (42.85 N/mm² failing stress in compression). Density is 2 340 kg/m³. Nevertheless, it is less hard than Roach, and can be worked with relative ease. Its remarkable resistance to weathering is evidenced by the loss of only 13 mm thickness from the most exposed parapets of St Paul's Cathedral, about 0.05 mm per annum. It is the bed most used for general building purposes, one use being the facing of the Royal Festival Hall, London.

Basebed which occurs below the Whitbed, is about 1 m high, is fine grained, even textured and has little shell. It works freely and is the best bed for carved work.

The three lower beds named, polish well and are very suitable for wall linings, the shellier varieties being particularly decorative.

Bath stone is the name given to a series of free working stones found in the Bath area. The city of Bath is largely built in the stone and Apsley House in London. Bath stones include:

Monks Park which is light cream in colour, of fine grain and fairly compact. Resistance to weathering and to pollution are good, but plinths and copings would be liable to attack by frost.

St Aldhelm's Box Ground which varies in colour from light brown to cream and has a coarser grain. Resistance to weathering, pollution and frost is very good.

Doulting stone, quarried in Somerset, varies in colour from light brown, to buff to cream. It is a fine grained stone and works well.

Clipsham stone from Rutland varies in colour from buff to cream and from oölitic to coarse shelly in texture. In the past quality has varied widely but the best stone has been excellent.

Guiting stone is a Cotswold oölite, pale to deep ochre in colour, It has a medium grain with minute shells and works easily.

Flints are nodules of silica which are found in chalk. Formerly they were knapped to reveal a glassy brown-black interior which is extremely durable. Today they are mainly crushed for concrete aggregate, for making calcium silicate bricks, etc. They are also used, mainly in the chalk districts, as cobbles set in mortar for facing walls and for paving.

Metamorphic stones

These consist of older stones which have been subjected to immense heat and pressure causing structural change. Thus, clay becomes slate, limestone marble and sandstone quartzite, see table 39, page 115. Features such as bedding planes, large crystals and fossils are lost.

Slates, marbles and quartzite are dealt with:

Slates

Slates were formed by immense earth pressures acting upon clays and forming *planes of cleavage*, distinct from and often almost at right angles to, the original bedding planes. Slates are split along their planes of cleavage. The resulting surfaces may not be perfectly flat.

Properties Good slate, such as that complying with BS 680:1971 *Roofing slates*, is one of the most durable building materials. In the past, slate was used for its acid and alkali resistance as laboratory bench tops, urinal slabs and lavatory basin tops.

Poor slate however, may begin to decay in a few months, especially in damp conditions in industrial areas. Sulphide of iron in slates may form hydrated iron oxide and sulphuric acid, which latter attacks any calcium carbonate present. Similar attack may take place by the action of sulphuric acid from the atmosphere, and some Lake District slates, which contain a high proportion of calcium carbonate, may fail on that account.

Slate is strong in tension and compression when compared in equal thicknesses and otherwise equal conditions with other stones, as evidenced by its use as cantilevered stair treads. Formerly, slate was used as a good electrical insulator, eg for switchboards. It has negligible absorption and can be considered to be impervious and, therefore, suitable for dpcs, provided that water is not under pressure so that it would penetrate the joints as may be the case at coping level.

Slate can be considered to have no moisture movement and its stability is valuable for snooker tables. Thermal movement must be allowed for, however, in dark coloured slate copings which are exposed to the sun.

Appearance Slate can be had with riven textures which vary from silky and smooth to very rough, or sawn, sanded or finely rubbed surfaces. Being very hard and denser than granite it also takes a good natural polish but does not retain it as well, in fact calcareous slates lose their polish very quickly.

Colours include grey to almost black, red, blue, purple, green and brown. Striped and variegated colorations sometimes occur.

When carved, the texture of chiselled surfaces contrasts with highly polished slabs and the difference in tone is very pronounced on the darker slates. The appearance of slate can be enhanced by wax, or by varnish which is more durable but very difficult to renew satisfactorily.

Sources: North Wales This is the chief source of slate in this country. This product is typically dark and light blue, purple, blue-grey, dark and light green and dark green and dark grey – with a smooth texture. The slate can be split to almost paper thickness. The so called *bests*, which are the thinnest roofing slates, are about 5 mm thick.

Cornwall The slate is grey and grey-green and has a characteristic lustre. Naturally stained red slates (*rustics*) are occasionally available. Roofing slates are of medium thickness. They are sold mainly in the smaller standard sizes and in random widths and lengths.

Lake District The slate from Westmorland is generally green and that from North Lancashire is dark blue. The riven surfaces are pleasantly irregular. These slates contain calcium carbonate which in the case of the green slates only is attacked by sulphur gases. This weakness, however, is not so important in thick units and fortunately the Lake District roofing slates are thicker than those from other sources.

The Continent Some very good slates have been imported from France and Portugal but others from Italy have failed to meet the requirements of BS 680 for Roofing slates.

Uses The use of slate as *blocks* for general wall masonry is mainly confined to the quarrying areas, but it is used widely for copings, window sills and surrounds.

As *slabs* it is used for flooring, external paving and wall cladding. In a thinner form slate can be used for roof coverings and damp-proof courses.

Slate is an excellent substrate for air-dried and stoved paint, being impervious, dimensionally stable and chemically inert.

Carved in low relief, slate is much used for memorial and other plaques.

Granules and chippings are used for surfacing bituminous felts (see *MBS: Finishes*) and as an aggregate in concrete blocks.

Ground slate is used as a filler in mastic asphalt (see page 244, plastics (see page 270) and paints.

Tests for slates Simple tests are as follows:
(a) a good slate has no odour when wet
(b) the texture should have a somewhat rough 'feel' and should not be greasy
(c) when holed or dressed the fracture should be clean and not flaky
(d) BRS Digest 'First Series' *The durability of roofing slates* stated that if a roofing slate decomposes to mud in a day or two when submerged in a solution of fresh battery acid poured carefully and slowly into an equal volume of water, it is unlikely to give satisfaction in use, even in county districts.

British standards for slate products are:
BS 680:1971 Roofing slates
This BS contains tests which will also serve for assessing the properties of slates used for damp-

proof courses and other purposes. Slates which will be subjected to only slight atmospheric pollution by sulphurous or other acid fumes are required to pass tests (1) and (2) only.

1 *Water absorption*
   After storage in boiling water for 48 hours water absorption is not to exceed 0.3 per cent.
2 *Wetting and drying test*
   The slate must not delaminate or split after 15 soaking and drying cycles.
3 *Sulphuric acid test*
   The slate must not soften, delaminate or split when immersed in a 20 per cent sulphuric acid solution for 10 days. (Unfortunately this test eliminates North Lancashire slates which are in fact known to be durable.)

BS 5642:1983 Sills and copings contains tests which are less severe than those for roofing slates. *BS743:1970* (amd. 1983) *Materials for damp-proof courses* requires slates to be at leat 4 mm thick and 230 mm long.

Marbles

True marbles are fully metamorphosed, but in the trade, the description encompasses the English 'marbles', travertine and other limestones which take a good natural polish.

Pure marble is very finely crystalline and ideal for crisp carved detail. It is white, but coloured minerals often occur as a general coloration, as veins or other markings. Marbles which have a consistent pattern throughout the thickness of a block lend themselves to 'matching' of adjacent slabs to form symmetrical patterns.

Marble is generally very hard, dense and resistant to abrasion. Veins are sometimes weak but satisfactory repairs can be made with modern adhesives. Like granite, marble takes an excellent self-polish which greatly enhances its appearance. Unlike granite, however, marble is attacked by acids and a polished surface is not retained for very long externally in a polluted atmosphere.

Sugaring sometimes occurs on exposed surfaces and highly coloured marbles tend to fade. Also, although marble is virtually impervious discolouration can result in damp conditions and the backs of slabs must be sealed with shellac to protect them from 'wet' constructions.

Types A very wide range of marbles is imported, some of the varieties currently available being:

Colour of polished stone	Name	Country
Black	Belgium black	Belgium
Dark grey-black	Belgium fossil	Belgium
Grey	Dove	Italy
White	Sicilian etc	Italy
	Pentelicon	Greece
Cream	Bianco de mare	Yugoslavia
Cream-fawn	Aurisina	Italy
Beige	Portuguese beige	Portugal
Beige-brown	Perlato	Sicily
Brown	Napoleon	France
Golden-yellow	Golden travertine	Iran, Spain and Yugoslavia
Green	Tinos	Greece (Isle of Tinos)
	Verde antico	Greece
	Serpentino	Italy
	Swedish green	Sweden
Red and pink	Norwegian rose	Norway
	Rosso levanto	Italy
	Rose aurore	Portugal

Table 42   Marbles

Ashburton stone from Devonshire is dark grey to black with white and red markings and Hadene stone from Derbyshire is buff and contains crinoids and other fossils. (Not true marbles).

In this country, although marble will not retain polished surfaces or colours externally and unprotected, white marble is very common in structures such as the Marble Arch, London, and in typical municipal cemeteries. Internally, marble – often coloured and polished – is often used as a 'prestige' material for flooring, wall linings and for fire-place surrounds. White marble is still the preferred material for counter tops where food is served.

Crushed marble is used as an aggregate in terrazzo and in resin-bound floorings (see *MBS: Finishes*, chapter 1).

Quartzites

Quartzite was formed from sandstone by intense heat and recrystallization. It comprises about 96 per cent silica, is harder even than granite and is extremely durable. Quartzites can often be split in all directions, and some of them retain traces of the original bedding planes, which permit splitting.

The riven surfaces are somewhat uneven and splintery and have a pleasant lustre. They are not very easy to keep clean but provide 'non-slip' properties for flooring.

Colours are grey, green and gold.

Sources Quartzite is imported from Norway, Sweden, South Africa and Italy.

Uses Wall 'tiles' (or slabs), flooring and external paving, see *MBS: Finishes*

MAINTENANCE OF STONEWORK

References

BRE Digests:177 *Decay and conservation of stone masonry*.
  125 *Colourless treatments for masonry*.
  280 *Cleaning external surfaces of buildings*.
  139 *Control of lichens, moulds and similar growths*.
BS 6270 *CP for cleaning and surface repair of buildings*, Part 1:1982 deals with natural and cast stones and mortars.

Information

The Society for the Protection of Ancient Buildings provides advice.

Causes of decay are dealt with on page 111. Here we are concerned with cleaning, preservation, water repellent treatments and restoration, all of which processes should be undertaken by, or with the advice of specialists.

Cleaning

To maintain good appearance, to minimize the likelihood of decay and to prevent defects becoming concealed by soot deposits, stonework which is exposed externally should be cleaned regularly and defective joints should be raked out and repointed.

BS 6270 refers to cleaning using water, chemicals, sand blasting and carborundum discs. The necessary frequency of cleaning depends upon the kind of stone, the degree of atmospheric pollution, the exposure and the required finish – varying from rock-faced to extremely smooth and glass-like. Limestone which is freely washed by rain is 'self cleansing' but in protected positions limestones and sandstones may require cleaning at 5 to 10 year intervals, while in polluted atmospheres polished marble may have to be cleaned every month.

Methods of cleaning various stones are outlined in table 43. Deposits on limestones can usually be softened by a mist spray of clean water, and sometimes marbles can be cleaned in this way, but rarely sandstones or granites. Water cleaning can be done within a few minutes, to several days assisted by a high pressure water lance, if required. All stones can be cleaned by mechanical means or by hydrofluoric acid. Caustic soda and soda ash are very damaging and must never be used on any stone. Incidentally, stones must be masked when adjacent or super-imposed metalwork is cleaned.

Preservation

Preservation is dealt with in BRED 280 and BS 6270.

Errors in design and workmanship should be detected and corrected before any attempt is made to preserve stone. For example, salts should not be sealed in. They should be removed by repeated wetting and sponging. Unfortunately, attempts to arrest decay by applicatiom of colourless surface treatments often do more harm than good. Some flake off and some hold soot and dirt. BRE Digest 125 *Colourless treatments for masonry* (OP) stated that 'Nothing will be lost by awaiting with patience the outcome of trials of any new stone preservative that may be offered.'

A silthane resin which penetrates and consolidates stone to a depth of at least 25 mm is very promising, but it is recommended that it should be used only with expert advice.

Stone	Method	Remarks
Limestone Marble	Clean water spray to soften deposits followed by light brushing. Marbles and hard, polished limestones can be washed with water containing mild detergent, rinsed with clean water, dried with wash leather and polished with soft cloth	Relatively slow. Not suitable for heavy encrustations, in frosty weather, or where buried ferrous metals or timber might be adversely affected and the method may cause brown stains on limestones. Marble is sometimes difficult to clean. To maintain the colour of dark, particularly green, marbles externally, after cleaning them, beeswax and natural turpentine should be applied and polished several times each year.
Granites Sandstones	Ammonium bifluoride	Acids, particularly hydrofluoric acid, are extremely dangerous in handling
All stones	Hydrofluoric acid (about 5 per cent concentration)[1]	Risk of severe damage to surrounding materials, particularly glass
	Grit-blast:[2] dry	Rapid, even with heavy encrustations No staining of stones Requires skill to avoid damage to soft stones Very dusty process, operatives must have independent breathing supply. Close screening is required but some dust escapes into atmosphere
	wet	Generally as above. Water reduces visible dust, but may give rise to the objections to the water spray process
	Mechanical: abrasive power and hand tools brushes	Rarely necessary for limestones

[1]Steam sometimes helps to remove deep seated soiling after acid cleaning
[2]Only non-siliceous grit should be used

Table 43   Cleaning of natural stones

Water repellents

These treatments which prevent capillary absorption without sealing surfaces, inhibit water movement inwards but allow surfaces to breathe. They are discussed under *Clear finishes* in *MBS: Finishes.* However, by creating microporous conditions water repellents can increase frost damage.

Restoration

Isolated repairs can be effected with carefully matched stones or by *plastic repairs* which although less costly, are liable to become disturbingly conspicuous with time. Where decay is general but superficial, the whole surface can be cut back to expose sound material. *The Department of the Environment, Ancient Monuments Branch* must be notified, and will give expert advice on the treatment of *Scheduled buildings.*

Vegetable growths

These are rarely destructive but are evidence of dampness. BRE Digest 139 *The control of lichens, moulds and similar growths.*

Ceramics are made from a mixture of mineral material (generally quartz sand) and a clay binder (hydrated aluminium silicate) with impurities such as chalk, dolomite and sulphates, plasticized with water. The mixture is shaped, dried to remove making and absorbed water, and fired. During firing hydrate water, carbon dioxide and other gases are driven off, recrystallization takes place and glass is formed producing a hard, insoluble material.

Ceramic products vary very widely in sophistication from hand moulded burnt clay bricks to man-made fibres for lining furnaces.

Ceramics also vary widely in degrees of the frost resistance which is essential for products exposed externally in the UK – see clay bricks, page 127.

Research is carried out by The British Ceramic Research Association Ltd at Queen's Road, Penkhill, Stoke-on-Trent, ST4 7LQ.

Products such as hand moulded bricks and tiles which are made from raw material which may contain more than 30 per cent water have considerable drying contraction and are rarely accurate in size. However wall and floor tiles moulded from dust with only 2 to 5 per cent moisture content have little drying contraction and very much greater accuracy is possible.

Increases in temperature in firing are accompanied by more complete recrystallization and an increase in the formation of glassy material promoting greater: density, hardness, strength, resistance to chemicals and to frost, and greater dimensional stability. After firing, irreversible expansion due to adsorption of water by clay products can be troublesome in bricks (see page 136) and in floor and wall tiling, particularly if the bricks or tiles are incorporated in the work during the first weeks after they are fired.

Ceramics may have an as-fired appearance, or if a clay-base *engobe* is applied before firing, a matt finish. Glazes in any colour result if the moulded product is coated with a suitable *slip* before it is initially fired or after firing to the *biscuit* stage.

A sound vitrified glaze is virtually impermeable if its composition is accurately matched to the properties of the body and thereby subsequent micro-cracking (*crazing*) will be avoided.

Glazes which are to be exposed externally must be frost resistant.

The main ceramics products used in building are listed in table 44 and they are considered below.

FIRED CLAYS AND SHALES

These products include ordinary bricks (see page 126), clay roof tiles and flooring quarries (see *MBS: Finishes*)

TERRACOTTA

Terracotta is made from yellow to brownish-red clays intermediate in uniformity and fineness between ordinary bricks and vitrified wall tiles. It was used in Victorian buildings as hollow blocks filled with concrete during construction, for walling and cornices, where otherwise natural stone would have been used. Well burnt blocks filled with concrete containing sound aggregate have proved to be durable in many buildings such as the Natural History Museum in London. Terracotta is used now for unglazed chimney pots, air bricks and copings.

Shrinkage in firing is about one twelfth, and joints must be wide enough to accommodate variations in size and shape. Terracotta which is glazed is generally called *faience*.

FAIENCE

Faience is a glazed form of terracotta or stone-

Ceramic product	BS	Usual type of body
Bricks	3921:1985	Fired shales and clays
Roof tiles	402:1979 (plain tiles)	
Floor quarries	6431[1]	
Pavers	6677 (3 parts:1986)	
Firebacks	1251:1970	Fireclay
Refractory bricks	1758:1966	
Flue linings and chimney pots	1181:1971	
Sinks and wash tubs	1206:1974[2]	
Drain pipes and fittings	65:1981	Stoneware
Some faience products		
Floor tiles	6431[1]	Terracotta
Terracotta wall facing blocks sills and copings	5642:1983	
Hollow wall, partition and floor blocks		
Flue linings and chimney pots	1181:1971	
Air bricks	493:1970	
Some faience products	—	
Glazed tiles for internal walls	6431[1]	Earthenware
Some sanitary appliances		
Sanitary appliances	3402:1969 (quality) 1188:1974 (basins and pedestals) 5503, and 5504 (wc pans) 5605 (bidets)	Vitreous china
Special products, eg electrical insulators	—	Porcelain

[1] BS 6431 (many parts 1983) *Ceramic floor and wall tiles*
[2] Dimensions and workmanship only

Table 44   Ceramic building products

ware. Much faience is first fired to the *biscuit* condition and then glazed before refiring. Alternatively, the unfired clay may be glazed and 'once fired' a process which improves resistance to crazing of the glaze while reducing the range of colour obtainable. Faience glazes tend to have an orange peel texture. They provide permanent colour and although easy to clean are by no means self-cleansing in polluted atmospheres. Crazing of inferior products collects grime and may be very unsightly. Chipped glazes are unsightly and faience should not be used where it is likely to receive heavy knocks. Rounded corners rather than sharp arrises are to be preferred.

Ceramic glazes are inherently durable but where water is able to enter behind the glaze, crystallization of salts, and sometimes frost action, cause failure. Choice of a faience with low water absorption is some guarantee of frost resistance.

Faience is usually made in slabs up to 300 × 400 × 30 mm but copings, sills, plaques and sculptural forms can be made to order.

FIRECLAY

Fireclay is a simple product used for grate backs and flue liners, in which a high kaolin content in the clay binder provides high fire resistance. However, some bricks, faience and hollow clay blocks are made in fireclay where fire resistance is not the primary requirement.

STONEWARE

Stoneware is similar in composition to fireclay but because it is fired at a higher temperature it contains a higher proportion of glass, is harder and less absorbent and is suitable for drainage goods without a glaze. Stoneware drainage goods used to be glazed, either by introducing salt into the kiln during firing so that the sodium combined with the silica in the clay, or by conventional spraying of a glaze frit.

EARTHENWARE

The raw earthenware materials, blended from different sources, may contain a considerable proportion of limestone. Earthenware is a finer product than stoneware and is used as the body for glazed wall tiles, and for ordinary quality table 'china' but water absorption of the fired product may be as high as 15 per cent or even more making it less suitable for sanitary ware than vitreous china, since any small cracks in the glaze permit water to penetrate into the body.

VITREOUS CHINA

Vitreous china has a higher glass content than earthenware and its water absorption is only about 0.5 per cent, so there is negligible penetration of the body by water should the glaze crack. It is also stronger than earthenware, and vitreous china is now used for most sanitary fittings.

PORCELAIN

Porcelain is similar in most respects to vitreous china but it is often made from purer materials under more strictly controlled conditions to give properties which are required for specific uses, such as electrical insulators.

6 Bricks and blocks

BS 3921:1985 *Specification for clay bricks and blocks* describes a *brick* as a walling unit designed to be laid in mortar and not more than 337.5 mm long, 225.0 mm wide and 112.5 mm high, as distinct from a *block* which is defined as a unit having one or more of these dimensions larger than those quoted.

BS 6100 *Glossary of building and civil engineering terms*, Part 5, Section 5.3:1984 defines terms relating to *Bricks and blocks*. BS 6677, Parts 1, 2 and 3:1986 deal with *Clay and calcium silicate pavers for flexible pavements*.

For economy, lengths of walls and piers should be multiples of format sizes, but if necessary, solid units can be cut, and the widths of joints can be varied.

The main uses of bricks and blocks are as units laid in mortar to form walls and piers. The uses of exposed brickwork both externally and internally, for its appearance and low maintenance costs, and of brick paving – mainly externally – are increasing. Unlike in situ concrete, no support is needed to vertical surfaces during construction and also complex shapes can be built, particularly with bricks, which in concrete would require equally complex and costly formwork.

Relatively small units provide visual 'scale' and some bricks and blocks have attractive textures and colours. The appearance of walling is much affected by that of the mortar. Crudely shaped and unsympathetically coloured jointing or pointing can spoil the appearance of any units. Some bricks and blocks are suitable to be painted and others are good substrates for plaster and renderings.

Bricks and blocks may vary in colour between batches and units and require to be selected and mixed on site to avoid local concentrations of colours in the finished work.

BRED 280 deals with *Cleaning external surfaces of buildings*.

The strength in compression of brickwork and blockwork can be very high, but, unless the work is reinforced, tensile strength is usually ignored in design. The relationship of unit, mortar and brickwork strengths is referred to in chapter 15 *Mortars for jointing*. Briefly, there is no strength advantage in using very strong mortars with weak bricks or blocks, in fact, mortar should generally be weaker than the units.

Durabilities of bricks and blocks vary widely depending on quality – a property which cannot be deduced from casual inspection, or from simple tests.

Ordinary brickwork, like dense concrete, has relatively high thermal conductivity. Nevertheless, walls comprising fair-faced 102 mm brickwork inner and outer leaves, with thermal insulating fills or batts in cavities, can have U values below (ie better than) the 0.6 W/m² deg C maximum required for dwellings.

In this volume blocks are dealt with on page 143, and mortars on pages 301 and 302.

BRICKS

References
Bricks and Brickwork CC Handisyde and BA Haseltine. The Brick Development Association.

The main British Standards for brickwork are:

CP 111:1970 *Structural recommendations for load-bearing walls* (under revision)

BS 5390:1976 *Code of practice for brick and block masonry.*

BS 5628 *CP for the structural use of masonry*

Part 1:1978 *CP for the structural use of unreinforced masonry*

Part 2:1985 *CP for the structural use of reinforced and prestressed masonry*

Part 3:1985 *CP for the use of masonry – Materials and components, design and workmanship* includes resistance to rain penetration

BS 6270: *CP for cleaning and surface repair of buildings*, Part 1:1982 includes *Clay and calcium silicate brick masonry*

Information is available from: The Brick Development Association Ltd (BDA), Woodside House, Winkfield, Windsor, Berkshire, SL4 2DX.

Strength and durability should be sufficient for the conditions in which a brick is to be used and, if it is to be seen as a facing brick, it should be of good appearance. Alternatively, a good base for decoration, or a keyed brick (figure 12) for rendering or plastering may be required.

The size and shape of bricks should be regular to facilitate bonding. In its most common form, a brick can be held in one hand, and its length is equal to two widths and three heights with 10 mm joints. The metric lengths and widths, being slightly smaller than the imperial sizes, bond with old brickwork, and the 65 mm height bonds with the old $2\frac{5}{8}$ in. bricks with $\frac{3}{8}$ in. joints. In the Midlands and the North of England some manufacturers make bricks 73 mm high to relate to the old $2\frac{7}{8}$ in. bricks which were common in that area, and in the South some manufacturers of handmade bricks make 50 mm high bricks to suit the old 2 in. height.

Figure 12 shows some typical *standard* bricks and figure 13 shows nine of about forty *special shapes* contained in BS 4729:1971 (amended to 1978) *Shapes and dimensions of special bricks*.

Most bricks have a *frog*, or depression, on one or both bed faces which serves to reduce cost in manufacture and in handling and also provides a mechanical key for mortar. Frogs are usually laid upwards and, with double-frogged bricks the deeper frog is laid upwards. CP 111 compressive stresses are for the 'frogs up' condition. Frogs may be placed downwards to reduce weight and the London Brick Company's cellular brick is designed to do this. If frogs are placed downwards tests are necessary to determine permissible compressive stresses. Perforations and cells also reduce weight, and incidentally make for more uniform firing of clay bricks.

Bricks are made in four materials:

1 Burnt or fired clay
2 Calcium silicate – see page 138.
3 Dense concrete – see page 142.
4 Lightweight concrete (for fixing) – see page 143.

CLAY BRICKS

Clay bricks were made in Britain by the Romans and then reintroduced from the Low Countries and Germany in the thirteenth century. Today, more than 8000 million bricks are produced annually in this country, and most of these are made from clay.

Research is carried out by the *British Ceramic Research Association Ltd* (BCRA).

Useful references include:

BRE Digests 164 and 165   *Clay brickwork 1 and 2*
Model specification for load-bearing clay brickwork The British Ceramic Research Association (BCRA) 1967.
BS 3921:1985 is a *Specification for clay bricks*. It provides definitions and aims to quantify properties which are important in use, ie:
Sizes
Voids in bricks
Water absorption
Crushing strength
Durability – ie frost resistance and soluble salt content
Efflorescence
Suction rate
Appearance

Procedures are given for sampling and testing and a check list for ordinary clay bricks is contained in Appendix G.

Sizes

In the past, large variations in size were characteristic of ordinary fired clay products, and in clay bricks they often caused serious problems in bonding – in particular – narrow piers.
The BS gives one standard *Coordinating size*:

	Overall measurement of 24 bricks mm	
	Maximum	*Minimum*
Length	5235	5085
Width	2505	2415
Height	1605	1515

$225 \times 112.5 \times 75$ mm, which includes 10 mm in each direction for joints and tolerances, and the related *Work size*: 215 (2 headers plus 1 joint) $\times 102.5 \times 65$ mm. The overall dimensions of 24 lengths, widths and heights must fall within stated limits and the length, width or height of no individual brick shall exceed the relevant coordinating size.

Modular bricks

Although full scale tests indicate that the use of modular bricks, rather than the standard brick, does not result in any significant saving in the cost of brickwork, BS 6649:1985. *Clay and calcium silicate modular bricks* describes a $200 \times 100 \times 75$ coordinating size and it is likely that this and other modular sizes will be included in a revised BS 4729.

Special shapes, other than rectangular prisms, can be made by sawing where the interior thereby exposed is visually acceptable. Suitable bricks can be carved or *rubbed*. Special shapes can also be specially ordered from manufacturers.

BS 4729:1971 *Standard special bricks* describes shapes and dimensions of bricks which are in general use, and may be held in stock – including bats, closers coping bricks and those shown in figure 13.

Other standard special shapes include *Calculon* bricks – see page 128 and figure 15.

Classification

Clay bricks by raw material, origin, normal use or by the presence of voids.

Voids in bricks – see figure 12.

BS 3921 classifies bricks as:

Solid – having no *holes*, *cavities* (ie holes closed at one end) or *depressions*.

Cellular – having no holes, but *cavities* may exceed 20 per cent of the gross volume.

Perforated – having holes up to 25 per cent of the gross volume. The area of any hole must not exceed 10 per cent of the area of a face and the aggregate thickness of solid material measured horizontally across the width and at right angles to the face, to be at least 30 per cent of the overall width.

Frogged – with depressions in one or more bed

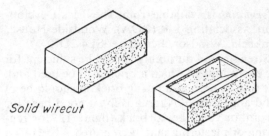

Solid wirecut

Pressed or hand moulded brick

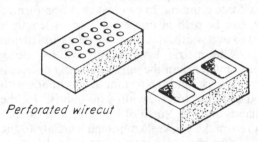

Perforated wirecut

Cellular pressed brick

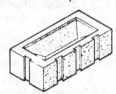

12   Standard bricks   Keyed pressed brick

Grade	Average compressive strength	Average weight	Perforation
	N/mm^2	kg	
A10	68.9	4.08	23 per cent
B7.5	51.7		
C5	34.5	3.95	'solid'

Calculon bricks

faces total not to exceed 20 per cent of the gross volume.

In addition to the BS classifications, clay bricks can be described by the raw material, eg a *gault*; method of manufacture, eg a *wirecut*; by colour, or by texture. *Stock* is a very commonly used term which in itself, like 'deal' in describing an undefined timber, has no precise meaning.

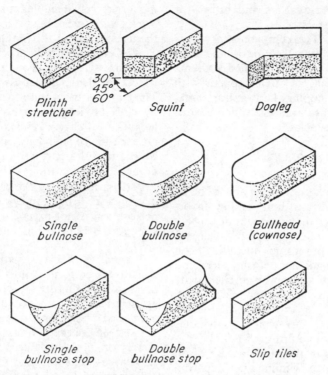

13   Typical 'Standard special' bricks – BS 4729:1971

The term is usually applied to hand or machine formed soft-mud bricks often described by geographical origin, eg *London stocks* and the *Sussex and Buckinghamshire red-purple multi-coloured bricks.*

Types by uses

Bricks which are used for general purposes and where appearance is not important are known as *Commons* Bricks which are required to have good appearance are called *Facings* which are either specially made or selected.
Engineering bricks have high strength and are typically, but not necessarily impervious, durable and costly.

Proprietary bricks

'Phorpres' cellular brick, see figure 12, has three deep depressions or cells. It is 25 per cent lighter than the standard Fletton brick, and brickwork is about 15 per cent lighter.

Strength is adequate for normal two storey construction when the cells are laid downwards as would be normal.
'Phorpres' keyed brick has dovetailed keys for plaster or renderings on one stretcher and one header face – see figure 12.

'Calculon' bricks (London Brick Co Ltd and Redland Bricks Ltd) are made in three strengths for highly stressed internal loadbearing cross walls and spine walls and can be used as an internal skin to cavity perimeter walls. Each type incorporates a hand hold and manufacturers' tests show a 30 per cent increase in the speed of laying and a saving of approximately 40 per cent mortar when compared with 228.6 mm solid brick walling. The *C5* type is smooth all round and the *A10* and *B75* types are available either smooth all round or keyed for plaster.

The size of type C5 differs from that of types A10 and B7.5. Actual dimensions (subject to normal tolerances) are given in figures 14 and 15. Standard specials for A10 and B7.7 type are

129

quarter, half and three-quarters and bricks for returns and for use in piers.

Ibstock coping bricks Two complementary units provide a cavity in which the dpc is laid over standard bricks. The interlock formed prevents bricks being pushed off, as often happens with normal brick-on-edge copings laid on horizontal dpcs. *Acoustic shapes* are available which absorb sound at specific frequencies.

Manufacture

The raw material, which must be capable of being shaped, dried and burnt without undue distortion, largely determines the properties of bricks.

Brickmaking clays are composed of silica and alumina and various impurities including iron compounds, lime, magnesia, potash, soda and sulphur, most of which are combined chemically into compounds such as feldspar and mica. Clays, such as that used in the manufacture of Fletton bricks, contain natural fuel which contributes subtantially to economy in firing.

The main processes in the manufacture of all ceramic products involve 'winning and preparing the raw material forming the shape required and drying and firing it.

'Winning' and preparation

Methods of *winning* the raw material affect the economy of manufacture but not the properties of the brick. Preparation involves the removal of stones, etc and in some cases the addition of fuel. Weathering by frost action is rare today. Grinding and in some cases the addition of water follow.

Forming

All ceramic products must be formed to dimensions which allow for shrinkage in firing later. For hand moulding a plastic clay is essential but a relatively stiff clay can be moulded by machine and the lower moisture content reduces shrinkage in drying.

Soft-mud moulding This is done exceptionally by hand in a *stock mould*, or by machine produc-

ing a relatively irregular shaped and dimensioned brick. An interesting surface texture results from the use of sand in the moulds as a release agent.

Wire-cutting In this process a moderately stiff clay mix is forced through a die, the mouth of which has the length and width of a green brick. The continuous column formed in this way is cut by wires into lengths corresponding to green brick heights. Ordinary *wire-cuts* are recognized by the absence of frogs and by marks on the bed faces caused by solid particles in the clay having been dragged by the cutting wires.

Perforation during extrusion saves clay, fuel in drying and firing, and weight in handling. If perforation is not excessive there is little loss of strength. BRED 273 is entitled *Perforated clay bricks*.

Pressing Stiff clays, without the addition of water, can be formed into bricks by mechanical pressure, and this obviates the need for drying the green bricks in a separate drier.

Pressed bricks have frogs on one bed face but pressed engineering bricks often have shallow panels on both bed faces. Cellular bricks with very large 'frogs' are also formed in this way.

Fletton bricks, first made in the village of that name, are made in Northamptonshire and Bedfordshire by the *semi-dry* process. In this method the clay is ground to a powder pressed automatically and dried and fired in modified Hoffman kilns. Engineering bricks are an example of bricks made by a *stiff-plastic* process.

Harder clays found in Scotland, the North of England, Staffordshire and South Wales are ground, either dry, or with water in edge-runner pans, mixed with water, extruded into slabs and pressed into bricks.

Drying and firing

'Green' bricks must be dried before they are fired. Natural drying in *hacks*, stacks about seven courses high, protected from rain, is rare today and most bricks are dried, either in a separate drier or, in the case of *semi-dry* and *stiff-plastic* processes, in the drying zone of a continuously fired kiln.

Setting, or stacking one green brick upon another in a chequerwork pattern, is done mechanically, but the skilled operation of setting for

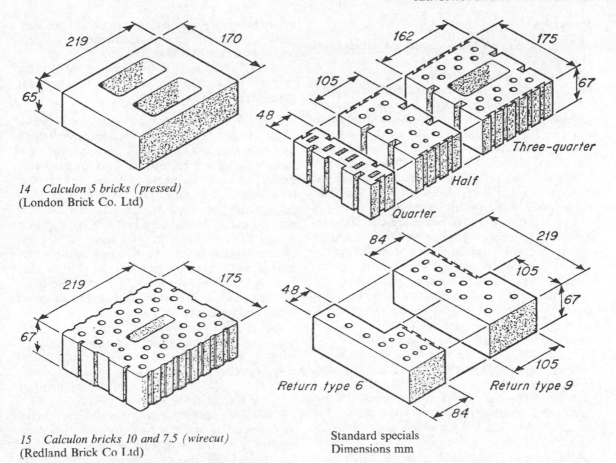

14   Calculon 5 bricks (pressed)
(London Brick Co. Ltd)

15   Calculon bricks 10 and 7.5 (wirecut)
(Redland Brick Co Ltd)

Three-quarter

Half

Quarter

Return type 6

Return type 9

Standard specials
Dimensions mm

firing is still done mainly by hand. Stacks of bricks are moved by forklift trucks.

The parts of bricks which are covered by other bricks during drying and firing, usually burn to the natural colour of the fired clay and show as *kiss marks*. Exposed parts are covered with 'scum' resulting either from salts which migrate to the surface of a brick or from a reaction between the kiln gases and lime in the clay.

Firing or *burning* of bricks in which fuel is incorporated in the clay is still carried out to a very small extent in *clamps* in which the green bricks are set in layers alternating with coke breeze, the whole being walled and roofed over with waste bricks from previous firings. The whole mass burns and when it has cooled, the bricks are sorted. Underburnt bricks are set in a further clamp and overburnt bricks or *'burrs'*, which are highly vitrified and often badly distorted, can be used for random walling.

Today, most clay bricks are fired in permanent kilns, either intermittent or continuous-burning.

Intermittent kilns are used only where special colour effects are required, as with Staffordshire blue bricks and some soft-mud bricks.

The great majority of bricks are fired in continuous kilns, which are economical for large scale production. They are of two types: (i) those in which the fire is moved around a circuit of interconnected chambers, eg the *Hoffman kiln*, and (ii) the *tunnel kiln* in which the bricks, usually predried, are moved through a stationary fire. In both cases, the green bricks are preheated by the outgoing hot gases from the firing zone and the fired bricks are cooled by the incoming air required for combustion.

Properties

Although the range of clay bricks available in this country is very wide, where bricks are to be used in constantly wet conditions, or where a particular appearance or very high strength are required the choice may be severely limited. Wherever possible, selection should be based on observations of the behaviour of bricks for a reasonsble period, in conditions approximating as closely as possible to those in which they will be used. BDA Practical Note no 3 *The selection of bricks for performance* and BRE Digests 164 and 165 *Clay brickwork 1 and 2* are useful references.

The properties of clay bricks are characteristically variable even within one delivery to a site, so that tests carried out on isolated samples can be very misleading. BS 3921 recommends that manufacturers should carry out frequent tests and make the results available in the form of control charts. Such records, by showing variations in properties with time, provide more information than can be deduced from isolated tests.

In general, good clay bricks have a compact texture, are reasonably free from cracks, lime, stones and pebbles and the harder varieties give a metallic ring when struck with a trowel. Good bricks are well-burnt, ie they have achieved a good ceramic bond. This is not generally indicated by simple tests although an expert who is familiar with a particular type of brick may recognize a well-fired sample by its colour and hardness.

The following properties are now considered in greater detail:

> Appearance
> Strength
> Water absorption
> Soluble salt content
> Chemical resistance
> Frost resistant
> Moisture movements
> Thermal movement
> Fire resistance
> Sound insulation
> Weight

Appearance

Formerly bricks of good appearance were obtained by careful selection from bricks which would otherwise have been regarded as *commons*. Now most *facings* are obtained by special processes, and care in manufacture.

BS 3921, Appendix F is a useful guide, if not a numerical classification, to the assessment of appearance of facing bricks.

Colour is either integral or superficial. *Integral colour* results directly from the characteristics of the various clays from which *common and some facings* are made. Natural colours of fired clay bricks include red, white, yellows, brown and blue, but not green. Many facing bricks are permanently coloured by the addition of inert mineral oxides to the clay body or by applying oxides to their faces before firing. Generally, colour darkens as the temperature of firing increases and where the normal colour of a clay brick is known to be dark, a light coloured brick (or *'salmon'*) suggests inadequate firing, with associated lower strength and durability.

Red bricks Red, caused by iron oxide, is perhaps the typical clay brick colour, produced in most parts of the country. Red-purple multicoloured stocks made from 'soft-mud' incorporating fuel in the mixture are formed by hand or machine mostly in the Home Counties.

White bricks are made from fireclay and chalky clays including *gault* clay found in Yorkshire, shaped by either wire cutting or by pressing. Gault clays can also cause primrose yellow and pink or red colours, depending on kiln temperature and atmosphere.

Yellow bricks include London Stocks which are made from the brickearth and chalk found in Kent and Essex. Fuel added to the clay before firing leaves characteristic voids and burnt particles in the fired brick.

Brown bricks in various shades from light to dark are now increasingly used.

Blue bricks result from a high content of iron compounds and firing in reducing (low oxygen) conditions as in the Staffordshire blue engineering brick.

Superficial colour results from a surface treatment such as *sand facing*, from oxide pigments applied before firing, or from control of the atmosphere in the kiln. Damage in handling, or at a later stage, exposes the body of the brick the appearance of which may be quite different from that of the thin facing.

Texture results from the method of forming, from mechanical treatment, or from sand or broken brick particles blasted onto the surfaces of the *green* bricks.

Glazes High cost, and fashion changes, have stopped production of glazed bricks in UK. The processes were:

Salt glaze The extremely hard and impervious transparent *salt glazes* were a standard finish on drainage goods. Suitable clays limited the colours to brown and red-brown but the process is simple: sodium introduced into the kiln during firing of the bricks reacts with the silica in the clay forming sodium silicate.

Applied glazes These enamel glazes are opaque and black, white or any colour and the bricks were very much more costly than salt glazed products. The *green* brick is first partially fired to the *biscuit* condition. It is then sprayed with, or dipped in a glazing *slip* which must match the body so that it is not crazed by the expansion of the latter during subsequent refiring in a kiln which is enclosed to avoid contact with the products of combustion.

Strength

Approximate ranges of compressive strengths of various bricks are shown in table 45.

Clay bricks may have strengths up to about 180 N/mm^2, and by taking into account their high strength and that of mortars, in accordance with the rules laid down in CP 111, economies can be effected in the thicknesses of multistorey walls. However, considerably lower strengths are adequate for the loadings which are usual in small buildings. The Building Regulations 1985 require only 2.8 N/mm^2 for inner walls of two-storey houses.

BRE Digest 65 emphasises that every clay brick is an individual with wide differences in strength which are not evened out when bricks are stacked on a truck. Quality control testing indicates how much uncertainty attaches to an isolated test result, and BS 3921 encourages its adoption for the benefit of users as well as manufacturers.

BS 3921 requires crushing strength to be assessed on the average minimum results of 10 bricks, selected and tested with frogs filled in accordance with clause 9 and Appendix D, respectively.

Type	Compressive strength N/mm^2
Calcium silicate bricks – range (BS 187)	7–50
Concrete bricks – range BS1180	7–40
CLAY BRICKS	
Clay bricks – range	4–180
Hand moulded	12–35
Diatomaceous earth	2.76–5.62
Red-purple, soft-mud, multi-coloured	12–23
London stocks	6–25
Flettons – solid (BS 3921)	20.5 and 27.5
– cellular	7.0
Facings	20–120
Loadbearing common bricks	7–60
Calculon bricks C 5	34.5
B 7.5	51.7
A 10	68.9
Engineering Class A	50.0
Class B (BS 3921)	70.0

Table 45   *Approximate compressive strength ranges of bricks*

Water absorption

The water absorption of clay bricks is expressed as percentage of the dry weight of a brick. It varies from less than 3 to about 30 per cent. BS 3921 requires the results of a 5 h boiling test to be expressed as the average percentage of the dry weight of 10 bricks selected and tested in accordance with Clause 9 and Appendix E, respectively. Neither water absorption nor *saturation coefficient* (the ratio of 24 hour cold absorption to maximum absorption) are reliable indices of durability.

For the design of laterally loaded walls (BCRA SP 90) use is made of the characteristic flexural strength of brickwork and this depends upon the water absorption of the brick used, the grade of mortar and the direction of bending of the brickwork.

BS 5628 *CP for use of masonry* Part 1 relates water absorption to flexural strength and BS 3921 limits the water absorption of Engineering and DPC classes of bricks – see table 46.

Designation	Class	Minimum average compressive strength[1] (N/mm²)	Maximum average water absorption after 5 h boiling[2] (percentage by weight)
Engineering bricks – BS 3921	A B	70.0 50.0	4.5 7.0
Bricks for damp proof courses – BS 3921	DPC 1[3] DPC 2	5.0 5.0	4.5 7.0
All others – BS 3921		5.0	no requirements
Building Regulations 1985 Load-bearing walls not exceeding 2.7 m ceiling–ceiling in 2 storey houses – external – internal		7.0 5.0	no requirements

[1] Average crushing strength of 10 bricks sampled and tested in accordance with clause 9 and Appendix D, BS 3921, respectively
[2] BS 3921 test
[3] Recommended for use in buildings by BS 5628 Part 3:1985

Table 46   Strength and water absorption requirements for clay bricks

Soluble salt content

All clay bricks contain some soluble salts varying in quantity from brick to brick even within one delivery. Soluble salts derive from the original clay or from its reaction with sulphur compounds from the fuel used for firing the bricks. Salts in bricks and, indeed, from other sources, sometimes cause staining, efflorescence, decay of certain bricks and, as sulphates, expansion and disintegration of mortars and renderings.

Ferrous sulphates often react with lime in fresh mortar and cause brown stains, and salts of vanadium and chromium sometimes cause yellow and pale green patches.

BS 3921 categories for maximum salt contents, by mass of soluble ions, are:

L – *low*	calcium	0.30 per cent
	potassium	0.030 per cent
	sodium	
	sulphate	0.50 per cent
N – *normal*		no limits

Efflorescence Soluble salts in bricks, in particular magnesium and sodium salts, and to a less extent calcium and potassium salts, together with lime compounds in fresh cement-lime mortars may be carried to the surface and crystallize as a white or near-white deposit. The extent to which this efflorescence appears depends upon the amount and solubility of salts present and upon wetting and drying conditions. Efflorescence is often seen, particularly in the spring, when bricks dry out after having been allowed to become wet in the stack on the site, or during laying. However, efflorescence is likely to show at any age if brickwork is saturated and subsequently dries out. Parapets and retaining walls often do this. Efflorescence on brickwork may also derive from mortars and adjacent materials. Whether efflorescence shows mainly either on the bricks or on the mortar is largely determined by their relative permeabilities and consequent rates of evaporation from the respective surfaces.

In crystallizing, salts expand, and if this occurs below surfaces (*crypto-efflorescence*), crumbling of underfired bricks may result. BS 3921 categorizes bricks by the worst display of salts on the surfaces of bricks in samples as follows:

Nil   –none perceptible

Slight   –up to 10 per cent of area of face and no powdering or flaking of surfaces.

Moderate –more than 10 per cent but not exceeding 50 per cent of area and no powdering or flaking of surfaces

Heavy   –more than 50 per cent and/ or flaking of surfaces. Consignments of bricks are not acceptable if any of them is in this category.

Clay bricks contain sulphates, more particularly if they are fired at lower temperatures.

Sulphate attack on mortars and renderings

In persistently wet conditions sulphates react slowly with tricalcium aluminate (a constituent of Portland cement and hydraulic lime) and the calcium sulpho-aluminate formed causes it to expand and later to soften and disintegrate. Sulphate attack is common on mortars in clay brickwork which remains very wet for long periods. BRE Digest 89, *Sulphate attack on brickwork*, states that a vertical expansion of 0.2 per cent is quite common in facing brickwork where sulphation of the mortar has occurred. Sometimes renderings are attacked.

Ideally, all brickwork should be kept dry and free from contamination, but where persistently wet conditions are unavoidable the sulphate content of clay bricks should not exceed 0.5 per cent. If bricks with such a low sulphate content are not available or if sulphate attack may result from soils, flue condensates or from gypsum plaster being accidentally or mistakenly mixed with Portland cement, sulphate resisting Portland cement should be used in mortar – see page 154.

It is strongly recommended that expansion joints should be provided in clay facing brickwork at not more than 12 m centres.

Chemical resistance

Decay of less durable bricks results from the drainage of water from the calcium sulphate deposits which form on limestone in polluted atmospheres. Generally, however, burnt clay products have high resistance to most chemicals. In particular engineering bricks with suitable mortars are satisfactory for sewerage, industrial chimneys and pickling tanks. BS 3679:1963 *Acid resisting bricks and tiles* gives requirements for four types (not including floor quarries) in respect of sizes, composition and texture. Apparent porosity of red, blue and refractory bricks and tiles must not exceed 12 per cent and a percentage to be agreed for chemical stoneware bricks and tiles. Tests are described in the Standard.

Frost resistance

As with other properties, clay bricks vary widely in their resistance to frost. Also exposure hazards differ greatly and materials which may be perfectly satisfactory for walling between damp-proof course and roof, may decay rapidly if they are frozen whilst they are saturated, as may occur in earth retaining walls and brick on edge copings.

Neither high strength, low water absorption, or saturation coefficient are satisfactory indices of frost resistance. Thus, some bricks which are included in the *engineering class* are not frost resistant and some bricks approaching 48 N/mm^2 in crushing strength decay rapidly when they are wet and subject to frost. Flettons with crushing strengths from 21.0 – 28.0 N/mm^2 should never be used as brick-on-edge copings. On the other hand, some weaker bricks such as London stocks 6.0 – 25.0 N/mm^2 may be frost resistant.

Where bricks may be frozen while they are saturated, bricks of *class F* are needed. At present the best evidence of frost resistance is provided by use in buildings for a number of years and BS 3921 requires bricks to have performed satisfactorily for three years under conditions of exposure at least as severe as those proposed in a building in the locality in which it is intended to use them. If no such building exists, sample panels built in an exposed position under the supervision of an independent authority will be acceptable.

Although no wholly satisfactory accelerated tests have been devised, BS 3921 describes *Classes F, M* and *O*.

Class F   —*frost resistant* bricks are not adversely affected when saturated and repeatedly frozen and thawed and when subjected to 100 cycles of freezing and thawing when incorporated in brickwork panels.

Class M   —*moderately frost resistant* bricks are durable where they are saturated but not subjected to repeated freezing.

Class O   —*not frost resistant* These may be suitable for internal use if they are protected from water during and after construction as recommended in BS 5628, Part 3.

Durability

BS 3921 gives 6 designations based on both *Frost resistance* and *Soluble salts* assessments, ie: *FL, FN, ML, OL* and *ON*
The standard points out that while a *FL* brick would be required for brick facings exposed to the weather an *ON* brick would suffice if the brickwork is rendered.

Moisture movements

The *reversible* moisture movement of clay products is less than that of concretes and calcium silicate bricks and is generally of no practical significance.

Material	Movements	
	Irreversible per cent	Reversible per cent
Clay bricks	0.10–0.20 Ex	negligible
Calcium silicate bricks	0.001–0.05	0.001–0.05
Lightweight concrete blocks	— 0.06 Sh	0.05
Concrete bricks	0.03	0.04–0.06

Ex – expansion   Sh – shrinkage
Table 47   Approximate moisture movement of typical bricks and blocks

On the other hand, *irreversible* expansion of clay products due to adsorption of moisture from the atmosphere, varying with types of clay and temperature of firing, may amount to 0.1, or exceptionally 0.2 per cent.

Figure 16 shows the rate of movement up to 100 days. BRE CP 16/73 states that at $7\frac{1}{2}$ years the expansion is about $1\frac{2}{3}$ times more than at 127 days. The further movement up to 60 years is expected to be only 'a tiny fraction' greater than at $7\frac{1}{2}$ years.

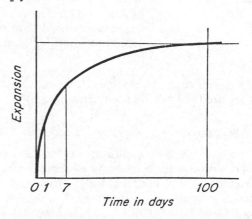

16   Expansion of clay bricks after firing
Graph from BRE Digest 65

Although the movement of brickwork is not directly related to the movement of unrestrained bricks, the expression of movements at 127 days as an *index* would be useful in design. In particular, a brick with a low index would be used where its expansion may be compounded by the drying, and creep shrinkages of a reinforced concrete structural frame.

Conversely, where bricks have a high index, or are recently fired, particular attention would be paid to the design and execution of movement joints in brickwork. Because about half of movement takes place in the first seven days after bricks begin to cool after firing, BRE Digest 65 recommends that at least a week should pass before bricks are used, particularly where they are to be laid in a strong mortar which will become rigid very quickly. (The *Model specification for loadbearing clay brickwork* issued by the British Ceramic Research Association recommends that 'no bricks shall be built into the work until two days have elapsed from the time of drawing from the kiln'.)

Movement joints CP 111 advocates the inclusion of vertical movement joints at 10–15 m intervals, and horizontal joints at every fourth storey or 12 m to allow for differential movements between external and internal leaves of walls.

Thermal movement

The coefficient of thermal expansion of typical clay bricks is $5–8 \times 10^{-6}$ deg C. Although small, thermal movements can be significant, for example between the inner leaf of a cavity wall and an outer leaf of dark coloured bricks exposed to solar radiation.

Diatomaceous bricks have negligible thermal expansion.

Thermal conductivity

Conductivity varies with density. Diatomaceous earth bricks which are very light, have a 'k' value of about 0.14 W/m deg C and can be used for lining flues carrying gases up to 870°C but ordinary clay bricks are poor insulators. A 'k' value of 1.15 W/M deg C is given for *common* bricks in the *IHVE Guide*. Conductivity also varies with moisture content and table 48 shows thermal conductivities for brickwork with various bricks and having moisture contents which may occur in inner and outer leaves of cavity walls.

Fire resistance

Having been fired at temperatures higher than those which normally arise in building fires, clay bricks provide excellent resistance to fire, and special refractory bricks are available. BS 1758:1966 covers *Fireclay refractories* (bricks and shapes).

Sound insulation

Well constructed brickwork provides insulation against airborne sound in ratio to its density and thickness although this is seriously reduced by bad bricklaying which leaves small even paths for sound through a wall.

Weight

Approximate average dry weights of clay bricks are:

	kg
Calculon 10 and 7.5	4.09
Calculon 5	3.95
Engineering	
Accrington red, pressed	3.86
Staffordshire blue, pressed	3.27
Fletton: common	2.27
cellular	1.82
London Stock	2.14–2.27
Diatomaceous earth brick	0.91

Brick type	Approximate density of brickwork (kg/m^3)	Approximate conductivities (k) of brickwork* (W/m deg C)	
		Protected (1 per cent ^+mc)	Exposed (5 per cent ^+mc)
London Stocks	1 440	0.44	0.60
Multi-coloured stocks (soft mud process)	1 540–1 600	0.50–0.54	0.68–0.73
*Flettons:*ˣ			
Cellular	1 470	0.46	0.62
Facings and commons (frog down)	1 700	0.62	0.84
Facings, commons and Calculons (frog up)	1 800	0.71	0.96
Perforated wirecuts	1 800–2 180	0.71–1.16	0.96–1.57
Solid wirecuts	2 050–2 100	0.98–1.05	1.33–1.42
Dense, high strength pressed or wirecut	2 200–2 300	1.18–1.33	1.60–1.80

*The thermal conductivities have been taken from BRE Digest 108 for corresponding bulk dry densities
$^+$Moisture content expressed as a percentage by volume
ˣLondon Brick Company products (k values are based on estimated maximum densities of brickwork)
Table 48   Approximate thermal conductivities of clay brickwork

CALCIUM SILICATE BRICKS

Calcium silicate bricks, commonly called either sandlimes or flintlimes according to the aggregate employed, were invented in 1866, and in 1967 some 5 per cent of all bricks produced in this country were of this type.

BS 187:1978 *Calcium silicate (Sandlime and Flintlime) bricks* describes six classes.

BRE Digest 157 deals with *Calcium silicate brickwork*.

The materials used, and the method of forming them by pressing followed by steaming, make for sharp arrises and remarkable uniformity in strength, size, shape, colour and texture.

In appropriate grades, the strength, hardness and durability of calcium silicate bricks is adequate for almost all purposes but they must not be exposed to acids or to strong salt solutions. Fire resistance can be taken as being equal to that of clay bricks. Values for sound and heat transmission through walls approximate to those for clay bricks of equal density. Calcium silicate bricks are free from soluble salts, the presence of which in clay bricks is referred to on page 133. The BS requires the bricks to be free from visible cracks, balls of clay or loam and particles of lime.

Unlike new clay bricks which tend to expand, calcium silicate bricks (and even more so concrete bricks) shrink when they dry out and this must be taken into account in the design of brickwork if cracking is to be avoided.

Pavers

BS 6677, Part 1:1986 deals with *Calcium silicate pavers for flexible pavements*. Type PA are suitable for light traffic and type PB for roads. Common sizes are: 200, 210 and 215 × 100, 102.5 and 105 × 50 and 65 mm

Manufacture

Controlled proportions of sand, crushed or uncrushed flint or combinations of such materials are mixed with hydrated lime or quicklime and water, to which stable and inert pigments may be added. Unlike clay bricks, calcium silicate bricks are not burnt. After pressing, the bricks are steamed at pressures up to 1.7 N/mm^2 for about eight hours in an autoclave, during which process the lime reacts with the surfaces of the silica particles forming hydrated calcium silicates and leaving practically no free lime.

When exposed to the atmosphere, the calcium silicate gradually reacts with carbon dioxide and forms calcium carbonate, the brick slowly gains in strength and hardness and finally resembles a natural calcareous sandstone.

Size

BS 187:1978 lays down *Coordinating sizes* of 225, 112.5 and 75 mm and the following *work sizes* (mm):

	Lower limit	Work size	Upper limit
Length	212	215	217
Width	101	102.5	105
Height	63	65	67

An alternative *work size* height of 73 mm may be available and *special shapes* to BS 4729 are made by most manufacturers.

Modular format sizes with *works sizes* of 290 × 90 × 190 × 90 × 90 and 65 mm are being made.

The smaller dimensional variations of calcium silicate bricks make for easier brick laying and more uniform joint widths than with clay bricks.

Dimensions are deemed to be acceptable where the deviations in the individual dimensions of ten bricks, sampled as described, are within stated limits.

Calcium silicate bricks are usually solid with frogs in one bed face.

Special shapes to BS 4729 are made by most manufacturers and calcium silicate bricks are now available in solid and cellular types as well as those with frogs.

Properties

Strength

Table 49 gives the strengths of the six classes in BS 187:1978. (Very few bricks with crushing strength below 14.0 N/mm^2 are now made, and class 1 is omitted.)

Use	Class	Minimum mean crushing strength of 10 bricks[1] N/mm^2	Minimum 'predicted lower limit of crushing strength'[2] N/mm^2	Colour marking
Loadbearing	7	48.5	40.5	Green
or	6	41.5	34.5	Blue
Facing	5	34.5	28.0	Yellow
	4	27.5	21.5	Red
	3	20.5	15.5	Black
Facing or Common	2	14.0	10.0	—

[1] Calculated on smaller bed face after deducting area of frogs
[2] The 'predicted lower limit of crushing strength' is defined in Appendix 'C' of BS 187

Table 49   Minimum strengths for uses of BS 187: 1978 Calcium silicate bricks

In addition, some makers can provide calcium silicate bricks with average crushing strengths up to 50 N/mm^2 or even higher, but the extremely high strengths of *Class A* clay *engineering* bricks cannot yet be obtained commercially. Flint-lime bricks with crushed stone aggregates show substantially higher strengths when tested with the frogs filled with mortar, but sandlime bricks shown no significant increase in strength where the frogs are filled.

Bricks are tested wet. Requirements for uniformity of strength, which would not be practicable for clay bricks, are included in the BS for calcium silicate bricks.

Hardness

Hardness (resistance to abrasion) varies with class and quality of brick, the stronger classes being suitable for lightly trafficked paving.

Moisture movement

The extent of moisture movement is intermediate between that of clay and concrete bricks and wet bricks built into long walls, particularly if a strong mortar is used, tend to crack when they dry.

Initial irreversible drying shrinkage is equal to subsequent reversible moisture movement, ie 0.001 to 0.05 per cent linear. The 1978 specification limit is 0.04 per cent for all classes of bricks except *class 2*. This maximum presumes that precautions, such as the following will be observed for calcium silicate brickwork above ground level:

1 If available, use bricks with a drying shrinkage substantially lower than 0.04 per cent.
2 Keep the bricks as dry as possible before and during construction and until construction is complete. If wetting is essential in very hot weather as little water as possible should be used.
3 Use a mortar which contains no more Portland cement than is required to suit loading, resist possible freezing during bricklaying and the exposure to be expected in service. A relatively weak mortar is able to accommodate the moisture movement of the bricks so that any hair cracks tend to form around individual bricks, and cracks through both joints and bricks are unlikely to occur.
4 Avoid restraint, where possible, and provide movement joints at susceptible places such as between the heads and sills of superimposed windows, at changes in thickness or height and at intervals of not more than 8 m in long walls. Movement joints are also required between calcium silicate and clay brickwork to allow for differential and possibly opposed movements of the different materials. Joints may be filled

139

with a resilient material to within 13 mm of the outside face and either sealed with mastics as the work proceeds, or left open until the construction water has evaporated from the brickwork before being pointed with mortar. Rigid fixings or renderings must not be carried over such contraction joints. Shrinkage of internal walls can usually be accommodated by joints including polythene, building paper or similar separators.

Brickwork			Minimum BS 187 class brick	Appropriate cement: lime:sand mortar	
				Risk of frost during construction	
				nil	possible
Internal	Inner leaf of cavity walls	plastered		1:3:10–12	
	Internal walls	unplastered			
External	Backing to external solid walls		2	1:2:8–9	
	External walls Outer-leaf of cavity walls Facing to solid walls	above damp proof course near to ground level			
		Below this dpc but more than 150 mm above ground level			
		within 150 mm of ground or below ground		1:1:5–6	
	Free-standing walls		3		
	Parapets with dpcs at top	unrendered			
		rendered		1:2:8–9	
	Brick sills and copings		4	1:$\frac{1}{2}$:4–4$\frac{1}{2}$	
	Earth-retaining walls back-filled with free-draining materials				

[1]Cement:sand with plasticizer, and masonry cement:sand mixes equivalent to cement:lime:sand mixes are:

cement:lime:sand	cement:sand with plasticizer	masonry cement:sand
1:0–$\frac{1}{4}$:3	—	—
1:$\frac{1}{2}$:4–4$\frac{1}{2}$	1:3–4	1:2$\frac{1}{2}$–3$\frac{1}{2}$
1:1:5–6	1:5–6	1:4–5
1:2:8–9	1:7–8	1:5$\frac{1}{2}$–6$\frac{1}{2}$
1:3:10–12	1:8	1:6–7

Notes: The proportions of lime are for lime putty.
Where sulphates are present in ground water, sulphate-resisting Portland cement may be required.
Higher-strength bricks and mortars may be necessary for calculated brickwork. This table should be read in conjunction with chapter 15 *Mortars for jointing*

Table 50 Minimum quality calcium–silicate bricks and mortars for durability

Durability

Calcium silicate bricks of the quality appropriate to the type of use have been found to have 'satisfactory durability' over a period of at least fifty years. Wetting and drying and repeated freezing and thawing have little significant effect on bricks of 14.0 N/mm^2 compressive strength and above. Such bricks can form external paving, BS 6677, where salt will not be used for defrosting, and flooring where the possibility of staining is not a matter for concern. Stronger bricks are recommended in very exposed conditions.

Sulphur dioxide acts chemically on the calcium silicate bonding agent and in severely polluted atmospheres surface erosion and blistering have occurred, but only after 25 years and in bricks of relatively poor quality. Calcium silicate bricks should not be used where they would come into direct contact with sewage, or like cement products, where they would be exposed to strong acid fumes or to splashing by acids. The bricks are also attacked by high concentrations of magnesium or ammonium sulphates eg in industrial wastes.

Sea air is not damaging, but calcium silicate bricks may deteriorate if they are repeatedly wetted with sea water or solutions of sodium chloride or calcium chloride, particularly in severe frost when they are saturated with such solutions. Sulphate salts which occur in some soils and ground waters do not harm calcium silicate bricks but sulphate-resisting cement should be used in the mortar. The higher strength grades should be used below damp-proof courses.

Appearance

Quality control in manufacture enables bricks of uniform colour and texture to be produced in different strengths.

Colour The natural colours are white, off-white, cream or pale pink according to the aggregates used. The tone darkens when the bricks are wet. By including pigments, any colour from pastel to dark tones is obtainable. Cost is increased, especially for sky blue or deep green colours. One manufacturer is producing multi-coloured bricks. New bricks may initially, show white *lime bloom*, but this is not harmful.

Texture Sandlime bricks are usually smooth and fine-textured, and flintlime bricks are rougher and coarser. Intentionally textured bricks are now made. Care must be taken to prevent arrises of calcium silicate bricks being damaged.

Fire resistance

Only in BS:5390 is any distinction in fire resistances made between clay and calcium silicate brickwork, where loadbearing, single leaf, unplastered calcium silicate brickwork walls are required to be thicker in order to achieve 3 and 4 hour resistances.

Calcium silicate bricks are suitable for chimneys, where they are not constantly subjected to red heat.

Absorption

The water absorption of calcium silicate bricks is generally higher than for most clay bricks but it is not a satisfactory indicator of durability of the permeability of brickwork. Weather resistance of brickwork is determined by the size, structure and distribution of pores, by the mortar and the effectiveness of the bond between bricks and mortar.

Sound insulation

This is as for clay bricks of the same density.

Thermal insulation

The k value of calcium silicate bricks is 0.5–0.7 W/m deg C for densities of 1 630–1 800 kg/m^3.

For practical purposes, brickwork can be taken to have the same k value as clay brickwork of the same density.

Thermal expansion

Thermal expansion is 0.014×10^{-3}°C. As with other bricks, BS 5390:1976 recommends that allowance should be made for possible stresses in long walls.

Weight

This is similar to that of average clay bricks, ie 2.7 to 3.4 kg. The density of calcium silicate brickwork is 1 620–2 400 kg/m^3.

Typical densities in kg/m³ being:

	Class 3	Class 5
Sandlimes	1 630	1 700
Flintlimes	1 800	2 000

Cost

For bricks having comparable properties and appearance the cost of calcium silicate bricks is often very competitive with that of clay bricks. Moreover the accuracy and uniformity of shape and size makes them easy to lay.

CONCRETE BRICKS

Concrete bricks are harder, more difficult to cut and less pleasant to handle than clay or calcium silicate bricks and are less common.

Drying shrinkage varies considerably, from 0.019 to 0.080 per cent of the length and is greater than that of calcium silicate bricks. As with those bricks, special care should be taken in selection, design and in building, to avoid cracking of brickwork after construction.

Concrete bricks are described in BS 6073 *Precast concrete masonry units,* Part 1:1981 *Specification* and Part 2:1981, *Method for specifying*. The specification includes units for fixing, but not for paving.

BS 6073 requirements

Some definitions differ from those in BS 3921.

Solid bricks may have small holes (not exceeding 20 mm wide or 500 mm² in area) which pass through or nearly through up to 25 per cent, or frogs up to 20 per cent of the total volume. Up to 3 holes each not exceeding 3250 mm are allowed as aids to handling.

Perforated bricks have small holes passing through exceeding 25 per cent of the volume and holes for handling as for solid bricks.

Hollow bricks may have holes passing through up to 25 per cent of the volume.

Cellular bricks have holes closed at one end up to 20 per cent of the whole volume.

Strength BS 6073 specifies only one crushing strength for ordinary bricks, ie:

7.0 N/mm² minimum average of ten bricks with a coefficient of variation not exceeding 20 per cent. Stronger bricks are 10,15, 20, 30 and 40 N/mm².

Special purpose bricks exceed 40 N/mm², and 350 kg/m³ cement content.

Durability Good quality concrete is resistant to frost and BS 6073 implies that bricks complying with this specification are suitable in all usual exposures.

Drying shrinkage must not exceed 0.06 per cent.

Sizes

BS 6073 specifies the following *works sizes* (mm)

Length	Height	Thickness[1]
290	90	90
215	65	103
190	90	90
190	65	90

[1] *BS 3921 refers to width*

It is unlikely that any one manufacturer produces all these or all the BS 4729 *Standard special bricks* sizes.

Maximum deviations (mm) are +4, −2 on lengths and +2, −2 on height and thicknesses.

Co-ordinating sizes are normally 10 mm greater than works length and height and 9.5 mm greater than thickness.

Information is provided by the Concrete Brick Manufacturers' Association, 60 Charles Street, Leicester, LE1 1FB.

Manufacture

The principles stated in chapter 8 for making good-quality concrete apply.

Standards are listed for binders, aggregates, additives and admixtures, including pigments. It is pointed out that sulphate-resisting cements are likely to be effective only if bricks are made as recommended by the cement manufacturers for particular conditions in use.

The *CBMA* states that 'brickwork should not be restrained against free movement by adjacent parts of a structure, including its foundations'. It is recommended that with a mortar not stronger than 1 Portland cement: 2 lime: 8–9 sand, movement joints should divide brickwork into rectangular panels 6–7.5 m long. Movement joints should occur on both sides of openings in walls, or bed-joint reinforcement may be provided above and below the opening to prevent cracks forming. Reinforcement can also increase the 'safe' lengths of panels.

Fixing bricks of lightweight concrete are required to be 'of a consistency to permit the easy driving, of, and to provide a good purchase for nails and screws'. They are often useful for 'coursing' purposes in conjunction with lightweight concrete blocks.

BS 6073 requires a minimum crushing strength of 2.8 N/mm^2.

BLOCKS

Blocks are larger than *bricks*, and whereas a brick can usually be lifted in one hand, a block usually requires two hands. *Slabs* are larger again.

CP 121 Walling, Part 1:1973 deals with *Brick and block masonry*. Blocks can be laid more quickly than bricks. Generally, blocks are intended to be plastered or rendered, but many products, more particularly concrete blocks, are sufficiently regular in shape and size and otherwise of good appearance without plaster for walling.

Block heights of 215 and 290 mm relate to three and four brick courses and a block 440 mm long relates to two brick stretchers, all with 10 mm joints.

Strength

The Building Regulations 1985 Approved Document A states minimum strengths for blocks in walls where the ceiling to ceiling height does not exceed 2.7 m for: *three-storey dwellings*: outer leaves to cavity walls throughout and inner leaves and internal walls to ground floor 7 N/mm^2.

CLAY BLOCKS

BS 3921:1974 dealt with both clay bricks and blocks, but there is now no BS for clay blocks, which are no longer manufactured in this country.

Clay blocks are generally extruded hollow units. After firing, ordinary clays are dense and brittle which presents problems in cutting and fixing. Blocks made from diatomaceous earth, however, are less dense, can be easily cut and they accept nails well.

Inner leaves and internal walls to floors 1 and 2 may be 2.8 N/mm^2.

Two-storey dwellings: all walls can be 2.8 N/mm^2. If ceiling to ceiling heights exceed 2.7 m, strength of blocks must be at least 7 N/mm^2. Mortar must not be weaker than 1:1:6 – see *Mortars*, chapter 15). Minimum thicknesses are 190 mm for solid external walls and 90 mm for each leaf of external cavity walls and for internal loadbearing walls. For internal loadbearing walls 90 mm minimum is recommended.

Often with economic advantage, structural stability requirements are satisfied by calculation in accordance with CP 111, or BS 5628 Part 1:1978. Preliminary design block strengths for CP 111 are: 2.8 N/mm^2 for one and two storey houses; 3.5 N/mm^2 for three storey houses flats and maisonettes and 5.0 – 7.0 N/mm^2 for four storey flats and maisonettes. The higher strengths are proposed where opening sizes are larger than normal.

Sizes

The former BS *Standard formats* for wall blocks are given in table 51 and for floor blocks in table 52.

Designation			Actual dimensions (mm)		
			length	width	height
300	62.5	225	290	62.5	215
300	75	225	290	75	215
300	100	225	290	100	215
300	150	225	290	150	215

Table 51   Formats of clay wall blocks

Figure 17 shows typical clay wall blocks and figure 18 shows a block coursing with bricks.

Figure 19 shows typical blocks for 'pot' floors with a 'filler tile' to give uniform suction for plaster.

Designation	Actual dimensions (mm)		
	length[1]	width	depth
200, 300	295	295	75
			100
			125
			150
			175
			200
			225
			250

(Designation column values: 75, 100, 125, 150, 175, 200, 225, 250)

[1]Length is measured along the direction which is normally parallel to the concrete reinforcement in the floor

Table 52   Part 2: Formats for clay floor blocks

Corner, closer, fixing blocks and blocks with recesses for conduits were also available.

Strength

The BS 3921:1974 minimum average compressive strengths for clay blocks tested in accordance with clause 43 of the BS were:

Non-loadbearing partitions	1.4 N/mm²
Facing and common blocks Ordinary and Special qualities Blocks for loadbearing internal walls	2.8 N/mm²
Hollow blocks for structural floors and roofs	14 N/mm²[1]

[1]This minimum strength differs from that given for blocks for walling because the methods of testing, and expressing the results were different.

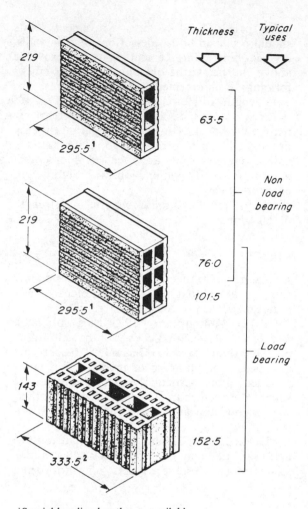

Non load bearing

Load bearing

[1]Special bonding lengths are available.
[2]These blocks can be easily cut to length.

17   Typical clay wall and partition blocks

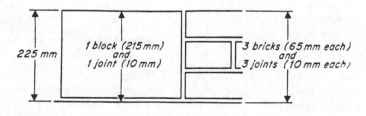

225 mm

1 block (215mm) and 1 joint (10mm)

3 bricks (65mm each) and 3 joints (10mm each)

18   A block coursing with bricks

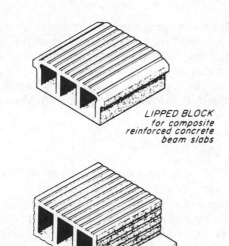

19 *Typical clay hollow blocks and filler tile for rc floors and roofs*

CONCRETE BLOCKS

References

BS 6073 *Precast concrete masonry units* (ie concrete bricks and blocks other than paving blocks).

Part 1:1981 *Specification for precast concrete masonry units.*

Part 2:1981 *Method for specifying precast concrete masonry units.* (Part 3 may contain information on appropriate uses.)

Concrete blocks, Michael Gage and TW Kirkbride, Architectural Press.

BS 6073 blocks used in their usual aspect, exceed the length, width or height specified for bricks, have heights not exceeding six times their thickness and do not exceed 650 mm in any size direction.

Concrete blocks are termed either *dense* or *lightweight*. See pages 180 and 196.

Density is not directly related to strength and is not specified in BS 6073.

So called *lightweight* blocks may be less than one third of the weight of *dense* blocks and can be laid more quickly than bricks. They have been used widely for the inner leaves, and are used increasingly for the outer leaves of cavity walls, for internal walls and non-loadbearing partitions.

Thermal insulation is provided by air in lightweight aggregates (see page 185), aerated mortars and in voids in blocks. Filling voids with polyurethane foam substantially improves the overall thermal resistance of blocks.

Typical lightweight blocks can be cut or chased with hand tools and hold nails and screws without plugs. However, as with calcium silicate bricks, drying shrinkage is significant and cracking of blockwork is particularly likely to occur in panels where distances between contraction joints exceeds 6 m or twice their height.

Manufacture

Dense and lightweight aggregate blocks are compacted by pressure or vibration, or both. Some are steam cured. Aerated concrete blocks are made with or without fine aggregate, preferably high pressure steam autoclaved – see page 239.

Types of blocks

Solid blocks have no formed holes or cavities, other than slots for cutting. *Cellular blocks* have one or more cavities which do not pass through and *Hollow blocks* have one or more holes which do pass through. The external wall thickness must be at least 15 mm or 1.75 × nominal minimum size of aggregate, whichever is the greater. The many special shapes which are available include: Half length blocks, cavity closers, quoin and sill blocks, also blocks with troughs to accommodate concrete with two or more 12 mm steel bars, which provide permanent formwork for tie beams or lintels.

Sizes

Table 53 gives the BS 6073 work sizes. Other available thicknesses include 110, 130 and 190 mm. Not all of the many special sizes and shapes are likely to be produced by any one manufacturer.

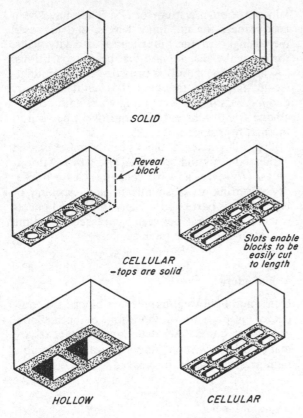

SOLID

Reveal block

CELLULAR
–tops are solid

Slots enable blocks to be easily cut to length

HOLLOW

CELLULAR

20 *Typical lightweight concrete wall and partition blocks*

Maximum permitted dimensional variations are $+3$, -5 mm on length and height and $+4$, -2 mm on thickness with a maximum average of ± 4 mm at any particular point.

Co-ordinating sizes Lengths and heights are increased by normal joint width, which is usually 10 mm.

Strength

BS 6073 requires average crushing strength for ten blocks not less than 75 mm thick, to be not less than 2.8 N/mm² and the strength for any individual block to be not less than 80 per cent of the average. The average transverse strength of five blocks less than 75 mm thick, which are deemed to be non-loadbearing, must be not less than 0.65 N/mm². Typical products have minimum crushing strengths of:

	Concrete density[1] (excluding voids) kg/m³	Strength N/mm²
Dense aggregate	1 500–2 100	2.8–35
Lightweight aggregate	700–1 500	2.8–10.5
Autoclaved aerated	400–900	2.8–7.0

[1] Weight of block divided by overall volume.

Length	Height[1]	Thickness[1]													
		60	75	90	100	115	125	140	150	175	190	200	215	220	225	250
390	190					115						200	/		/	
440	140		75				/		/						225	/
	190	60				/		140			/			220	/	
	215					115	125	140	150	175	190	200	215		225	250
	290			90					150		190	200	215			
	140				100	/	/		/						/	/
	190	/														
	215						125			175	/				225	250

[1] Vertical dimension where block is in its normal aspect   [2] Termed 'width' in BS 3921 for clay bricks and blocks

Table 53 *BS 6073 Work sizes for manufacture of concrete blocks (mm)*

Very low density blocks may require loads from joist hangers and similar point loads to be distributed.

Strength of concrete blocks and blockwork are dealt with by Gage and Kirkbride, and in BRE Digest 61, *Strength of blockwork and concrete walls*.

Drying shrinkage

BS 6073 Part 1:1981 limits drying shrinkage to 0.06 per cent generally, and to 0.09 per cent for autoclaved aerated concrete blocks.

Shrinkage cracks in walls should be avoided by the use of blocks with low drying shrinkage, (products are available with values as low as 0.03 per cent), and by taking the precautions described for calcium-silicate brickwork on page 138.

Thermal movement

Thermal movement which is greater in gravel aggregate blocks than in lightweight aggregate and aerated blocks is not normally critical, except in conditions such as partitions below poorly insulated flat roofs.

Thermal insulation

The thermal conductivity of blocks varies mainly with density. Thus, k values (W/m deg C) are 1.10–1.75 for dense aggregate blocks 0.21–0.60 for lightweight aggregate blocks and 0.15–0.27 for autoclaved aerated concrete blocks.

The Building Regulations 1985 Approved Document L require a maximum U value for walls of dwellings of 0.6 W/m^2 deg C.

This standard can be met, even where one leaf of brickwork is employed, by using thicker and lighter blocks and by supplementary insulation outside, in the cavity and/or internally.

Expanded polystyrene filling in voids in blocks assists insulation. Lightweight concrete 'bricks' are now supplied for use instead of cutting blocks to make up coursing, and for other infill work. Bricks cause 'cold bridges' and must not be used for such purposes.

Sound insulation

As with imperforate walls of brickwork or con-crete the mass law applies – see *MBS E and S,* chapter 6. A well built two 100 mm leaf, 75 mm cavity wall with two coats of plaster each side can satisfy the Building Regulations 1985 for a party wall between houses.

Fire resistance

Fire resistance is good. Blocks with class 1 aggregates provide 2 hours fire resistance in unplastered 100 mm walls.

Mortars

For external walls mortar should not be stronger than needed for strength or durability. Mortar should bond well to blocks to avoid rain penetrating through micro cracks at the mortar/block interface.

Durability

Durability of factory made concrete products is generally very high.

Blocks made with dense aggregates complying with BS 882 or 1047, or having an average compressive strength of at least 7.0 N/mm^2, or blocks for which the manufacturers provide evidence of suitability, can be used for the external leaves of cavity walls above and below ground floor damp proof courses.

An open textured block is less likely to suffer from frost attack than smooth blocks which may develop micro cracks.

Finishes and appearance

The appearance of blockwork depends upon both the choice of blocks and mortar in terms of colours and textures, including the form of mortar joint. For good appearance, the bonding pattern and quality of workmanship are also important.

A wide range of surface textures include 'as cast', sawn, and split creating a rock-like appearance. Various profiled blocks are available.

Thus blocks of good appearance need not be plastered or even painted, and the sound absorption of exposed open-textured blocks is good.

The manufacturer's advice should be obtained

as to the best procedure for plastering and rendering proprietary lightweight concrete blocks.

A proprietary product has a coloured facing of thermo-setting resin with a sand filler, the edge of which projects as a lip, giving an accurate and narrow joint.

Limes

Limes can be broadly classified as *non-hydraulic* or hydraulic binders. High calcium and magnesian limes are termed non-hydraulic because they will not harden without air being present, (for example under water).

BS 890:1972 *Building limes* which covers hydrated lime (powder), quicklime and lime putty of non-hydraulic and semi-hydraulic limes will be superseded by BS 6463: *Quicklime, hydrated lime and natural calcium carbonate*. Strength development being small and slow in non-hydraulic limes and the strength of hydraulic limes being less than that of Portland cement, today limes are rarely used as the sole cementitious ingredient in mortars, renderings or plasters, and never in ordinary concretes.

BS 6100 *Glossary of building and civil engineering terms,* Part 6 *Concrete and plaster*, Section 6.1:1984 deals with *Binders*.

HIGH CALCIUM LIMES (*pure* or *fat* limes)

These are produced by burning a fairly pure limestone, essentially calcium carbonate, so as to drive off the carbon dioxide leaving calcium oxide or *quicklime*. When water is added to quicklime considerable heat is evolved, there is considerble expansion, and the resulting product is calcium hydroxide. If the operation is carefully controlled, as it can be in a factory, so that just sufficient water is added to hydrate the quicklime, the lumps break down into a dry powder known as *dry hydrate*. Where lime is hydrated on the building site, or in a builder's yard (which is rare today), an excess of water is added and the resulting *slaked lime* should be passed through a fine sieve to remove slow slaking particles and then left to mature for at least three weeks.

Although they are unlikely to be present in hydrated lime which complies with BS 890, unslaked particles tend to slake and expand after lime has been used, causing localized *popping* and *pitting* of plaster, or expansion of brickwork. The tendency of limes and cements to expand is expressed as *soundness*.

High calcium limes are mainly of use in building because they are *fat*, ie they make for workable mortar, rendering and plaster mixes. Fatness improves with prolonged maturing of slaked lime (no harm is done thereby) and although 'dry hydrate' can be used immediately after mixing with water, its plasticity is greatly improved by *soaking overnight*, ie for at least 12 hours.

High calcium limes also retain water even when they are applied to absorptive backgrounds. Initial stiffening depends on loss of water – by evaporation or to absorptive materials such as bricks. Hardening depends on combination with carbon dioxide from the air (*carbonation*) with reformation of the original calcium carbonate. Because hardening is necessarily from the outside, the interior of a mass hardens more slowly, even where a mix includes sand which makes access of air to the interior somewhat easier.

In addition to its use in mortars, renderings and plasters, hydrated high calcium lime is used in the manufacture of calcium silicate bricks, see page 138.

MAGNESIUM LIMES

These non-hydraulic limes are made from limestones which contain more than 5 per cent, and usually more than 35 per cent, of magnesium oxide. They are less easily slaked than high calcium limes usually by slaking to powder under a heat-conserving thick layer of sand. After screening through a larger mesh than that used for high calcium lime to enable the sand particles to pass through, some unhydrated magnesium oxide usually remains and this carbonates and gives the mortar greater strength than high calcium limes. If the latter process is delayed, as may happen where plaster is painted at an early age, delayed hydration of this oxide may lead to expansion of the finished work.

HYDRAULIC LIMES

These limes which harden to some extent by an internal reaction are made by burning chalk or limestone which contain clay, producing compounds similar to those present in Portland cement. They are stronger but less fat or plastic than non-hydraulic limes.

Semi-hydraulic lime is derived mainly from the grey chalk of the southern counties of England, hence the term *greystone lime*.

BS 890 requires the combined content of calcium and magnesium oxides to be not less than 60 per cent for hydrated lime or 70 per cent for quicklime and lays down minimum hydraulic strengths.

Eminently hydraulic lime is not clearly distinguished from semi-hydraulic lime.

The term *blue lias* is applied to eminently hydraulic limes made from the liassic limestone found in Somerset, and elsewhere.

Table 55 shows that eminently hydraulic lime is weaker than Portland cement and as it does not improve plasticity there is no point in mixing them together.

	Crushing strength N/mm^2	
	7 days	6 months
Hydraulic lime	0.69	3.44–6.89
Portland cement	17.23	27.58

Table 54   Strengths of typical Hydraulic lime and Portland cement mortars (1:3 mixes by volume) from BRS Digest 46 (First series) *Building limes* revised 1966

Hydraulic limes present some difficulty in slaking. While like all limes they must be thoroughly slaked, excess water would lead to premature hardening and the exact amount of water required can only be determined by experience with the particular lime concerned. Clearly,

hydraulic lime cannot be soaked overnight to improve its workability.

Some eminently hydraulic lime is supplied as quicklime and this should be slaked in the manner described for magnesium lime.

POZZOLANAS

Pozzolanic materials containing silica, alumina and some iron oxide combine with slaked lime to form a hydraulic cement.

Natural pozzolanas include volcanic ashes such as that from Pozzuoli near Mount Vesuvius, trass from the Upper Rhine and santourin from Greece, while certain sands derived from igneous rocks can be substituted for ordinary sands in mortars.

Artificial pozzolanas include crushed burnt clay bricks and tiles, granulated blast furnace slag and pulverized fuel ash.

Cements

The term 'cement' is often applied to certain adhesives (see chapter 14), plasters, and to magnesium oxychloride binders eg in door cores and magnesite flooring – see *MBS: Finishes*, chapter 1.

This chapter is concerned with cements used mainly in concretes (chapter 8), fibre reinforced composites (chapter 10), mortars (chapter 15) and in screeds and external renderings (*MBS: Finishes*, chapter 3). The cements described here are Portland cements, supersulphated cement and high alumina cement, of which Portland cements are by far the most widely used – see BRED 237 *Materials for concrete*.

BS 6100 Section 6.1:1984 defines *Binders*.

These cements are *hydraulic*, ie they depend upon water rather than air for strength development. When water is added to cement a chemical reaction begins immediately and continues while water is still present. Hydraulic cements stiffen at first and later develop strength. Only a small quantity of water is required to hydrate cement and additional water evaporates leaving voids, which reduce the density and therefore strength and durability of products. Thus, strength is related to the water : cement ratio as shown in figure 23, page 163. Excess water also increases shrinkage in drying.

Strength development ceases at about freezing point and at higher temperatures its rate is related to temperature and age, see figure 29. In the process of hydration, cements, in particular high alumina cement, evolve sufficient heat to be useful in maintaining the temperature of concrete in cold weather. In hot weather, however, exothermic heat may lead to differential stresses and resulting cracking of massive structures. In drying, cement pastes shrink, although by re-wetting about half of the movement is reversible – see figure 21, and page 162.

Normally, no two Portland, supersulphated or high alumina cements must be mixed together, and all plant and tools must be carefully cleaned to prevent this happening.

Tables 55 and 56 give information comparing the properties of the more important cements.

SETTING AND HARDENING

In broad terms, *setting* means stiffening only, and *hardening* means useful strength development, which are different phases in the overall process of the hydration of a cement. Quite soon after water has been added cement pastes begin to stiffen and for convenience in use this must be neither too slow nor too rapid.

Type of cement	BS no.	Setting times[1]		Compressive strengths[2] (minimum) N/mm^2			
		Initial (minimum)	Final (maximum) hours	24 hrs	3 days[3]	7 days[3]	28 days[3]
Portland – ordinary – rapid hardening	12 12	45 mins 45 mins	10 10	— —	13.0 18.0	— —	29.0 33.0
Portland blast furnace – low heat	146 4246	45 mins 1 hour	20 15	— —	8.0 3.0	14.0 7.0	22.0 14.0
Low-heat Portland	1370	1 hour	10	—	5.0	—	19.0
Sulphate-resisting Portland	4027	45 mins	10	—	10.0	—	27.0
High alumina	915	Not less than 2 hours and not more than 6 hours	Not more than 2 hours after initial set	41.4	48.3	—	—
Supersulphated	4248	45 mins	10	—	7.0	17.0	26.0

[1] See BS 4550 *Methods of testing cement*, Part 3, Section 3.6 1978 *Physical tests*
[2] Results of tests carried out as method BS 4550 Clause 1, Part 3, Section 3.4: 1978
[3] Must be greater than earlier strengths
[4] BS 12 App. D test

Table 55   BS requirements for setting times and compressive strengths of cements

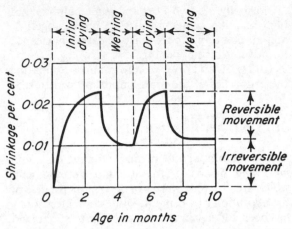

21 *Moisture movement of concrete under typical practical conditions. From "Concrete Practice' Volume II R H Elvery, C R Books*

The setting times of Portland cements are controlled by the addition of small proportions of gypsum in manufacture.

The relevant British Standards describe tests by which two degrees of setting of cement pastes known as *Initial* and *Final* sets can be determined.

During setting cement pastes may expand and British Standards include a *soundness* test for assessing, and limits for, setting expansion.

The finer the cement particles, the larger the superficial area which is available for hydration by water and the more rapid are setting and early hardening of a given type of cement. Thus, in BS 12:1978 *Portland cement* the minimum specific surfaces are given for ordinary and rapid-hardening Portland cements respectively.

Table 55 compares setting times and strengths of six cements, and table 56 gives other properties

	British Standard No.	Rate of strength development	Rate of heat evolution	Drying shrinkage	Resistance to shrinkage cracking	Inherent resistance to chemical attack		
						sulphates[1]	weak acids	alkalis
Main types of Portland cement								
Ordinary	12	Medium	Medium	Medium	Medium	Low		
Rapid-hardening	12	High	High	Medium	Low	Low		
Low-heat	1370	Low	Low	Above medium	High	Medium–high		
Sulphate-resisting	4027	Low–medium	Low–medium	Medium	Medium	High–very high		
							Low[2]	Good
Other types of Portland cement								
Ultra-high early strength	—	High–very high	High–very high					
Water-repellent	—							
Hydrophobic	—	Properties similar to those of ordinary Portland cement						
White	—							
Cements containing blastfurnace slag								
Portland blastfurnace	146	Low–medium	Low–medium	Medium	Medium	Low–medium	Medium	Good
Supersulphated	4248	Medium	Low	Medium	Medium	High–very high	Very high	Good
High-alumina cement	915	Very high	Very high	Medium	Low	Very high	High	Low
Pozzolanic cements	6610 6588	Low	Low–medium	Above medium	High	High	Good	Good

[1]See BRE Digest 89   [2]BS 8110 states that Portland cements should not be used in acidic conditions of pH 5.5 or less

Table 56   *Types of cements and their properties.* Based on BRS Digest *150 Concrete materials*

of cements. Figure 22 shows the strength development of concretes made with various cements, up to 12 months.

During initial drying out cement paste shrinks, and if drying occurs before sufficient strength has been developed the cement paste is bound to crack. A proportion of the initial drying shrinkage is irreversible but subsequently, with increase and decrease of water content, cements expand and contract and these reversible movements are about half as much as the initial shrinkage – see figure 21.

TYPES OF CEMENTS

The types of cement are:
Portland cements
    Ordinary
    Rapid-hardening
    Ultra high early strength
    Sulphate resisting
    White
    Low-heat

Cements based on Portland cement
    Water repellent
    Masonry
    Portland blastfurnace cement
    Hydrophobic
    Pozzolanic

Supersulphated cement

High alumina cement.

The main properties of most of these cements and BS requirements for setting times and compressive strengths are given in table 55.

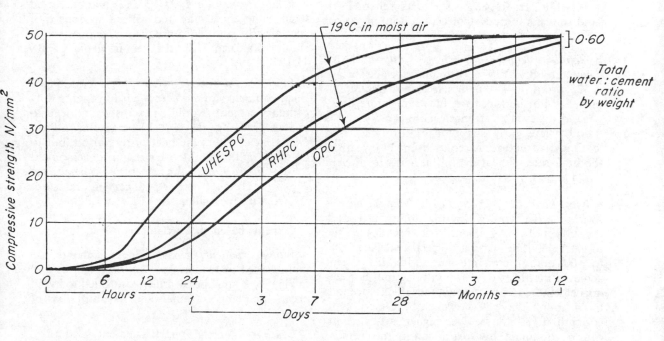

Notes
Types of Portland cements (PC)
    UHESPC   ultra high early strength
    RHPC   rapid hardening
    OPC   white, coloured and sulphate resisting. Ordinary Portland cements have similar strength properties
    Higher strengths in thoroughly compacted concretes are obtained at all ages with lower water:cement rations – see page 163

22   *Strength development of concretes made with Portland cements* (*approx. average cube strengths*)

Portland cements

In 1824 Joseph Aspdin patented his improvement upon hydraulic lime and called the product Portland cement, because it somewhat resembled Portland stone.

Portland cements are made by calcining a slurry of clay (silica, alumina and iron oxide) with limestone (calcium carbonate), in a rotating furnace, up to 180 m in length. The resulting clinker, with a small proportion of gypsum to retard the setting of the cement, is ground to a powder, the fineness of which influences the properties of the product.

Ordinary Portland cement   This, the least expensive and by far the most widely used type of cement, is suitable for normal purposes. It should conform with the requirements of BS 12:1978 *Portland cement (ordinary and rapid-hardening)* minimum specific surface is 225 m^2/kg.

BS 8110 Part 1:1985 does not recommend Portland cements in persistent acid conditions of pH 5.5 or less.

Rapid-hardening Portland cement   Also covered by BS 12:1978. Minimum specific surface is 325 m^2/kg. This cement sets at the same rate as ordinary Portland cement but being more finely ground it develops strength more rapidly and is useful where early stripping of formwork and loading of structures is required. In cold weather, the high rate of heat evolution helps to prevent damage by frosts.

Ultra-high early strength Portland cement – Swiftcrete (Cement Marketing Company Ltd) is the subject of an Agrément Board certificate. Due to extremely fine particles (specific surface 7 000 to 9 000 cm^2/g), concrete made with this cement achieves the three-day strength of rapid hardening cement in 16 hours and its seven-day strength in 24 hours. There is, however, little increase in strength after 28 days. Even higher early strength can be obtained by steam curing this cement without the danger of corrosion of reinforcement which would arise from the inclusion of calcium chloride in Portland cement.

The early heat of hydration is greater than that of rapid hardening cement.

Creep of concrete made with ultra high early strength cement is greater than that of other Portland cements but shrinkage is similar to that of ordinary Portland cement.

The advice of the cement manufacturer should be obtained before using admixtures.

Sulphate resisting Portland cement   This cement has a reduced tricalcium aluminate content (not exceeding 3.5 per cent) and is therefore better able to withstand chemical attack arising in wet conditions from sulphates in industrial wastes, clay bricks and flue condensates. It is suitable for use in concrete which is dense and otherwise of good quality in ground water having a concentration of sulphur trioxide up to 0.1 per cent and in subsoil containing up to 0.5 per cent. It is described by BS 4027:1980 *Sulphate-resisting Portland cement*.

BRE Digest 250 discusses *Concrete in sulphate bearing soils and ground waters*, and BRE Digest 89 *Sulphate attack on brickwork*.

White Portland cement   This is manufactured from white china clay and white limestone in a special kiln and is consequently expensive. It should conform with the requirements of BS 12:1978.

Low-heat Portland cement   This is required for thick-concrete work, where the heat generated by ordinary cements would be excessive and lead to serious cracking. BS 1370:1979. *Low heat Portland cement*, limits the heat of hydration, but to ensure a sufficient rate of strength development the *specific surface* of the cement must be at least 275 m^2/kg. It is a true Portland cement and is not diluted with inert fillers.

Cements based on Portland cement include:

Water repellent cement   These cements are mainly used in renderings. Mix proportions and mixing time must be carefully controlled to avoid excess air being entrained with resulting loss of strength.

Masonry cement   Portland cement: sand mixes are not ideal for bricklaying and external renderings. They are usually too strong, lack the requisite plasticity and water retention, and tend to crack. Masonry cement which contains a fine inert filler and a plasticizing agent overcomes these difficulties, while avoiding the possible expansion of a lime, used as a plasticizer, which is not thoroughly slaked. BS 5224:1976 *Masonry cement* requires compressive strengths of 4.0 N/mm² and 6.0 N/mm² at 7 and 28 days respectively. See chapter 15 *Mortars*.

Coloured Portland cements These cements made by adding pigments to ordinary grey or to white cements, are no longer produced in the UK.

Portland-blast-furnace cement This is made by adding generally about 30 and not more than 35 per cent by weight blast furnace slag to ordinary Portland cement clinker before grinding. It should comply with BS 146:1973 *Portland blast-furnace cement*.

The requirements for fineness, setting times, soundness and strength are as for ordinary Portland cement, but the rate of hardening in the first 28 days and the amount of heat evolved is less, so the cement is less suitable for use at low curing temperatures.

However, the strength of mature concrete is about the same as for concrete made with ordinary Portland cement. Portland blast-furnace cement has good resistance to dilute acids and sulphates and can be used for construction in sea water.

Low heat blast-furnace Portland cement This cement, which is suitable for use in hot places, is described by BS 4246, Part 2:1974. The heat of hydration is not more than 250 kJ/kg at seven days and 290 kJ/kg at 28 days. Setting is slower, ie 'initial set' not less than 1 hour and 'final set' not more than 15 hours. Crushing strengths at 3, 7 and 28 days are less than for Portland blast-furnace cement.

Hydrophobic cement The particles of this cement are coated with a water-repellent film and it can be stored in damp conditions for a long time without deterioration. When the cement is mixed with aggregate the film is rubbed off and hydration then takes place normally. Hydrophobic cement has better workability than ordinary Portland cement and improved waterproofing properties. It complies with the main requirements in BS 12:1978.

Pozzolanic cements Pozzolanas (see pages 150 and 156), are included in the manufacture of the following cements:

which contains 15–30 per cent pfa, with a possible addition of chalk, can be used in concrete for most purposes.

BS 6610:1985 *Pozzolanic cement with pfa* containing 35–50 per cent pfa will generally be used as a *low-heat cement* with improved resistance to the action of sulphates and weak acids.

The setting times for both cements are as for ordinary Portland cement.

Admixtures for Portland cement

Of the many products intended to modify the properties of concrete, some are of doubtful value, none of them has any useful effect upon concrete which is inherently poor in quality and some have 'side effects' which can be undesirable. Concrete admixtures BS 5075 Part 1:1982 describes *Accelerating, retarding and water-reducing admixtures*.

They should be used as recommended by the manufacturers and under strict control.

The following admixtures are suitable for use with ordinary or rapid-hardening Portland cements but not generally with other types of cement.

Workability aids

These admixtures enable less water to be used.
The main types are:

1 *Water-reducing, set-retarding agents* Water-reducing admixtures, some of which entrain up to 3 per cent air, generally increase workability, cohesiveness of the mix and hence the strength of concrete, which latter they do to a greater extent than would be expected from the reduction in the amount of water added to the mix. Some water-reducing admixtures also retard setting of the cement for 2–6 hours.

2 *Fine powders* Extremely finely ground mineral powders (eg pulverized fuel ash, chalk, lime, kaolin and diatomaceous earth) act as lubricants and improve cohesiveness by filling the pores in an ordinary cement mix. However, if too much of these powders are added the water:cement ratio must be increased, so that strength is reduced, and there is a greater tendency to shrinkage cracking.

3 *Surface active agents* (Plasticizers) 'Wetting agents' improve workability of mixes by reducing the surface tension of water so that surfaces are wetted more easily, and, more importantly, by entraining very small air bubbles. Most such admixtures reduce strength.

Superplasticizers can extend the performance of normal plasticizers, reducing labour in placing and finishing, it is expected, without reducing the durability of concrete.

However, like so many sophisticated aids, they are very costly and have limited applications. They are very little used in the UK. The self-levelling tendency precludes their use for surfaces which slope more than about 3° to the horizontal. Mixes must be specially designed, and the proportions of all ingredients, and the processes, must be strictly controlled – see BS 5075 Part 3:1985.

Air-entraining agents

By entraining 4 to 5 per cent of minute, discontinuous and uniformly distributed air bubbles into concrete its workability, and durability, particularly in resistance to frost, are increased. Although air entrainment allows a lower water: cement ratio to be used, the reduction in density causes loss of strength of up to 15 per cent in richer mixes – see BS 5075:Part 2:1982.

Accelerators

Accelerators increase the rate of setting and strength development of ordinary and rapid-hardening Portland cements up to about 28 days. Since accelerators increase the rate of setting of cement and may have other 'side effects' they should be used only in cold weather where the usual precautions (see page 181) are taken. In such cases the increased rate of heat evolution offsets the effects of an atmospheric temperature a few degrees below freezing point. Calcium chloride ($CaCl_2$) is no longer recommended to be used in reinforced concrete, excessive concentrations having led to corrosion of reinforcement. In mortars calcium chloride can corrode metal ties and cause efflorescence.

Formodac (BP Chemicals Ltd) can be used in various types of cement in winter, or to enable rapid release of precast concrete from moulds. (It is acidic, and very short contact with the eyes causes severe damage.)

Damp-proofing and permeability-reducing admixtures

Damp-proofing admixtures prevent water movement by capillary action, whereas permeability-reducing admixtures prevent its passage under pressure. No admixture, however, entirely prevents the passage of water vapour and the description 'waterproofing' admixtures is wrongly applied.

Retarders

Admixtures based on sugars, starches, zinc oxide or boric oxide retard the setting of cement without significantly affecting workability and strength. They may be used in very large masses of concrete in hot weather, exceptionally in ready-mixed concrete and as surface retarders – see *Integral finishes on concrete, MBS: Finishes*.

Polymers

Polymers can strengthen, reduce permeability and increase durability of concretes – see page 172.

Pigments

BS 1014:1975 describes *Pigments for cement, magnesium oxychloride and concrete*.

Except for small quantities of concrete or to obtain special colours, if they are available, it is generally preferable to use cements to which pigments have been added under controlled conditions at the cement works.

Pozzolanic materials

Pozzolanas (see page 150) react with lime which is liberated during the hydration of Portland cement. A pozzolana – usually pulverized fuel ash (pfa) – may be added to Portland cement in manufacture page 156 or on site. Early strength development is slower, and less heat is evolved, which is beneficial in mass concrete structures such as dams.

Concrete containing about 20 per cent pfa by mass, at one year gives the same strength as OPC concrete.

More than 35 per cent by mass of pfa can improve resistance to the alkali-silica reaction – see page 163.

Lower water contents can be used in pfa concrete for given workability, resulting in improved durability.

BS 3892 is *Pulverized fuel ash*, and Part 1:1982 PFA specifies its use as a *Cementitious component in structural concrete*.

Supersulphated cement

BS 4248:1974 describes properties for this cement. It is made by grinding together 85 to 90 per cent of blast furnace slag, 10 to 15 per cent sulphate and 1 to 5 per cent of an activator such as Portland cement clinker. This cement differs in many respects from other cements and must not be mixed with them or with lime.

Calcium sulphoaluminate hydrate – the material formed when Portland cement is attacked by sulphates – is formed during hardening and it, together with a very low calcium hydroxide and tri-calcium aluminate content, accounts for the much higher resistance of supersulphated cement to sulphates.

Resistance to acids is high, the cement has been successfully used in an environment of pH 3.5, and, unlike high alumina cement it is resistant to solutions of caustic alkalis (sodium and potassium hydroxides) and is not subject to conversion. Early strength development is slow, particularly in cold weather, but at later ages strengths are at least equal to those of comparable Portland cement mixes – see table 55. Low temperature steam curing, say at 49°C, increases the rate of early strength development but higher temperatures are not desirable. The rate of heat evolution is low, it is a *low heat* cement suitable for mass concrete and for work in hot climates. Supersulphated cement is generally used where its special properties are required. Particular attention must be paid to good concreting practice. Thus water and aggregates must be clean. Mixes should be designed for thorough compaction in proportions not leaner than 1:6 by weight for concrete or 1:2 by volume for mortar, with total water : cement

ratios not exceeding 0.5. Concrete mixes should be batched by weight.

Moist curing should continue for at least three days.

High alumina cement

BS 915:Part 2 1972 (amended to 1983) *High Alumina Cement* requires an alumina content at least 32 per cent by weight. Bauxite and limestone are fused continuously in reverbatory furnaces (not clinkered like Portland cement) and cast into *pigs*. These are cooled and the extremely hard product (sometimes used as an aggregate for special concrete) is broken and ground to cement.

The cement is grey-black, and its properties differ from those of Portland cement.

The hydrated cement comprises calcium aluminate hydrates together with alumina hydroxide gel, in contrast to the calcium silicate hydrates and calcium hydroxide gel formed in Portland cement.

Heat evolution is rapid and extremely high strengths can be obtained, but it is not quick setting (although a *flash set* can be induced by mixing it with a proportion of Portland cement when loss of strength results).

While sound, high alumina cement is resistant to sugar, oils and fats, fertilizers, vinegar beer and peaty waters. Resistance to acids and sulphates is superior to that of Portland cement, although resistance to caustic alkalies is low.

Following serious structural failures it has become evident that in certain adverse conditions of concrete manufacture and service, high alumina cement undergoes *conversion*, which is a crystal change accompanied by serious loss of strength and porosity. The cement is then attacked by chemicals such as calcium sulphate (gypsum) in plaster.

Conversion results from a high water:cement ratio and high temperature during curing and high temperature and/or humidity thereafter. Rapid heat evolution enables work to continue in cold weather but concrete must return below 26°C within 24 hours of being placed, which may require cooling with running water.

The Building Regulations 1985 allow the use of high alumina cement only at high temperatures. With crushed firebrick aggregate it resists temperatures up to 1300°C, and a *refractory* white cement up to 1800°C.

BS 8110:1985 *The structural use of concrete* does not permit the use of high alumina cement.

BS915 amendment 1983 advises that guidance on correct use should be obtained from the manufacturers and specialised publications.

8 Concretes

CI/SfB E

Concrete is generally understood to be a mixture of cement, water and aggregate which takes the shape and texture of its mould, or *formwork* on site.[1] When cured at a suitable temperature and humidity it hardens. Cements and admixtures have been described in the last chapter. Formwork and reinforcement which provides tensile strength in reinforced concrete are considered on page 178.

Types of concretes with their main properties and uses are shown in table 57. Those exceeding 200 kg/m³ are classified here as *Dense concretes*. Less dense concretes, made by aerating the mix (*cellular concrete*), by using lightweight aggregates or by omitting the fine aggregate (*no-fines*) concrete, are classified as *lightweight concretes* – see page 184.

Tensile strength is given to concretes by reinforcement – usually steel – which can be pre- or post-tensioned.

General references:

BS 6100 *Glossary of terms for building and civil engineering,* Part 6, *Concrete and plaster.*
BS 8110 *The structural use of concrete*
　　Part 1:1985 *CP for design and construction*
　　Part 2:1985 *CP for special circumstances*
　　Part 3:1985 contains design charts
This BS is an *Approved Document* of the *Building Regulations* 1985.
CP 114:1969 *Structural use of reinforced concrete in buildings*
CP 115:1969 *Structural use of prestressed concrete in buildings*
CP 116:1969 *The structural use of precast concrete*
BRE Digest 13 *Concrete mix proportioning and control*
BRE Digest 237 *Materials for concrete*
BS 5328:1981 *Methods for specifying concrete.*

---

[1]The term 'concrete' is now sometimes also applied to polymer/aggregate products see page 156, and 'Bituminous concrete' is referred to in the USA.

Introduction to concrete and Concrete Practice D E Shirley, BSc, MIChem E, published by The Cement and Concrete Association Advisory Service which has offices in all major cities
Concrete materials and practice LJ Murdock and G F Blackledge, Edward Arnold Ltd
Concrete Technology Vols 1 and 2, D F Orchard Contractors Record Ltd

Appearance

Variations in the appearance of concrete surfaces result from:
　(i) materials eg grey, white or coloured cement. Colour, shape, texture and grading of aggregates.
　(ii) formwork profiles and textures.
　　　(iii) work to surfaces after casting, varying from light spraying and brushing of freshly cast concrete to bush- hammering, and grinding and polishing of hardened concrete.
　(iv) exposure to atmosphere.
　(v) abrasion
See also M BS *Finishes.*

It should be noted that glossy surfaces on concrete derived from formwork or after casting from mechanical smoothing, eg of terrazzo, soon become rough where they are exposed externally.

Obtaining good appearance on concrete requires special techniques and control, and therefore involves extra cost – see *MBS: Finishes,* chapter 5, *Integral surface finishes on concrete.* Moreover, unlike clay bricks and natural stones, appearance rarely improves with weathering – see BRE Digest 126, *Changes in the appearance of concrete on exposure.*

Factory casting facilitates control, and some processes can be used to give textures which are not possible on building sites. High quality smooth surfaces are difficult to achieve, and generally, more interesting appearance results where aggregates are exposed, eg, by scraping 'green' concrete, by bush-hammering or by grinding hardened concrete as in terrazzo.

159

Changes in appearance

Unlike some clay bricks and natural stones appearance of concretes rarely improves with weathering – see BRED 126 *Changes in the appearance of concrete on exposure*.

Smooth surfaces, as-cast from glossy formwork or achieved mechanically after casting, are soon roughened where they are exposed externally. Variations in appearance will also result from differences in absorption of moisture and dirt. Concrete shrinks, and smooth faced concrete tends to *craze* into networks of 'hair cracks' in which dirt collects.

Hydrated lime, leached from Portland cement, reacting with atmospheric carbon di-oxide, often causes ugly calcium carbonate stains, and sometimes stalactites on buildings.

Other changes in appearance externally result from atmospheric pollution and from vegetable growths including algae and mosses.

BS 6270 *CP for cleaning and surface repair of buildings*, Part 2: 1985 deals with *Concrete and precast concrete masonry*, Part 1:1982 includes *cast stone*

DENSE CONCRETES

Potentially, the best concrete products can be made in factories, however, concrete is one of the few materials which are often made on the building site. In practice its quality varies considerably and it is important to understand the factors which make for good and consistent quality. In short, these are: suitable cement, aggregate and water, thoroughly mixed in proportions which make possible the lowest water:cement and cement:aggregate ratios consistent with thorough compaction. Drying must be prevented, temporary support and a sufficient temperature maintained until the required strength is attained.

Very broadly, for any given type of aggregate, high density in concrete is associated with high strength, hardness, durability, imperviousness, frost resistance and thermal conductivity.

Dense concretes are dealt with under the following headings:

Properties of hardened concrete
Materials for dense concretes, page 162
Water:cement ratio, page 163

Properties of hardened concrete

Strength properties

Characteristic crushing strengths of the more usual structural grades of Portland cement concretes are shown in table 58.

The direct tensile strength of concrete varies between 1/8 and 1/14 of its compressive strength but the tensile strength measured in bending is usually about 50 per cent greater.

Higher early strengths are obtained by using special cements, or by steam curing Portland cement concrete, and provided concrete is fully compacted, strength at all ages increases as the water:cement ratio of the mix is reduced.

Figure 22 page 153 and table 59 compare the strength development of concretes made with other cements and up to 12 months.

CP 115:1969 *The structural use of prestressed concrete* gives values for moduli of elasticity 'E' of gravel concretes of varying strengths as in table 60. *Creep*, which is plastic deformation caused by a constant load, occurs more rapidly at first but slowly approaches a limit after about five years (see BS 8110:1985). The extent of creep is roughly in proportion to the load applied and is greater with weaker and less mature concretes.

For concretes of 26–35 N/mm^2 at 28 days and with normal design stresses the average value is 30×10^{-6} per mm N/mm^2.

Permeability

Concrete which is made with a low water:cement ratio and is very thoroughly compacted has good resistance to the absorption of water. Admixtures can sometimes contribute to impermeability (see page 155) but no concrete is completely impervious to water vapour.

Chemical resistance

The chemical resistance of cements is considered

Type	Aggregate	Density of aggregate kg/m³	Density of concrete kg/m³	Compressive strength at 28 days N/mm²	Modulus of elasticity N/mm²	Drying shrinkage per cent	Thermal conductivity W/m deg C	
Dense concretes	Iron shot	4005–4561	5286	up to 69			—	Radiation shielding
	Gravel	1360–1760	2240–2480	14.0–70.0 / 41.4–69 (special purpose)	20 700–34 500 / 34 500–44 800	0.03–0.04	1.4–1.8	Fire resistance Class 2
	Crushed limestone	1360–1600	2160–2400	24.1–34.5				Fire resistance Class 1
	Crushed brick		1680–2160	13.8–27.6				
Lightweight concretes	No-fines — Gravel	1360–1600	1600–1950			0.016–0.028	0.08–0.94[2]	Structural
	No-fines — Clinker	720–1040	880–1440	2.76–6.89		0.033–0.040		
	Lightweight aggregate — Foamed slag, sintered PFA or expanded clay with some natural sand	320–1040	720–2000	2.0–62.0 (structural concrete minimum 15.0)	6 890–20 700[4]	0.030–0.070	0.24–0.93	Superior thermal insulation and fire resistance Class 1
	Various, see table 66	(64)[3] 480–1040	(400)[3] 560–1760	(0.48)[3] 1.40–27.5	1450–3120	0.03–0.09	0.25–0.35 / 0.16–0.91	
	Aerated	—	400–1440	1.38–10.35	autoclaved	0.22 air cured / 0.06 autoclaved	0.08–0.26	

[1] 2.07–4.82 N/mm² concrete
[2] For concrete 1762–1842 kg/m³
[3] These values are for exfoliated vermiculite and perlite and are not typical of the ranges
[4] Generally $\frac{1}{3}$ – $\frac{2}{3}$ of corresponding gravel concrete
(Information from BRE Digests 150 (out of print) and 123 and other sources)

Table 57 Types of concrete

Grade	Characteristic[1] crushing strength 28 days N/mm²	Lowest grade for use in:
C 7.5	7.5	Plain mass concrete
C 15	15.0	Reinforced concrete with lightweight aggregate
C 20	20.0	Reinforced concrete with dense aggregate
C 30	30.0	Concrete with post-tensioned tendons
C 40	40.0	Concrete with pre-tensioned tendons

[1]Characteristic strength is that below which not more than 5 per cent of the test results are allowed to fall.

Table 58 Grades of concrete for uses – based on BRED 244

	Average crushing strength			
Age at test	Ordinary Portland cement		Rapid-hardening Portland cement	
	Storage in air 18°C 65% RH N/mm²	Storage in water N/mm²	Storage in air 18°C, 65% RH N/mm²	Storage in water N/mm²
1 day	5.5	—	6.9	—
3 days	15.0	15.2	17.2	17.2
7 days	22.0	22.7	24.1	24.8
28 days	31.0	34.5	33.1	37.2
3 months	37.2	44.1	38.6	45.5

(1 cement: 6 aggregate, by weight; 0.60 water:cement ratio.) From BRS Digest 14 (First series)

Table 59 Typical strength development of concrete

Cube strength in compression N/mm²	Modulus of elasticity N/mm²
20.7	20 700
27.6	27 600
34.5	31 000
41.3	34 500
55.1	41 300
68.9	44 800

Table 60 Moduli of elasticity of dense concretes

in table 56. The chemical resistance of ordinary Portland cement concretes increases with crushing strength, but special cements and sometimes special aggregates are needed where conditions are severe. However, CP 23/77 Chemical resistance of concrete BRE, points out the need for high quality concrete irrespective of the choice of materials and any protective measures.

Frost resistance

Concrete may be damaged by expansion of ice crystals, which are most likely to form in capillary pores or cracks, resulting initially from mixing water which was surplus to that required to hydrate the cement. Air-entrainment admixtures (see page 155), form discontinuous pores which improve resistance to frost.

Resistance to abrasion

This depends upon the hardness of the aggregate particles and on the ability of the mortar matrix to retain them.

Resistance of concrete to fire

Reference to this property was made on page 40. Up to about 120°C the strength of ordinary concrete increases, but there is a serious loss of strength at higher temperatures, flexural strength being more affected than compressive strength.

Loss of strength is less with leaner mixes and with Portland blastfurnace cement. High alumina cements with crushed firebrick can be classed as refractory.

Loss of strength is considerably less where aggregates which do not contain free silica, eg limestone and furnace formed aggregates are used – see table 61.

Low density in cellular and lightweight aggregate concretes improves fire resistance. Fire resistance classifications of aggregates are given on page 41.

The survival of reinforced concrete in fire depends upon the protection afforded to the steel reinforcement by the concrete cover. Wire reinforcement helps to retain this but once the cover has spalled off, the steel conducts heat readily and failure is rapid.

Thermal movement

The coefficient of thermal expansion of concretes varies from 6 to 13×10^{-6} deg C according to mix proportions, type of aggregate and curing conditions. The average value is about 10×10^{-6} deg C so that a concrete member 30 m long expands about 12 mm with a 40 deg C rise in temperature. Limestones and broken brick aggregate concretes suffer about half the movement of ballast concrete.

Moisture movement

Concrete shrinks when it dries and expands when it is wetted, the greater part of the initial drying shrinkage being irreversible – see figure 21, page 152.

Movement increases with the richness of a mix, with water:cement ratio, and where rigid aggregate is not used, ie in lightweight aggregate and aerated concretes. On the other hand, moisture movement can be halved by high pressure (not low pressure) steam curing – see page 182.

If the stresses induced by shrinkage exceed the tensile strength of concrete, cracks tend to occur and this is particularly likely where concrete dries out before it has had time to develop much

Temperature °C	Permanent loss of crushing strength per cent	Colour change in aggregate[1]
250	5	—
300	18	} pink or grey
600	64	} grey
900	85	}
1 200		} buff
above 1 200		yellow

[1] Igneous rocks do not change in colour

Table 61 Loss of strength in fire

strength, or where concrete elements are fixed rigidly at their ends. Some relief is given by the ability of concrete to creep and reinforcement at close centres restrains moisture movement, so that cracks will not normally be visible in first quality reinforced concrete.

Uneven drying causes distortion, as in floor screeds which dry only from their upper surfaces and curl upwards at their edges – see *MBS: Finishes*, chapter 1.

Materials for dense concrete

Ordinary plain concretes require cement, water and aggregates. Cements are dealt with on page 150 and admixtures, including polymers, on page 155.

Glass, steel and polymer fibre reinforcements are dealt with on page 236.

Steel bar and wire reinforcement is referred to on page 178. Water, aggregates and polymers in concrete are now considered.

BRE Digest 237 deals with: *Materials for concrete*.

Water

Water for concrete should be reasonably free from impurities such as suspended solids, organic matter and salts, which may adversely affect the setting, hardening and durability of the concrete. This requirement is usually satisfied by using water which is fit for drinking, or where the quality of water is in doubt, it may be assessed by comparing the setting times of cement pastes and the compressive strengths of concretes made with it, and with distilled water, respectively. BS 3148:1980 *Methods of test for water for making concrete*, includes notes on the suitability of water.

Sea water does not normally reduce the strength of Portland cement concrete and can safely be used for plain concrete. However, efflorescence may occur, and because salt promotes the corrosion of steel sea water must not be used for reinforced concrete.

Water:cement ratio

The amount of water required to hydrate cement

is very small – about 4.7 litres per 50 kg Portland cement. Water is also needed to enable concrete to be fully compacted but further water creates voids, every 1 per cent of which reduces the strength of concrete by about 5 per cent.

Thus, the density and strength of concrete reduces as the water content increases above that required for full compaction of concrete. *A low water:cement ratio* is also necessary for: imperviousness, resistance to frost, chemicals and abrasion, and to minimise drying shrinkage.

The water:cement ratio is expressed as:

$$\frac{total\ weight\ of\ water}{weight\ of\ cement}$$

or as

a *free water:cement ratio*. Free water being the total weight of water in concrete less the weight of water which is absorbed by aggregates.

Figure 23 shows the relationship between water:cement ratios and crushing strengths for a typical concrete and table 62, from BS 8110, relates water:cement ratios and cement contents to grades of concretes and exposures.

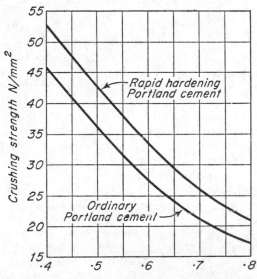

Note: mechanical compaction is required to obtain the higher strengths.

23   Relationship between crushing strength and water:cement ratio of fully compacted Portland cement concrete cubes at 28 days

Exposure	Maximum free water: cement ratio	Minimum cement content kg/m³	Lowest grade of concrete
Mild	0.8	180	C20
Moderate	0.65	275	C30
Severe	0.60	300	C35
Very severe	0.55	325	C35[1]
Extreme	0.50	350	C45

[1] Air entrained

Table 62   Water:cement ratios, cement contents and grades for unreinforced concretes with 20 mm nominal maximum size aggregates in various exposures

Aggregates

BS 6100, Part 6, Section 6.3:1984 defines *terms for aggregates*. BS 882:1983 specifies *Aggregates from natural sources for concrete*

Aggregate must be sufficiently strong, free from constituents which can react harmfully with the cement, be well graded into particles of optimum sizes, and have minimal or no moisture movement. Shape and texture affect workability of concrete mixes, and hence strength and durability of the hardened concrete. Also, in hardened concrete, weather resistance, hardness, thermal conductivity and appearance of aggregates can be important.

Strength   The strength of aggregates is rarely a limiting factor. In fact, aggregates of moderate and low stength reduce the stress in the cement paste and can increase the durability of concrete, see *lightweight structural concrete*, page 187. However, BS 882 states requirements for heavy duty floors and pavings.

Density   The bulk density of natural dense aggregates varies from 1 450 to 1 600 kg/m³. Very heavy aggregates, such as barytes, have been used for biological shielding from radiation. BS 4619:1970 describes: *Heavy aggregates for concrete and gypsum plaster.*

Shrinkage   Certain aggregates such as dolerites and some sedimentary rocks, especially greywacke and mudstone, are shrinkable. BRED

35 states that the shrinkage of concretes made with them can be up to four times that of concretes made with non-shrinkable aggregates.

Cleanliness Aggregates should be free from significant quantities of substances which:

1 are chemically incompatible with cement, eg sulphates and organic material. Certain aggregates which contain reactive silica cause *alkali-aggregate reaction* (AAR) in concrete was first noted in the UK in 1971. Most commonly, siliceous aggregates form a calcium alkali silica gel (ASR), which imbibes water and expands. Resulting cracks make concrete vulnerable to frost attack and corrosion of reinforcement. Essential conditions are sufficient alkali solution and water and a susceptible aggregate. Where the latter cannot be avoided, concrete should have a low water:cement ratio and a low alkali Portland cement with not less than 30 per cent pulverized fuel ash or 50 per cent blast furnace slag. The concrete should then be kept dry – see BRED 258,
2 reduce bond with aggregate, eg clay and oil coatings,
3 expand, eg bituminous coal,
4 decompose, eg organic matter,
5 attract moisture, eg salt in excess of 2 per cent by weight of cement. Chlorides accelerate the rusting of reinforcement and impair the sulphate resistance of sulphate resisting cement,
6 cause staining, eg pyrites.

Gravels and sands should be washed by the suppliers to remove soluble matter and silt. BS 812:1975 describes a site test for organic impurities. Part 4:1976 deals with chemical properties.

BS 882 controls clay, silt, dust and shell content and flakiness for concrete grades and it suggests maximum chloride contents for concrete types.

Crushed clay brick, sometimes used as aggregate for its fire-resisting properties, or for its appearance when it is exposed, must not contain more than 1 per cent of sulphates expressed as sulphuric anhydride.

Specific surface The larger the superficial area of the aggregate particles by reason of: angularity of the aggregate, rough texture or a high proportion of small particles, the less workable the concrete will be. On the other hand, angularity and rough texture allow a greater adhesive force to develop.

Grading

In a well graded aggregate the various sizes of particles interlock, leaving the minimum volume of voids to be filled with the more costly cement. The particles also flow together readily, ie the aggregate is *workable*, reducing labour in placing concrete and enabling a lower water:cement ratio to be used.

The proportions of the different sizes of particles is known as the *grading* of an aggregate which is usually expressed as percentages by weight passing various sieves conforming to BS 410:1976 *Test sieves* ie 50, 37.5, 20, 14, 10, 5, 2.36 and 1.18 mm, 600, 300 and 150 μm.

Conventionally, aggregate which is mainly retained on a 5 mm BS sieve (eg natural gravel and crushed gravel and stones) is called *coarse aggregate* and aggregate which mainly passes through a 5 mm sieve is called *fine aggregate*.

Generally, the largest particle of coarse aggregate should be as large as can be placed without 'bridging' between reinforcement or between reinforcement and an external face. For heavily reinforced members the largest particle should be 7 mm less than the space between the main reinforcement bars and it should never exceed the minimum cover to the reinforcement less 7 mm. 20 mm is usually the nominal maximum size of aggregate for ordinary reinforced concrete work. Larger sizes could sometimes be used and for mass concrete the maximum size can be 40.0 mm or greater provided it does not exceed one-quarter of the minimum thickness of the member.

Aggregates are often described by their maximum size with the qualification '*down*', meaning that they contain appropriate proportions of the smaller sizes.

It is important to note that either *coarse* or *fine* aggregates as defined above can also be described as being *coarsely* graded if they contain a high proportion of the larger particles or *finely* graded if they contain a high proportion of the smaller particles.

All-in aggregate contains particles of all sizes required. It must be remembered that the volume of a mixture of coarse and fine particles is less than the sum of the volumes of the separate parts. Approximate equivalents are:

Separate volumes		Equivalent volume of 'all-in' aggregate (approx.)
Fine aggregate	Coarse aggregate	
1	2	$2\frac{1}{2}$
$2\frac{1}{2}$	3	$3\frac{3}{4}$
2	4	5
$2\frac{1}{2}$	5	$6\frac{1}{3}$
3	6	$7\frac{3}{4}$

Table 63   Equivalent volumes of materials

Materials

As aggregate forms the bulk of hardened concrete and transport is costly, it is usually desirable to use local material.

For ordinary concrete, the larger particles, known as *coarse aggregate* are typically 'natural' or crushed gravel, or rock such as granite, basalt, hard limestone or sandstone. The small particles or *fine aggregate* are sand or a gravel or rock crushed to provide the required sizes.

BS 882:1983 *Aggregates from natural sources for concrete* specifies materials, including those suitable for heavy duty wearing surfaces of floors and pavings.

BS 1047:1983 specifies *Air-cooled blast furnace slag* for use as coarse aggregate in concrete.

Alternative aggregate combinations for concrete

In practice, aggregates are delivered to sites in one of four combinations of graded aggregates:

Fine aggregate only
Fine and coarse aggregate
Single maximum sizes
All-in aggregate

1 *For members less than 75 mm thick* – including mortars, renderings and floorings – *Fine aggregate*, typically a sand, most of which passes through a 5 mm BS sieve graded to comply with the appropriate BS.

2 *For members more than 75 mm thick*

(*a*) *Fine aggregate* as above and *Coarse aggregate* most of which is retained on a 5 mm BS sieve and graded in accordance with BS 882. This is the most common combination for ordinary reinforced concrete.

(*b*) *Single sizes* each graded down from 40, 20, 14 and 10 mm. Precise proportions of each size can be measured and mixed together on site to give the particular overall gradings required for high quality 'designed concretes'.

(*c*) *All-in aggregate* should contain proportions of *fine* and *coarse* aggregate gradings to give a satisfactory overall grading. However, it is not easy to prevent the larger and smaller particles separating, with consequent loss of the specified grading. All-in aggregate is, therefore, commonly used for mass unreinforced concrete, in grades C7.5, C10 and C15.

BS 882:1983 gives grading limits for particle sizes, table 64 for coarse aggregates, and table 65 for fine aggregates in three overlapping zones of varying 'coarseness'. The BS also contains a table for the grading of *all-in aggregate* of 40, 20 and 10 mm nominal sizes. Fine aggregate must comply with the overall limits shown, and not more than one in ten consecutive samples must fall outside the limits for one of the gradings: C, M or F. However, aggregate which does not comply with the table can be used if it can be shown to produce concrete of the required quality.

Fine aggregate for heavy duty floor finishes must comply with C or M.

However, there is no ideal grading which is applicable to all aggregates, and for important work, tests should be carried out to determine the grading which gives maximum workability in the wet concrete, and density and strength in the hardened concrete.

A ratio of 1 fine aggregate: $1\frac{1}{2}$ to 3 coarse aggregate is generally satisfactory, but the most advantageous ratio depends upon the type of aggregates, the maximum size of coarse aggregate and the grading of the fine aggregate. The finer the grading of fine aggregate the smaller the proportion required to give the most satisfactory result.

BS 812 *Methods of sampling and testing of mineral aggregates, sands and fillers* has four parts. It includes methods of determining physical, mechanical and chemical properties. Also methods of determining particle size and shape, and shell content of coarse aggregate.

The procedure for determining the grading of

Sieve size mm	Graded aggregate			Single-sized aggregate			
	40 mm to 5 mm	20 mm to 5 mm	14 mm to 5 mm	40 mm	20 mm	14 mm	10 mm
50.0	100	—	—	100	—	—	—
37.5	90–100	100	—	85–100	100	—	—
20.0	35–70	90–100	100	0–25	85–100	100	—
14.0	—	—	90–100	—	—	85–100	100
10.0	10–40	30–60	50–85	0–5	0–25	0–50	85–100
5.0	0–5	0–10	0–10	—	0–5	0–10	0–25
2.36	—	—	—	—	—	—	0–5

Table 64   BS 882:1983 Grading limits for coarse aggregates for concrete – percentages by mass passing BS sieves

BS sieve mm	Overall limits per cent	Additional limits for grading per cent		
		C	M	F
10.0	100	—	—	—
5.0	89–100	—	—	—
2.36	60–100	60–100	65–100	80–100
1.18	30–100	30–90	45–100	70–100
600	15 100	15–54	25–80	55–100
300	5–70	5–40	5–48	5–70
150	0–15[1]	—	—	—

[1]20 per cent for crushed rock fines, except for heavy duty floors

Table 65   BS 882:1983 Grading limits for fine aggregates for concrete

an aggregate is given in BS 812:1975. So that samples for sieving truly represent the bulk of an aggregate, at least ten small samples are made into a main sample and this is reduced to a specified size by repeated *quartering*.

Silt, clay and dust

For various aggregates BS 882 gives the maximum percentages of clay, silt (ie very fine sand) and dust as:

	Per cent by mass
Uncrushed, partially or crushed gravel	1
Crushed rock	3
Uncrushed or partially crushed sand or crushed gravel fines	3
Crushed rock fines	15
Crushed rock fines for heavy duty floor finishes	8
Gravel all-in aggregate	2
Crushed rock all-in aggregate	10

Alternatively, BS 882 gives maximum volumes of clay and silt based on the BS 812:Part 1 field settling test, and where required additionally on the decantation test.

BS 812 *Field settling test* is simple to carry out but acts only as a guide to the silt and sand content of a sample. Results should be correlated with those obtained by the decantation test. It is not applicable to crushed stone sands or to coarse aggregates. 5 ml of a solution of two tea spoons of sodium chloride in 1 litre of water is put in a 250 ml cylinder. Sand is added gradually up to about 100 ml and the salt solution to 150 ml. After shaking vigorously, the cylinder is stood on a level surface and tapped to level the upper surface. After three hours the height of silt and clay is expressed as a percentage of that of the sand below.

If the volume of silt and clay does not exceed eight per cent it can be assumed that it will not exceed 3 per cent by weight.

167

Fibres in concrete

Polymers, glass and steel fibres in concrete are discussed on page 235.

Polymers in concrete

Polymers in concrete are discussed on page 172.

Workability

The term *workability* is used to describe the ease with which concrete mixes can be compacted and the highest possible workability must be aimed at so that concrete will be as completely compacted as possible while using the lowest possible water: cement ratio.

Workability should be obtained by the use of a well graded aggregate and one which has the largest maximum particle size which will pass readily between and around the reinforcement, see page 166, rather than by increasing the cement:aggregate ratio which increases the shrinkage, and cost of concrete.

The use of smooth and rounded, rather than irregularly shaped aggregate also increases workability but in high strength concretes there may be no overall increase in strength, because with equal water:cement ratios irregularly shaped aggregate produces the stronger concrete.

Air-entraining admixtures improve the workability of mixes (and improve the frost resistance of hardened concrete) but the reduction in density of the concrete is accompanied by a loss of strength up to about 15 per cent.

On most building sites a rough indication of workability is obtained by the *slump test*. See page 177.

Quality control

Owing to the inevitable variation in the quality of concrete which is made from time to time, in order to ensure that very few samples fall below a stated strength, most of the concrete which is produced must be stronger and therefore more costly than it is required to be.

Figure 24 shows how a *mean strength* is derived from a *characteristic strength* (below which not more than 5 per cent samples may fall) and the variation in quality. It is evident that the less

efficient the *control* in the measurement of ingredients and in the process of manufacture, the higher the mean strength must be to maintain the specified *characteristic strength*.

BRS Digest 13 stated that for concretes having 21.0 N/mm^2 crushing strength, almost twice as much cement was needed in volume-batched concrete with poor control, as in weight-batched concrete with constant supervision. Apart from saving cement, good control reduces honeycombing, laitence and permeability – factors which adversely affect durability and appearance.

For design purposes the required mean crushing strength can be obtained by multiplying the *characteristic strength* by the control factor given in table 66.

A useful reference is BRED 244 *Concrete mixes: specification, design and quality control.*

Control	Control factor
Good Cement and aggregates weight-batched Regular checking of water content of aggregates Competent supervision	1·33
Moderate Cement and aggregates weight-batched Water content of aggregate checked Average supervision	1·66
Poor Aggregate batched by volume Inexperienced supervision	2·00

Table 66   Control factors

Specification of concrete mixes

BRE Digest 244 mentioned above, should be studied.

BS 5328:1981 *Methods for specifying concrete* provides means of specifying, essentially either by materials, mix proportions and methods (*Prescribed mixes*), or by performance specification (*Designed mixes*).

The BS also provides proformae for specifying, and means of checking compliance for both structural and non-structural concretes. Appendix 'A' *Guidance on the specification of concrete* should be studied.

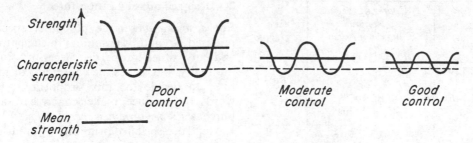

24   *Variations in strengths with control*

1   **Prescribed mixes** (sections two and three)

The *purchaser*[1] specifies the mix proportions and accepts responsibility for their being able to provide the desired properties in the concrete, so that although strength tests are useful as an indication of quality, low strength results are not a means of rejecting *prescribed mixes*. The proportions of the constituents of *Ordinary prescribed mixes* are taken from tables in the BS but for *Special prescribed mixes* they are designed by the purchaser's engineer.

(a)   *Ordinary prescribed mixes* (section two)

These mixes, which must not include admixtures, are likely to suffice for all but the most advanced engineering work. Proportions for mixes designated by their expected 28 day *characteristic compressive strength*[2] in N/mm² as C7.5P, C10P, C15P, C20P, C25P or C30P are given in the BS as weight of total dry aggregate per 100 kg of cement (table 69) and percentages by weight of fine aggregate to total aggregate (table 70). The BS advises that where crushed stone or gravel sands are used the purchaser should be satisfied that they will provide the desired properties in the concrete.

Aggregates for grades C7.5P, C10P and C15P may be measured by volume, but for grade C20P, C25P and C30P they must be measured by weight. Cement must be measured either by whole bags or in a weighing device. Water contained in aggregates and added water may be measured by either volume or weight. Requirements are also given for sampling, testing and tolerances.

For nominal maximum aggregate sizes of 40, 20, 14 and 10 mm table 69 gives the weights of dry aggregate to be used with 100 kg of cement and table 70 gives percentages by weight of fine aggregates of varying coarseness: total aggregates.

(b)   *Special prescribed mixes* (section three)

Deliberate design of concrete is economical for large contracts. In this category of mixes the purchaser's engineer, taking into account all the relevant factors, designs the mix and specifies the type of cement and minimum cement content (see table 68), the type and nominal maximum size of aggregate, and admixtures. The engineer must be satisfied that the specification can provide the required strength in the concrete, but as strength tests are not a means of judging compliance supervision of manufacture is critical.

The engineer also states the proportions of the constituents and 'where necessary' specifies:

Workability, eg for 'ready-mix' concrete
Maximum free water/cement ratio
Air content of fresh fully compacted concrete
Maximum and/or minimum temperature of fresh concrete
Admixtures.

Other items dealt with by section three include:
Inspection, sampling and testing
Batching and mixing
Information to be provided
Tolerances.

[1]The BS differentiates between the *purchaser* and *producer* of concrete.

[2]*Characteristic strength* is defined as 'that value of strength below which 5 per cent of the population of all possible strength measurements of the specified concrete are expected to fall'.

Grade	Cement	Coarse aggregate Nominal maximum sizes 10, 14, 20 or 40 mm	Fine aggregate
C7.5P C10P C15P	BS 12 or BS 146	BS 882 or BS 1047 40 or 20 max	BS 882
		or BS 882 all-in or reconstituted aggregate – preferably with the higher fine aggregate proportions given in table 65	
		The aggregate may be measured by volume	
C20P C25P C30P	BS 12 BS 146 or BS4027	BS 882 or BS 1047 40, 20, 14 or 10 max	BS 882 natural sand and/or crushed stone or gravel
		Not reconstituted or all-in aggregate	
		The aggregate must be measured by weight	

Table 67 BS 5328:1981 Requirements for ordinary prescribed mixes

Concretes	Concrete grade	Minimum cement content kg/m³
Prestressed	all*	300
Reinforced	all*	240
Concretes	C20	220
containing no	C15	180
embedded	C10	150
metal	C7.5	120

*See the appropriate code for minimum permitted grades, CPs 114, 115 or 116

Table 68   Minimum cement contents
(Table 6 from BS 5328)

2 Designed mixes (section three)

For *designed mixes* the purchaser specifies the required strength, the minimum cement content, and any other properties necessary to ensure durability, but leaves the contractor or manufacturer to decide the mix proportions. In other words, the producer makes concrete to satisfy the purchaser's *performance specification*.

Mixes are specified primarily by grade designations:

C7.5, C10, C15, C20, C25, C30, C35, C40, C45, C50, C55, C60, F3, F4, F5, IT2, IT2.5 or IT3 which relate to *characteristic compressive* (C), *flexural* (F) and *indirect tensile* (IT) strengths respectively, all in N/mm^2 (MPa) at 28 days.

The general requirements are as for *Special prescribed mixes* except that strength testing is, of course, essential in order to prove compliance of *designed mixes* with the BS.

Design of concrete mixes

The process in the design of high strength concretes are complex but the design of mixes for most purposes is illustrated by figure 25 and described in detail by: *Design of normal concrete mixes*, DC Teychenné, R E Franklin and H C Erntroy, HMSO.

This publication divides concrete mix design into five main stages, ie determination of:

1 *Target water/cement ratio* – taking into accouunt the maximum free water/cement ratio for durability, the type of cement, the required characteristic strength at a given age, the site control and the target mean strength.
2 *Free water content* – taking into account workability and the maximum aggregate size.
3 *Cement content* – from 1 and 2 as limited by any specified maximum or minimum values giving a modified water/cement ratio.
4 *Total aggregate content* – taking into account the relative density of the aggregate.
5 *Fine and coarse aggregate contents* – the proportions depending on the maximum size of aggregate, workability, the grading zone of the fine aggregate and the free water cement ratio.

Another reference is: *Concrete mix design* F D Lydon, Applied Science Publishers Ltd.

Grade of concrete	Nominal maximum size of aggregate (mm)	40		20		14		10	
	Workability	Medium	High	Medium	High	Medium	High	Medium	High
	Range of slump (mm)	50–100	80–170	25–75	65–135	5–55	50–100	0–45	15–65
		kg	kg	kg	kg	kg	kg	kg	kg
C7.5P		1080	920	900	780	N/A	N/A	N/A	N/A
C10P		900	800	770	690	N/A	N/A	N/A	N/A
C15P	Total aggregate	790	690	680	580	N/A	N/A	N/A	N/A
C20P		660	600	600	530	560	470	510	420
C25P		560	510	510	460	490	410	450	370
C30P		510	460	460	400	410	360	380	320

N/A not applicable   Slumps are for standard samples

Table 69   Weights of dry aggregate to be used with 100 kg of cement for ordinary prescribed mixes (Table 1 from BS 5328)

Grade of concrete[1]	Nominal maximum size of aggregate (mm)	40		20		14		10	
	Workability	Medium	High	Medium	High	Medium	High	Medium	High
C7.5P C10P C15P		30–45[2]		35–50[2]		N/A		N/A	
	Grading zone 1	35	40	40	45	45	50	50	55
C20P	2	30	35	35	40	40	45	45	50
C25P	3	30	30	30	35	35	40	40	45
C30P	4	25	25	25	30	30	35	35	40

N/A not applicable

[1] The proportions given in the tables will normally provide concrete of the strength in N/mm² indicated by the grade except where poor control is allied with the use of poor materials

[2] For grades C7.5P, C10P and C15P a range of fine-aggregate percentages is given; the lower percentage is applicable to finer materials such as zone 4 sand and the higher percentage to coarser materials such as zone 1 sand

[3] For all grades, small adjustments in the percentage of fine aggregate may be required depending on the properties of the particular aggregates being used

For grades C20P, C25P and C30P, and where high workability is required, it is advisable to check that the percentage of fine aggregate stated will produce satisfactory concrete if the grading of the fine aggregate approaches the coarser limits of zone C or the finer limits of zone F

Table 70   Percentage by weight of fine aggregate to total aggregate for ordinary prescribed mixes (Table 2 from BS 5328)

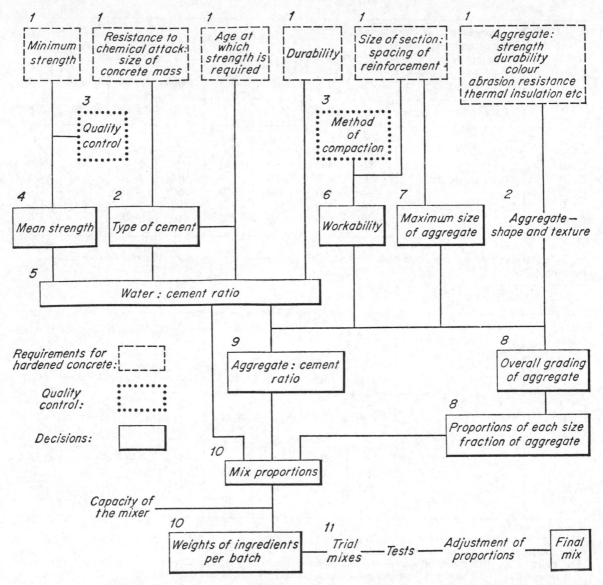

25   *Factors and steps in mix design based on diagram in 'Properties of Concrete' by A M Neville, Pitman*

Polymers in concretes (PIC)

Polymers in concrete can significantly increase flexural strength and improve resistance to wear. They reduce permeability and water absorption while increasing durability in aggressive atmospheres. Polymers are introduced either by premixing or by impregnation of hardened concrete.

Natural rubber or styrene butadiene elastomers, or suitable thermoplastics can be added during mixing of concretes as emulsions or as dispersed phases. Generally, polymers added to the mix retard setting and early strength development, but by increasing workability, polymers enable less water to be used, with corresponding increase in flexural and to a less extent, compressive strength. Also, the modulus of elasticity is reduced.

A well established use for premixed polymer concrete is as screeds and levelling compounds, with improved strength and adhesion to bases so they can be reduced to feather-edge thickness. Precast external wall panels, can be thinner and lighter than conventional concrete panels, and may weather more attractively.

For factory production and for remedial work impregnation of hardened concrete is a promising development. After drying a monomer is absorbed into the concrete and then converted into a polymer by a thermal-catalytic process, or by irradiation with gamma rays. This method reduces permeability more than the premix system and in this case there is a distinct increase in compressive strength. Poor concrete can be treated in-situ with up to 25mm penetration. On the other hand, high strength, and therefore non-porous concrete, cannot be further improved by impregnation. Consequently the high costs of materials and labour incurred in treating porous concrete must be compared with the costs of providing high quality conventional concrete.

Proprietary polyester concrete surface water channels and gullies are claimed to be 'three times stronger and substantially more chemical resistant than ordinary concrete'.

MANUFACTURE OF CONCRETE

Concrete can either be made wholly on the site, or the potential advantages of factory production can be partially secured by the use of *ready-mixed concrete*, or wholly secured by the use of pre-cast products. Table 72 suggests production methods for varying qualities of concrete on small, medium-sized and large sites

The delivery to building sites of BS 5328:1981 *Ready-mixed concrete* is particularly useful where continuous site production is not possible or where space for storage of materials and for mixing is restricted, and it makes high quality concrete and special mixes available on the smallest sites. Normal site tests should be carried out on the mixed concrete.

Apparatus for the manufacture of concrete is described in *MBS: S and F, Part 2*.

The processes of manufacture are:

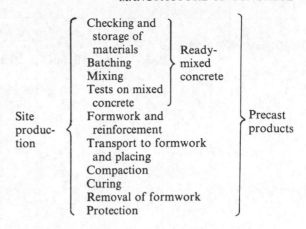

Storage of materials

Storage of materials must prevent deterioration of cement and contamination and segregation of aggregates.

Cement must be kept dry. Paper bags cannot be relied upon to prevent *air-setting* and resulting lumpiness. Particular care should be taken in storing extra rapid hardening and ultra high early strength Portland cements, and supersulphated cement. Exceptionally, where it is not certain that cement can be stored in dry conditions or that it can be used soon after delivery it may be advantageous to use hydrophobic Portland cement – see page 155.

Paper bags should not be stacked more than four or five feet high to avoid *warehouse set* caused by compaction. Cement should be used in the order in which it was received.

Aggregates The various sizes should be kept separately on clean, hard surfaces, sloping for drainage, and not directly on the ground. Wherever possible stock piles should be duplicated so that deliveries can drain for at least twelve hours before use.

Batching

Accurate batching of cement, aggregates and water make for savings in cost of designed mixes by enabling a lower control factor to be employed. It used to be customary to specify and to batch cement and aggregates in proportions by volume, as so called *nominal mixes* but, volume batching tends to be inaccurate because both

Workability	Water:cement ratio	Gravel aggregate			Crushed rock aggregate		
		10 mm	20 mm	37.5 mm	10 mm	20 m	37.5 m
Very low	0.4	3.5	4.5	5	3	4	4.5
	0.5	5.5	6.5	7.5	4.5	5.5	6.5
	0.6	7.5	—	—	6	7	—
	0.7	—	—	—	—	—	—
Low	0.4	3	4	4.5	—	3.7	4
	0.5	4.5	5.5	6.5	4	5	5.5
	0.6	6	7	7.5	5	6	7
	0.7	7	8	—	6	7	8
Medium	0.4	—	3.5	4	—	3	3.5
	0.5	4	5	5.5	3.5	4	5
	0.6	5	6	7	4.5	5	6
	0.7	6	7	8	5.5	6	7
High	0.4	—	3	3.5	—	3	3
	0.5	3.5	4	5	3	4	4.5
	0.6	4.5	5	6.5	4	4.5	5.5
	0.7	5.5	6	7.5	5	5.5	6.5

Table 71 Total aggregate:cement ratio (by weight) to give various degrees of workability with given water:cement ratios

cement and sand are subject to *bulking* and coarse aggregate is difficult to measure accurately by volume. Today, except for very small sites, cement is batched by weight and normally, and preferably, the aggregate also. Pre-packed cement/dry 20 and 10 mm 'down' aggregate mixes can be convenient on repair work – see BS 5838:Part 1, 1980.

Cement Cement varies in bulk density from about 1 120–1 600 kg/m³ according to the way in which the container is filled. Where a weighing device is not available, the bag can be used as a unit. Only exceptionally should a bag be split.

Sand Dry and wet sands have the same volume, but damp sand has a greater volume and if sand is measured by volume and allowance is not made for *bulking* concrete mixes may be seriously undersanded. Figure 26 shows the *bulking* of a typical sand. It is greater for the finer sands and may be as much as 40 per cent. In volume batching of damp sand a rule of thumb addition of 25 per cent by volume is often made, but an accurate assessment can easily be made by inundating a sample of the sand and noting the reduction in volume which results. In either case checking of concrete mixes by eye is always useful to avoid either 'hungry looking' or unduly 'fatty' mixes.

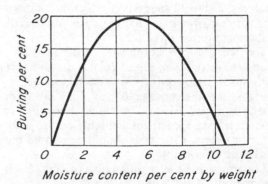

26 *Bulking of a typical fine sand*

Size of site	Ordinary concrete	High quality concrete (designed)
Small	Production of good quality site-mixed concrete may be difficult to achieve on small sites but is aided by use of *ready-batched BS 5838:1980 Dry packaged mixes or by ready-mixed concrete*	Requisite standard of batching, mixing placing, compacting and curing not practicable on small sites
		Ready mixed concrete; can be of high quality but requisite standard of placing, compacting and curing unlikely to be economical
Medium	Production of good quality site-mixed concrete is practicable	Requisite standard of batching, mixing, placing, compacting and curing of site-mixed concrete may not be economical
	Ready-mixed concrete may be economical	Ready-mixed concrete; placing, compacting and curing requires special care
Large	Production of good quality concrete is economical	Production on site is economical
	Ready-mixed concrete may be economical particularly for small quantities	

Note: Factory pre-cast concrete of ordinary or high quality can be used on all sites

Table 72   Production methods for sizes of sites and qualities of concrete

In weight batching the water content of sand must be determined from time to time and the weight of batches adjusted accordingly.

For example, in a batch of concrete containing 400 kg of sand having an average moisture content of 5 per cent the corrected batch weight is

$$\frac{400 \times 105}{100} = 420 \text{ kg sand}$$

Coarse aggregate   Deep and narrow gauging boxes reduce error in volume batching, but the method is laborious. Properly maintained weight-batching machines are very accurate and easy to use.

Conversion from volume to weight proportions. It is usually assumed that 50 kg of cement occupies 0.04 m³. The equivalent proportions for a nominal 1:2:4 mix, therefore, are 50 kg cement: 2×0.04 m³ fine aggregate: 4×0.04 m³ coarse aggregate, ie 50 kg cement: 0.08 m³ fine aggregate: 0.16 m³ coarse aggregate. Similarly a nominal 1:5 all-in mix becomes 50 kg cement: 0.20 m³ all-in aggregate.

	Typical loose dry densities kg/m³	Proportions by volume	Weight per volume of cement kg/m³	Proportions by weight
Cement	1 440	1	1 440	1
Sand	1 600	2	3 200	2.2
Gravel	1 360	4	5 440	3.8

Table 73   Conversion from volume to weight proportions

Table 72 shows how a 1:2:4 nominal mix can be converted into a mix by weight when the usual assumptions are made for the bulk densities of cement, sand and gravel.

For important work the weights of batches must be adjusted to allow for the water contained in coarse aggregate. For example the corrected batch weight for 1 000 kg of coarse aggregate with an average moisture content of 2 per cent would be:

$$\frac{1\,000 \times 102}{100} = 1\,020 \text{ kg coarse aggregate}$$

Water As the water:cement ratio determines the strength and durability of concrete, see page 163, the amount of water contained in each batch is critical. The gross weight of water (kg) per batch is water:cement ratio × weight of cement (kg).

The tanks fitted to the larger mixers have gauges which enable a measured quantity of water to be added to each batch. This must be adjusted from time to time to allow for the water contained in the aggregate.

The following approximations may suffice:

Condition of aggregate	Fine aggregate l/m^3	Coarse aggregate l/m^3
Very wet	120	60
Moderately wet	80	40
Moist	40	20

Table 74 *Approximate moisture content of aggregates*

For more important work the water contained in the aggregate can be determined either: by weighing a sample, drying it and noting the loss in weight, or by one of the displacement methods.

During the progress of work if changes in the moisture contents of aggregates are small, provided the quantities of cement and aggregates and the type of aggregates remain the same, the quantity of added water can be adjusted so as to maintain the workability indicated by a slump test on the first batch.

Mixing

Concrete may be mixed on the site, or 'at works' for pre-cast concrete or for delivery to the site as ready-mixed concrete.

On-site mixers The most commonly used type are batch mixers of the single-compartment drum type. BS 1305:1974 describes them as:

Tilting (T) – having a drum with one opening which rotates on an inclinable axis; and

Non-tilting (NT) – hving a drum with two openings which rotates on a horizontal axis.

Reversing drum (R) – having a drum rotating on a horizontal axis.

Forced action type (P) – a pan with blades or paddles.

The letter follows the numbers which represent capacity in litres and precedes numbers which represent capacity in cubic metres, the smallest capacity being 1.5 cubic metres and the largest 1 000 cubic litres.

Truck mixers are described in BS 4251:1974.

Some mixers incorporate weight batching equipment and attachments for hand scrapers to assist in loading the hoppers and normally, 200 litre and larger mixers can measure volumes of water.

So that water is evenly distributed, it should enter the mixer before or at the same time as the other materials. The proportion of coarse aggregate should be reduced for the first batch or two each day to compensate for the loss of mortar which sticks to the blades and inside the drum.

The time required for thorough mixing varies according to the characteristics of the mix and of the mixer. With batch type mixers it has been customary to require mixing for at least two minutes although some machines mix thoroughly in shorter periods. In all cases the product should be uniform in colour and consistency.

When the concrete has been mixed the complete contents of the drum should be discharged in one operation to avoid segregation of the larger stones.

Mixers should be thoroughly washed out and cleaned daily, and even after short stoppages, to prevent 'caking' with hardened concrete which reduces the machine's efficiency and they should be cleaned out when the type of cement is changed.

Tests on fresh concrete

Site tests and the apparatus required for these are described in BS 1881: *Testing concrete*.

Part 101:1983 is *Methods of testing concrete on site*

Part 102:1983 *Determination of slump*

Part 103:1983 *Determination of compacting factor*

Parts 106 and 107:1983 deal with *Measurement of air content*, and *Density of fresh concrete respectively*.

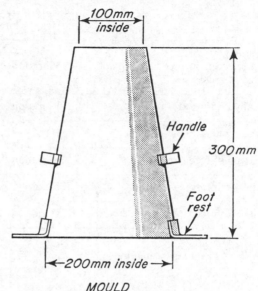

TAMPING ROD 600mm long x 16mm dia.

1 Take representative sample
   of concrete, and within
   2 minutes fill mould
   in 4 equal layers

2 Tamp each layer 25
   times with steel rod

3 Strike off top level and
   clean off any leakage
   around base of mould,
   and without delay:

4 Raise mould carefully
   in vertical direction

5 Measure slump to nearest
   6mm. If shear or collapse
   slumps occur, repeat test
   with another sample

6 Record results

7 Clean and dry mould.
   Do not oil

100mm
inside

Handle

300mm

Foot
rest

←200mm inside→

MOULD

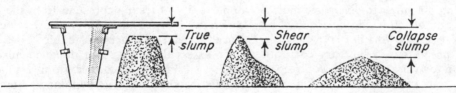

True slump   Shear slump   Collapse slump

27   The slump test

Consistency of manufacture The slump test, which is easy to carry out, indicates variations in the shape or grading of aggregate, or in the proportion of water being used. See figure 27.

Workability The *slump test* gives an approximate indication of the workability of Portland cement mixes which are neither too stiff nor too plastic. The *compaction factor* test is more accurate, but neither test is suitable where the maximum size of aggregate exceeds 40 mm.

Tests on hardened concrete

BS 1881 includes Parts 108, 109 and 110:1983 *Methods of making test cubes, beams and cylinders*, respectively. Part 111:1983 deals with *Methods of curing at 20°C* and Part 112:1983 with *Accelerated curing at higher temperatures.*

Other parts of the BS include:

Part 114:1983 *Determination of the density of hardened concrete*

Part 115:1983 *Compression testing machines*

Part 116:1983 *Determination of compression strength*

Part 117:1983 *Tensile splitting strength*

Parts 118–121:1983 *Determination of strength properties.*

Part 122:1983 *Water absorption of hardened concrete.*

The assessment of strength of existing structures is referred to on page 183.

Compression tests Cubes made before and

Type of concrete	Means of compaction	Compaction factor	Slump mm
Very high strength prestressed	Pressure	less than 0.70 approx.	0
	Heavy vibration	0.70–0.80	
High strength prestressed, reinforced paving and mass concrete	Vibration	0.78–0.85	0–25
Normally reinforced	Vibration	0.85–0.92	25–50
Mass concrete	Hand		
Heavily reinforced	Vibration	0.92–0.95	50–100
Normally reinforced	Hand		
Heavily reinforced Complex shapes	Hand	0.95	100–150

Table 75   Guide to workabilities

during the placing of concrete on the site are tested in crushing machines to give some indication of the strength which would be acquired by the actual work – see BS 1881:Part 116:1983.

Preliminary cube tests   Preliminary compression tests require very accurate control of materials and test conditions. The materials intended to be used are mixed in the laboratory in the proportions to be used in the work.

Works cube tests   Samples of the concrete which is being placed in each part of the work should be made into cubes in accordance with Part 108 of the BS. Specimens to be tested should be kept free from vibration and under damp sacks for 24 hours ± ½ hour before removing them from the moulds. They should be marked and stored in water at a temperature between 10 and 21°C. They should be covered with damp material to be taken to a laboratory where they must be stored in water again for 24 hours before being tested.

A typical specification requires that if a cube fails at 7 days to attain the strength specified for that age, another cube be made from the same concrete may be tested at 28 days. If the second cube fails to attain the strength specified for 28 days, the specification may give the contractor the opportunity of testing cores cut from the placed concrete to prove that the concrete which was placed provides the strength required.

To assess the strength of designed mixes, BS 5328:1981 requires a sample from a randomly selected batch of concrete by taking a number of increments at the point of delivery from the mixer, or ready-mix concrete delivery vehicle, in accordance with BS 1881. Normally the result is taken as the average of tests on two specimens after 28 days curing.

Formwork

Formwork provides the shape and surface texture of concrete members and supports them during setting and hardening. It must be grout-tight, true in line, level, face and profile, and strong enough to accept all constructional loads including those resulting from mechanical compaction. Formwork is best constructed in units for easy erection, striking without damaging the concrete, and so that it can be reused. It should allow easy removal of sides without disturbing soffits and, ideally, removal of soffits while retaining props.

The faces of formwork should be treated with mould oil to give a clean release but avoiding excess oil which stains concrete and which may interfere with the bond for plaster.

Reinforcement

Reinforcement should comply with the following Standards:

BS 4449:1978 *Hot rolled steel bars for reinforcement of concrete.*

BS 4482:1982:1969 *Hard drawn mild steel wire for the reinforcement of concrete.*

BS 4483:1969 *Steel fabric for the reinforcement of concrete.*

BS 4486:1980 *Hot rolled, and hot rolled and processed high tensile alloys steel bars for the prestressing of concrete.*

BS 4661:1978 *Cold worked steel bars for the reinforcement of concrete.*

BS 5896: *High tensile steel wire strand for prestressed concrete.*

BS 4757: 1971 *Nineteen wire steel strand for prestressed concrete.*

Reinforcement should be free from mill scale, loose rust, oil or grease.

Cover   Steel must be protected from corrosion and fire.

BRED 263, *The durability of steel in concrete* and Durability of Steel in Concrete, K W J Treadway, BRE, are references.

Early failures are common where corrosion products form and expand, cracking and spalling concrete cover.

Hydrated cement raises the pH of a matrix to 12.6–13.5 in which steel remains passive. Unfortunately, alkalinity is reduced by reactions with moist acidic atmospheres. This *carbonation* increases with permeability of the concrete and most rapidly at 50–75 per cent RH. Carbonation increases with time and corrosion occurs where it reaches reinforcement. The depth of concrete cover, therefore, should be increased in polluted atmospheres.

Chlorides stimulate corrosion, even in highly alkaline conditions, and depth of cover must be increased in marine conditions or where concrete will be exposed to de-icing salts. Chlorides can also derive from aggregates and from certain admixtures which are no longer recommended in reinforced concrete – see page 157. 95 per cent of tests on concrete mixes should show a chloride content not exceeding 0.35 per cent by weight of cement, and no results should exceed 0.5 per cent.

BRED 263 Part 3, which deals with repair to reinforced concrete, refers to the use of admixtures, including epoxy and polyester resins in mortars and grouts.

Table 4.8 of BS 8110, Part 1, should be studied. for example, for normal weight 20 mm aggregate in grade C45 concrete, having maximum free water:cement ratio and at least 350 kg cement/m³, cover must be at least 20, 30 and 60 mm in *mild, severe,* and *extreme* exposures respectively. Zinc coatings for 200 to 700 g/m² (0.03 to 0.10 mm) allow the depth of concrete cover to be reduced without increasing the risk of corrosion, and the depths of cover recommended for dense aggregate concretes to be used with lightweight aggregate concrete – see *Corrosion* chapter 9, and BRE Digest 109 *Zinc-coated reinforcement for concrete.*

For fire resistance should be minimum thicknesses of cover on main reinforcement from 12.5 mm for ½ hour, up to 63 mm for 4 hours.

Reinforcement should be placed in the exact positions shown on the drawings and the specified cover ensured, eg by spacers fixed to the reinforcement. Great care should be taken to avoid damage or disturbance to formwork when positioning reinforcement. Prefabrication of reinforcement helps in this respect.

Loose ends of wire ties, usually 1.6 mm soft iron, used to secure intersections in the reinforcement, must be kept away from formwork faces, particularly those which will be exposed to the weather.

Transport to formwork and placing

Whether concrete is moved from the mixer by lorries, barrows, dumpers, mechanical skips or pipeline it is important that the process does not cause segregation of the mix.

All plant, chutes etc, should be thoroughly cleaned after use without allowing the waste water to enter formwork.

'Wet' mixes are particularly likely to segregate

and, where possible, these should not be dropped into position. Chutes should be arranged so that a continuous flow is discharged at the lower end. Like mixers, chutes accumulate mortar and where they are used the first mixes should contain less coarse aggregate. Where concrete is discharged from long chutes it can first enter a hopper or bunker in which it is partially remixed before being placed in the forms.

A continuous supply of suitable mixes can be transported through pipe-lines 100–200 mm in diameter, by force pump or by pneumatic pressure. Force pumps can move concrete up to about 30 m vertically, or 300 m horizontally.

The mix should be of uniform consistency, not leaner than 1:6 by weight with a compacting factor in the range of 0.92–0.95 and 50–100 mm slump. The pipe-line must be lubricated with mortar before pumping of concrete starts, and when pumping stops it must be cleaned out, usually with compressed air and water aided by a dolly. See BRE Digest 133.

Immediately before concrete is placed, whether 'at works' or on-site, formwork should be thoroughly cleaned out and formwork and reinforcement should be re-checked.

Where concrete is placed on the ground, concrete *blinding*, or on good ground polythene sheeting, should be provided to prevent loss of fines.

'Flowing' in the formwork should be prevented by placing the concrete in layers of uniform thickness.

Compaction

Trapped air which should not exceed about 2 per cent when concrete is placed must be released if the maximum density with associated resistance to chemicals, water vapour, frost and abrasion is to be obtained. Figure 28 shows that 1 per cent of entrapped air causes a loss of about 5 per cent strength. Thorough compaction is also very important wherever concrete faces are to be exposed to view. Air is very liable to be trapped against form faces and at joints between hardened and newly placed concrete. Compaction should commence as soon as possible once water has been added to concrete although so long as it remains possible to fully compact concrete by the means

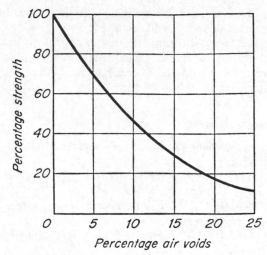

28 *Loss of strength due to air voids*

available, delay in doing so may not be serious up to perhaps two hours even in cool weather.

Minimum degrees of workability, expressed as slumps and compacting factors, which permit full compaction by hand or by machine of various Portland cement concretes are given in table 75.

Hand compaction Concrete which is sufficiently workable can be worked into place with spades. Air must be released from the surface by *slicing tools* to obtain a good finish.

By machine The stiffer mixes can only be thoroughly compacted by vibrators, rollers, presses or by centrifugal 'casting'.

Vibrators CP 114:1969 *Structural use of reinforced concrete in buildings* strongly recommends the use of vibrators provided that reduced water:cement ratios are adopted. Over-vibration leads to segregation. Vibration should stop when air bubbles no longer come to the surface. Vibrators are of the following types:

Immersion vibrators – or *poker* vibrators These should be operated by experienced persons. They should be inserted vertically at points not more than 450 mm apart. They should not be brought too close to the formwork or reinforcement and never drawn along horizontally. Vibrators should be withdrawn very slowly so that a hole is not left in the concrete.

External vibrators These are necessary where

the reinforcement does not allow space for poker vibrators. They must be firmly fixed to the formwork or hand-held mechanical hammers can be used. Formwork must be sufficiently rigid to transmit the vibration without being damaged or dislodged.

Beam vibrators Tamping beams with vibrators fixed to them are suitable for compacting the upper surfaces of horizontal slabs.

Table vibrators These are suitable for vibrating pre-cast goods. The moulds in which they are cast must be rigidly clamped to the vibrating table top.

Pan vibrators These consist of a steel-bottomed pan about 0.75^2 m in area with a vibrator fixed to the upper surface. They are suitable for awkwardly shaped horizontal surfaces but do not leave a truly flat surface.

Presses Hydraulic actuated presses are used for compressing paving flags and tiles in factory production.

Curing

In order to obtain the desired strength, compacted concrete must be free from physical disturbance, (a) water must be retained in the concrete and (b) temperature must be controlled.

(a) *Retention of water in the mix* Water is essential for hydration of cement and concrete ceases to develop strength if it dries. (There is no conflict here with the statement that a low water:-cement ratio makes for high strength in thoroughly compacted concrete.) If concrete dries before it has developed sufficient strength, cracking (due to shrinkage) results – see page 162. Loss of water can be prevented in the laboratory or factory by covering concrete with damp sacks at first and later by immersing it in water. Steam curing guarantees that no water will be lost from the concrete.

In-situ concrete can be protected from sun and drying winds by sheets of waterproof paper or plastics. clean sand can be laid on horizontal surfaces and kept wet for at least seven days. Alternatively, surfaces can be sprayed with an impervious composition which prevents evaporation.

Drying out should be prevented for the following periods:

Type of cement	Ordinary concrete (days)	Concrete subject to abrasion (days)
Ordinary Portland cement	4	7
Rapid hardening Portland cement	3	4
High alumina cement	1	—
Supersulphated cement	4	—

(b) *Control of temperature* At about freezing point cements cease to develop strength, and if the water is frozen concrete is permanently damaged. At temperatures up to about 40°C strength development is a function of temperature and time (or *maturity*). Figure 29 shows the 28 day compressive strength of a concrete cured at 3°C to be 85 per cent of the same mix cured at 17°C.

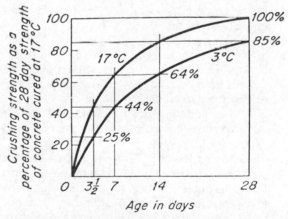

29 *Effect of temperature upon the rate of hardening of a typical ordinary Portland cement concrete. Based on graph from Report on 'Concrete Practice' Volume I (ISE and CCA)*

Thus, temperature during curing may require to be artificially raised to prevent freezing and to ensure a sufficient rate of strength development. On the other hand excessive rise in temperature naturally or artificially induced can lead to a *flash set* of cement and to cracking of concrete due to differential thermal stresses. Specific problems arise in cold and hot weather:

Cold weather The temperature of concrete should never fall below 5°C before and during placing, or below 4°C until it has hardened.

When the atmospheric temperature falls below about 4°C steps such as the following are normally essential to prevent freezing:

Increase heat evolved by cement:
 (a) Use rapid-hardening Portland cement or ultra-high early strength Portland cement.
 (b) Add an accelerator, which must not contain calcium chloride for reinforced or prestressed concrete or concrete containing embedded metal.
2 Heat ingredients:
 (a) Water-temperature should not exceed about 82°C
 (b) Frozen aggregate must *never* be used. Heating can be by steam pipes carried through stock piles.
To avoid a *flash set* of the cement, the aggregates and mixing water should be mixed before the cement is added so that their temperature is unlikely to exceed about 30°C.
3 Conserve heat:
Surfaces of concrete can be covered with strawboards or other insulating materials, and screens keep off cold wind. Thick timber formwork provides useful insulation.
4 Heat the building. Hot air blowers are available but great care, however, must be taken to avoid drying fresh concrete.
5 Heat the formwork. Concrete must never be placed in frost covered formwork or against frozen ground. Formwork can be heated by low pressure wet steam or hot air with a fine water spray introduced below protective housings, or by electric blankets.

Circumstances on building sites vary so widely that it is not possible to recommend specific measures. However, in frosty weather, when the day temperatures rise above 0°C it may be sufficient to use rapid-hardening Portland cement, to add an accelerator to ordinary Portland cement, to heat the mixing water and/or to insulate the formwork.

Steam curing High temperature during the setting of cements reduces their ultimate strength and all concrete should be cured for several hours at room temperature. Subsequently however, steam curing, which combines high temperature with conditions which prevent loss of water, enables very high early strengths to be obtained in Portland cement concretes (including most Portland cement concretes which include calcium chloride). Final strength, however, is not improved. Low pressure steam curing can be carried out inexpensively, even on a building site. High pressure steam curing, usually at about $2.0 \, N/mm^2$, requires an autoclave but has several advantages. Thus the 28 day strength of air-cured concrete can be attained in 24 hours, drying shrinkage, moisture movement and efflorescence are reduced and resistance to sulphate attack is improved.

Hot weather In particular, slabs must be protected from sunlight and drying winds by covering with damp sand, impervious sheets or water. Special measures, eg the use of low heat Portland cement, see page 154, may be necessary to avoid excessive temperature gain in large masses of Portland cement concrete and to prevent cracking which can result from differential thermal stresses.

Removal of formwork

Formwork must be left in position, and the supports maintained, until concrete is sufficiently strong to safely support its own weight and any loads which may be put on it. Concrete should have a cube strength at least twice the stress to which the concrete is likely to be subjected at the time of striking.

The times which should elapse before formwork is removed vary considerably according to the cement used, temperature of the concrete during curing and other factors. Minimum times are recommended in table 76. Longer times, based on surface temperatures, are given by CP 110 for slabs and beams.

Times for intermediate temperatures may be interpolated pro rata and for Portland cements one day should be added for each day the temperature is at, or below, freezing point. Times must be increased if loads such as building materials are being carried by new structures or if stronger concrete is required.

| | Type of cement | | | |
| | Ordinary and sulphate-resisting Portland cements | | Rapid-hardening Portland cement | |
	3°C	16°C	3°C	16°C
Walls and sides of beams and columns	5	1	3	1
Slabs: props left under	7	3½	4	2
props to slabs	14	7	8	4
Beam soffits: props left under	14	7	8	4
props to soffits	28	14	16	8

Table 76   Minimum number of days before removing formwork from concrete carrying its own weight only

The removal of supports may be delayed to obtain at least 10 per cent increase on the 28 day strength at three months and 20 per cent at one year, for all the five CP 110 grade concrete mixes.

Supports should be eased away uniformly and very slowly, so that the load is not suddenly imposed on partly hardened concrete. Formwork must be stripped carefully to avoid damage to arrises and projections, especially where vertical surfaces are exposed within twelve hours of casting.

Protection

After stripping formwork it may be necessary to protect concrete from damage by knocks, shock and vibration; from drying in hot weather and from loss of heat in cold weather – see page 181.

Construction joints

Whenever concreting is interrupted the construction joints which are inevitably formed are potentially weak. They may allow water to enter and they are always visible, particularly after a period of weathering. The positions and design of construction joints should, therefore, be decided at an early stage.

Joints should be straight, either vertical or horizontal, and in walls in positions related to window openings and other features. Generally, in columns construction joints are made as near as possible to the beam haunching and in beams and slabs within the middle third of the span. Vertical joints should be formed against temporary but rigid stop boards which must be designed to allow reinforcement to pass through.

To ensure the best possible bond between new and 'old' concrete lifts, *laitance* (a scum of cement and very fine material) must not be allowed to form on horizontal surfaces, the use of a drier mix at the end of each day helps in this respect. Preferably, within an hour or so after placing the lower lift, its surface should be sprayed with water and brushed to expose the coarse aggregate. If these measures are delayed it may be necessary to use a stiff brush to roughen the surface, or if it has hardened, to hack the surface but without damaging the aggregate. Then, although difficulties of access often make such operations difficult, the roughened surface should always be thoroughly cleaned and loose matter removed, preferably without re-wetting.

New concrete is satisfactorily placed directly against freshly cleaned and relatively dry concrete only if the newly placed concrete is very thoroughly compacted so that some of the water and fine material is absorbed into the surface pores of the hardened concrete. Alternatively:
(a) The clean surface is wetted sufficiently only to reduce the absorption of the hardened concrete.
(b) A thin grout of cement is brushed over the surface.
(c) Mortar about 3 or 7 mm thick, of the same materials and not weaker than the cement and sand in the new concrete, is laid immediately.
(d) The new concrete must be placed within 30 minutes. If the mortar is allowed to dry a weak joint will result.

Avoidance of segregation, and thorough compaction of newly placed concrete are particularly important along the joint plane.

Post casting

Assessment of strength of concrete in existing structures

Strength cannot be accurately deduced from any single in-situ test but BS 4408 (5 parts) gives *Recommendations for non-destructive methods of test for concrete* and BS 6089:1981, a *Guide to the assessment of concrete strength in existing structures* shows relationships between strengths of standard test specimens and those in structures.

Impregnation

This treatment to improve the properties of dry concrete is described on page 172.

Surface treatments

These aim to modify the appearance of hardened concrete – see page 158 or to provide a key for plaster, renderings or new concrete.

LIGHTWEIGHT CONCRETES

Lightweight concretes, ie *aerated concretes, lightweight aggregate concretes* and *'no-fines' concretes*, are defined as those weighing less than $2\,000\,\text{kg/m}^3$ and are made in densities down to about $160\,\text{kg/m}^3$. The subject is dealt with very thoroughly in *Lightweight concrete* by A Short MSc, AMICE, MI Struct E and W Kinniburgh FRIC (Maclaren and Son Ltd) and in the two volumes of *The First International Congress on Lightweight Concrete,* May 1968, Cement and Concrete Association and in the BRE Digests referred to later.

The 44 m diameter dome of the Pantheon Rome was largely constructed of pumice lightweight concrete in the second century AD. Today, advantages in the use of lightweight rather than dense concrete include: savings in the costs of handling materials and of supporting structures, superior thermal insulation and fire resistance; superior sound absorption of unplastered surfaces, some of which offer a better key for plaster. Lightweight concretes are usually easier

to cut, chase and nail into. On the other hand: compressive strength and the modulus of elasticity are reduced (although the latter reduction may improve resistance to mechanical damage). The moisture movement of aerated and lightweight aggregate concretes is high. Reversible moisture expansion is usually as great as the initial drying shrinkage. Protection of reinforcement against corrosion may reduce, and sound insulation reduces as density of the concrete decreases. Lightweight concretes are made in three main ways:

1 *Aerated or cellular concrete* Minute and non-communicating cells are formed by introducing air or gas into a matrix of cement (or in the case of autoclaved concretes, sometimes lime) with, in all but the lightest non-structural concretes, ground sand, pulverized-fuel ash or other fine siliceous material as fine aggregate.
2 *Lightweight aggregate concrete* Made by incorporating a cellular coarse aggregate.
3 *No-fines concrete* Made by omitting the fine aggregate and the smaller particles of coarse aggregate so as to leave voids.

Some properties of lightweight aggregate and aerated concretes are compared with those of dense concrete in table 77.

Aerated concretes

Concretes of this type have the lowest densities, thermal conductivities and strengths. Like timber they can be sawn, screwed and nailed, but they are non-combustible.

BRE Digest 178 and BS 8110 Part 2:1985 describe *Autoclaved aerated concrete.*

For work in situ the usual methods of aeration are by mixing in a stabilized foam or by whipping air in with the aid of an air entraining agent. In this country precast products are usually made by the addition of about 0.2 per cent aluminium powder to the mix which reacts with alkaline substances in the binder forming hydrogen bubbles. Air-cured aerated concrete is used where little strength is required, eg for roof screeds and pipe lagging, in densities as low as $160\,\text{kg/m}^3$. Full strength development depends upon the reaction of lime with the siliceous aggregate, and for equal densities the strength of high pressure steam-cured concrete is about twice that of air-cured

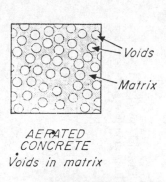

AERATED
CONCRETE
Voids in matrix

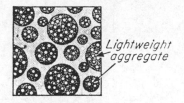

LIGHTWEIGHT
AGGREGATE CONCRETE
Voids in aggregate

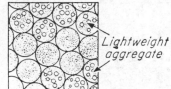

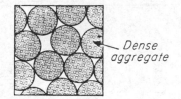

NO-FINES CONCRETE
Voids between aggregate

30   Types of lightweight concretes

concrete, and shrinkage is only one third, or less. No further curing is required after autoclaving. Blocks are usually cut at works to the required size from larger units.

Aerated autoclaved 'blocks' are now made in brick sizes for infill work.

Strength

Strengths sufficient for structural work are obtainable but the modulus of elasticity of aerated concrete is about one tenth that of dense concrete. Creep at working loads is not thought to be greater.

Moisture movement

The moisture movement of cement not being restrained by rigid aggregate, air-cured aerated concrete has very high drying shrinkage and without frequent shrinkage joints, this concrete if placed in situ would crack.

Precast products which are steam cured shrink only 20–25 per cent as much as air-cured concrete, but even where this movement takes place before precast units are incorporated in buildings the following precautions may be necessary:

(a) Specify BS 6073 Part 1:1981 *Precast masonry blocks* which stipulates maximum drying shrinkages of 0.05 to 0.09 per cent according to use, compressive strength and block density

(b) Keep units as dry as possible

(c) Use weak mortar – see chapter 15

(d) Provide shrinkage joints

(e) Before plastering or rendering allow concrete to dry fully and limit the areas – see *MBS: Finishes* chapters 2 and 3.

Weather resistance

Experience indicates that unprotected single leaf aerated concrete block walls have good resistance to rain penetration and to frost. However, for densities of 825 and 497 kg/m^3 water absorptions are about four times and eight times greater than that of dense concrete and external rendering is desirable wherever reinforcement is present.

Thermal insulation

Thermal conductivities of 0.084 W/m deg C and less are obtainable in dry concrete. External surfaces should be rendered or otherwise protected to avoid serious loss of thermal insulation due to absorption of water.

185

Aggregate	Density of aggregate kg/m^3	Dry Density of concrete kg/m^3	Compressive strength N/mm^2	Modulus of elasticity N/mm^2	Drying shrinkage per cent	Thermal conductivity $W/m\ deg\ C$	Main uses
Dense concrete							
Gravel aggregate	1360–1760	2240–2480	14.0–70.0	20 700–34 500–	0.03–0.04	1.4–1.8	Structural, anti-corrosion and fire protection (Class 2)
Lightweight aggregate concretes							
Clinker	720–1040	1040–1520	2.0–7.0	6890–20 700	0.04–0.08	0.35–0.67	Structural reinforced concrete[4] / Wall blocks[2] / Fire protection (Class 1)
Foamed blast furnace slag	320–880	960–2000	2.0–24.0		0.03–0.07	0.24–0.93 (960–1610)[1]	
Expanded clay or shale	320	720–1760	2.0–62.0	short duration loads	0.04–0.07	0.24–0.91 (720–1520)[1]	
Sintered pulverized-fuel ash	640	960–1760	2.8–55.0		0.04–0.07	0.32–0.91 (960–1520)[1]	
Pumice	500	640–1440	2.0–14.0		0.04–0.08	0.21–0.60	
Exfoliated vermiculite	65–130	400–800	0.7–3.5		0.25–0.35	0.16–0.26	Thermal insulation
Expanded perlite	80–240	400–1120	0.5–7.0		0.20–0.30	0.16–0.39	
Aerated concretes (air-cured)	—[3]	352–833	0.48–3.45	—	0.33–0.44	0.86–1.89	Thermal insulation / Wall[2] blocks / Fire protection (Class 1)
	—	721–1217	2.60–5.15		0.18–0.36	2.02–4.03	
(autoclaved)	—	960				0.26	—
	—	800	4.83	3100	0.06	0.20	
	—	640	3.45	2270	—	0.14	
	—	480	2.07	1445	0.07	0.11	
	—	320			—	0.08	

[1] Density ranges corresponding with thermal conductivity ranges.   *Oven dry densities
[2] See chapter 6 page 140 for minimum strengths of loadbearing and non-loadbearing blocks.
[3] 1 cement: 1–3 pfa and ground sand.
[4] See page 188 and BRE Digest 111.

Table 77 Properties of lightweight aggregate and aerated concretes (Information from BRE Digests and other sources)
No-fines concrete see page 189.

Fire resistance

Fire resistance as defined by BS 476:Part 8:1972 tests, is good, for example for walls without finishes:

102 mm loadbearing wall	2 hours
102 mm non-loadbearing wall	4 hours
142 mm non-loadbearing wall	6 hours

Hardness

Aerated concrete is much softer than dense concrete and requires protection from abrasion in the lower parts of walls and in similar positions. It can be easily sawn, worked with simple tools and nailed into. Retention of nails is better with cut nails than wire nails, and with the denser concretes.

Lightweight aggregate concretes

These concretes are dealt with in BRE Digest 123 *Lightweight aggregate concretes.*

Typical values for density, strength, thermal conductivity and drying shrinkage are given in table 77.

The least dense lightweight aggregate concretes, ie exfoliated vermiculite and expanded perlite concretes have strengths and thermal conductivities comparable to aerated concretes of the same density and are used for lagging pipes and for thermal insulating screeds. Intermediate densities are suitable for some building blocks.

Strength

Structural Lightweight Aggregate Concrete J K Nesbit, Concrete Publications Ltd, BS 8110 Part 2:1985 and BRE Digest 111 *Lightweight aggregate concretes* deal with structural applications.

Foamed slag, expanded clay, expanded slate and sintered pulverized-fuel ash concretes are suitable for reinforced concrete structures with strengths in compression up to 62 N/mm², and with densities 30–40 per cent, and thermal conductivities 50 per cent or more, less than those of gravel concretes.

As with dense aggregate concretes, the strength properties of *lightweight aggregate concretes* depend upon:

(i) type of aggregate
(ii) grading of aggregate
(iii) cement: aggregate ratio
(iv) water:cement ratio
(v) the degree of compaction.

Compared with ordinary dense concretes the following assumptions can be made:

Creep: 100 per cent greater
Permissible shear strength: 20 to 25 per cent less
Bond strength: 50 per cent less (except for vertically cast members)
Span: depth ratio: at least 10 per cent less
Elastic modulus: generally 1/3 to 2/3 of values for corresponding dense aggregate mixes increasing with compressive strength and density.

The GLC required mixes to be not leaner than 1 cement: 6 aggregate by volume. Because lightweight concrete is more porous and more likely to crack than dense concrete, discounting clinker which should not be used in reinforced concrete, protection of reinforcement against corrosion depends much more upon the quality of the concrete than on the type of aggregate used. Complete compaction is particularly important. The BRE recommend: 'where reinforced lightweight concrete is exposed to the weather, the concrete cover over the reinforcement should not be less than 51 mm and the maximum aggregate size should not exceed 13 mm'.

Moisture movement

Drying shrinkage is generally about twice that of dense concretes. The poor workability of some lightweight aggregates should be compensated for by the addition of sand or an air entraining agent rather than by using a richer mix which would increase drying shrinkage. Although the proneness of lightweight concrete to shrink and crack may be largely offset by its lower modulus of elasticity, the precautions advised for aerated concrete should be taken.

Fire resistance

In concretes the fire resistance of elements of structure (see page 37) depends largely upon the properties of the aggregate. Because furnace-formed lightweight aggregates are in *Class 1* (see page 41) reinforced concrete elements can be thinner than would be required if they were constructed with Class 2 aggregates.

Lightweight aggregates

Table 77 lists the aggregates in common use with their densities and some properties of typical concretes made with them. BS 3797:Part 2:1976 *Lightweight aggregates for concrete* includes requirements for exfoliated vermiculite, expanded perlite, expanded clay, expanded shale, sintered PFA and pumice.

BS 3681 Part 2:1973 describes: *Methods for sampling and testing lightweight aggregates for concrete*. Some notes follow:

Clinker (not to be confused with *breeze* ie small coke, which is no longer used as aggregate) consists of furnace residues sintered into lumps. It must be thoroughly burnt to avoid serious expansion later. BS 1165:1985 *Clinker and furnace bottom ash aggregates for concrete* limits combustible constituents to 10 per cent for general purposes and to 25 per cent for work which will not be in damp conditions. Some clinkers contain particles of iron which can cause stains and others may contain materials, particularly hard burnt lime, which expand when they are wet. The ill effects of the latter can be avoided by weathering the clinker before it is used. Tests for soundness and limits for sulphates are contained in the BS.

Foamed slag (BS 877:1973) In this country foamed slag is, after clinker, the most widely used lightweight aggregate. It is produced by directing jets of steam and compressed air onto molten blast furnace slag and then crushing and grinding the product.

Foamed slag concrete provides better thermal insulation than clinker concrete. It is used mainly for wall blocks and for insulating roof screeds but unlike clinker concrete it is suitable for reinforced concrete.

Expanded clay Certain clays which evolve gas and bloat when they are heated are formed as rounded pellets with a dense skin, and honeycomb interior in a rotary kiln, or into angular particles by crushing the porous cake from a sinter strand. Expanded clay provides the highest strength lightweight aggregate strructural concrete and is used for the same purposes as foamed slag concrete. *Aglite* and *Leca* are proprietary products.

Sintered pulverized-fuel ash (BS 3892:1965). Ash from powdered coal which has been burnt in electric power stations (PFA or *fly ash*), about 10 million tonnes of which are produced in the UK annually, is moistened and formed into pellets which are fired at about 1 200°C. The resulting spherical particles are graded into large and medium sizes or crushed to form fine aggregate. Concrete made with sintered PFA has drying shrinkage and moisture movement similar to those of expanded clay concrete. Uses are wall blocks, screeds and high strength structural reinforced concrete. *Lytag* is a proprietary product. Gradings are given for coarse aggregate supplied either *graded* (from 20, 14 and 10 mm down) or in *single sizes* (20, 14, 10 and 6 mm). Fine aggregate must fall into one of the two grading zones.

Exfoliated vermiculite Vermiculite is a mineral resembling mica which exfoliates when it is heated rapidly, producing the lightest aggregate (not exceeding 130 kg/m^3) apart from expanded polystyrene beads. Exfoliated vermiculite concrete has low thermal conductivity, but also low strength and its moisture movement is high. Its main use is for insulating screeds on flat roofs.

Expanded perlite This very light (not exceeding 240 kg/m^3) cellular aggregate is made by rapidly heating a glassy volcanic rock. Expanded perlite concrete is capable of higher crushing strengths, but in other respects it is similar to exfoliated vermiculite concrete. Its use in this country is small but increasing.

Other lightweight aggregates These include expanded slate (*Solite*) which is now being made in this country, imported pumice, and sawdust chemically treated to prevent adverse reactions with cement. The moisture movement of sawdust concrete tends to be high. At present synthetic materials such as expanded polystyrene beads are used only where superior thermal insulation justifies their higher cost, and the difficulties in handling such a lightweight aggregate. In fires the beads collapse and leave a weak concrete.

No-fines concrete

This description is applied to concretes which contain only a single-size 10 to 20 mm coarse aggregate (either a dense aggregate or a light-weight aggregate such as sintered PFA) with sufficient cement to join the particles while leaving voids between them. The density is about two thirds to three quarters that of dense concretes made with the same aggregates.

No-fines concrete is almost always cast in situ mainly as loadbearing and non-loadbearing walls including in-filling walls, in framed structures, but sometimes as filling below solid ground floors, and for roof screeds.

Walls

The surfaces of no-fines concrete provide an excellent key for external rendering and internal plaster finishes, which are essential to prevent air movement through walls with loss of thermal and sound insulation. Any rain which penetrates external renderings will travel inwards only 20 to 50 mm or so, but damp courses and construction joints should be designed to throw such water outwards.

In Germany no-fines dense aggregate concrete has been used in loadbearing walls up to twenty storeys. In this country walls 305 mm thick have been built up to ten storeys using a 1 cement:6 gravel aggregate mix at the lowest level reducing in richness to 1:10 at the top floor. Lightweight aggregates such as foamed blast furnace slag, clinker, sintered pulverized-fuel ash and expanded clay provide concretes which are less strong than those made with dense aggregates, but thermal insulation is superior and fixing and chasing are easier.

CP 111:1970 *Structural recommendations for load-bearing walls* requires a minimum crushing strength at 28 days of 2.76 N/mm². Short and Kinniburgh give average results of tests on 152.4 mm cubes made with 1:8 (the most commonly used proportions) with a water:cement ratio of 0.40 as follows:

Aggregate	Compressive strength at 28 days N/mm²
Rounded quartzite gravel	8.62
Irregular flint gravel	4.83
Crushed limestone	6.89
Crushed granite	7.58

The CP gives permissible stresses related to mix proportions and the crushing strengths of no-fines cubes.

Drying shrinkage

Aerated and lightweight aggregate concretes have high drying shrinkage but that of no-fines concrete is usually less even than that of dense concrete made with the same aggregate. Also because no-fines concrete shrinks more rapidly than dense concrete, plasters and renderings are less likely to crack.

Thermal insulation

The thermal conductivity (k) of no-fines gravel aggregate concrete is comparable to that of typical brickwork. Thermal transmittances for walls are as follows:

279 mm brick cavity wall, plastered inside
305 mm no-fines dense aggregate concrete, rendered and plastered
203 mm no-fines clinker aggregate concrete rendered and plastered
$\Big\}$ $U = 1.7 \text{W/m}^2$ deg C

305 mm no-fines clinker aggregate concrete rendered and plastered
$\Big\}$ $U = 1.3 \text{W/m}^2$ deg C

Sound insulation

The sound insulation of plastered no-fines concrete walls is slightly inferior to that of solid brick walls of comparable thickness.

Mixing

Aggregate should be damped before being placed in the mixer. Cement and then sufficient water should be added so that particles of aggregate are coated with cement without it bridging between them.

Formwork

Because no-fines concrete exerts only about one third of the pressure exerted by ordinary concrete, formwork can be of light construction. It does not require to be grout-tight and if expanded metal is used the mix can be seen as it is placed.

Reinforcement

Light reinforcement is advisable across the angles at openings. A coating of cement grout reduces the likelihood of corrosion.

Placing

Mixes should pour freely. Some gentle rodding may be needed but vibration should never be resorted to.

The concrete should be placed evenly in horizontal layers. As no-fines concrete does not segregate horizontal joints can be at three storey intervals; Cement slurry should be brushed on immediately before placing new concrete.

Fixings

Lightweight aggregate concretes may accept nails but plugs should be built into walls made with dense aggregates.

References include:
Metals in the service of man, W Alexander and A Street, Pelican Ltd
New Science of Strong materials J E Gordon, Pelican Ltd
Metals, Volumes 1 and 2, Design Engineering Series Morgan – Grampian (Publishers) Ltd
British Standards as listed later.

Metals can be defined as being elements which readily form positive ions and which are characterized by their opacity and high thermal and electrical conductivities. Electrical conductivity decreases with increasing temperature. The lightest metal in common use, aluminium, has about the same density as granite, about $2640 \, kg/m^3$, the densest, lead, weighs about $11\,400 \, kg/m^3$.

In the pure form metals are often very soft, eg lead, aluminium and iron, so that most metals used in building are alloys containing controlled proportions of different metals. Generally the melting point of alloys is lower, in other respects their properties are not directly related to those of the parent metals.

Metals must be selected to provide the required properties, including high strength, toughness at low temperatures, corrosion resistance, heat resistance and properties such as ductility to suit the intended method of forming and joining. Deformation and heat treatment affect hardness, strength, fatigue-strength ductility and malleability. Appearance may be important, while both the *initial* cost of products and their *cost in-use* must be considered.

Metals are described as either *ferrous*, containing a substantial proportion of iron (Fe) or as *non-ferrous*. Figures 31 and 32 show in broad outline the principal metals and alloys used in building.

Definitions

Stress is the force carried by unit area, expressed as N/mm^2.

Tenacity is strength in tension, usually expressed as: *tensile stress* at failure ie:

$$\frac{\text{maximum load}}{\text{original cross-sectional area}}$$

Strain is the deformation caused by a force. Tensile and compressive strains are expressed as ratios:

$$\frac{\text{increase (or decrease) in length}}{\text{original length}}$$

Strains also result from shear and torsional stresses.

Modulus of elasticity (E) or Young's modulus. Within the elastic range stress is proportional to strain so that E = stress/strain, see figure 33. Of the metals commonly used in buildings, steel has the highest E value, approximately 3 times greater than that of aluminium alloys, hence steel is far *stiffer* or more *rigid* – see figure 34.

The elastic limit is the point at which deformation of a stressed material ceases to be *elastic* and becomes *plastic*.

Yield is an increase of strain without any increase of stress. The *upper yield stress* is where yield begins and the *lower yield stress* is the lowest stress at which it proceeds.

Proof stress Non-ferrous metals and ultra-high strength steels have no clearly defined yield point and it is necessary to use a *proof stress* for specification and design purposes. Proof stress is the stress required to produce a specified amount, usually taken as either 0.1 or 0.2 per cent of non-proportional permanent strain – see figure 35.

Design stress often termed the *permissible* or *allowable* stress ensures structural safety in all conditions allowing for the possibility of defects in materials, bad workmanship, errors in design and for overloading in service. For example, the design stress in bending for mild steel beams is commonly taken as $165 \, N/mm^2$ which is about two thirds of the yield point stress of $247 \, N/mm^2$.

Ductility is the ability of a material to undergo plastic deformation before tensile failure. In

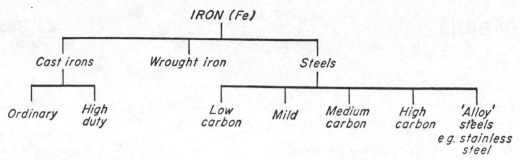

31   *Ferrous metals and alloys*

uniaxial stress, eg in the stretching of a wire, ductility can be expressed as the percentage elongation or reduction in cross sectional area. For triaxial stress conditions, eg in thick material subjected to flexure, ductility is determined mainly by the *Charpy V-notch test* and expressed as joules of energy required to cause failure. In the test a notched specimen supported at its ends is struck in the middle.

Brittleness is the opposite of ductility, fracture occurs with no plastic deformation. It is not related to tensile strength, thus high tensile steel is more brittle (less ductile) than mild steel. With steels lowering of temperature to the *transition* *temperature* suddenly changes the mode of failure from a ductile and fibrous to a brittle crystalline fracture. Compliance with BS 4360 for steel safeguards against brittle fracture, which, however, could only occur where there is a severe notch or crack in a structure.

Metals such as ordinary cast iron which are brittle at normal temperatures are called *cold short*, and those such as alpha brass which are brittle at elevated temperatures are called *hot short*.

Malleability is ability to be forged into required shapes.

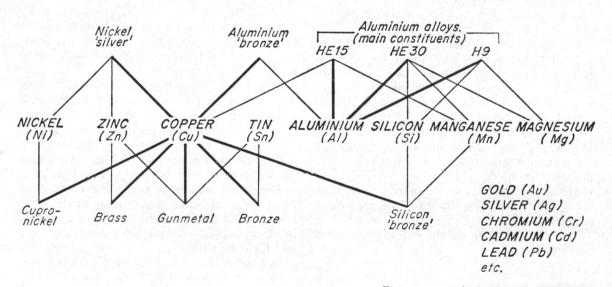

Elements are in bold type.
The thick lines indicate the predominant ingredients in alloys

32   *Non-ferrous metals and alloys*

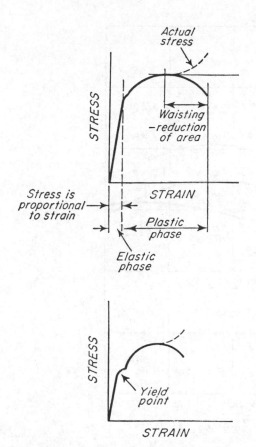

33 Stress-strain curves for steel

Toughness is a combination of strength and ductility which enables a material to withstand shock loadings.

Hardness, ie resistance to abrasion and penetration, is determined for metals by the Brinell, Rockwell and Vickers tests. The higher the number the greater the hardness. Hard metals include tungsten and cast irons. The expression *temper* is often used to describe hardness, a soft and ductile metal being said to be of 'low temper'.

Fatigue is a term used to describe the loss of strength resulting from repeated applications of a force which is less than would cause failure with a single application.

Creep is slow plastic deformation under a constant stress which becomes significant in concrete stressing tendons. It is more rapid at higher temperatures – particularly in metals such as lead.

Work hardening is an increase in strength and hardness of metals when rolled, forged or otherwise manipulated at normal temperature. Unlike hot working in which the grain structure during manipulation is constantly refined, cold working distorts the grain structure thus increasing strength and hardness. Work hardening can be either an advantage, eg steel sheet is strengthened by cold rolling while retaining sufficient ductility for subsequent manipulation, or, a disadvantage, eg when ductility has been reduced to a level which will not allow further manipulation without heat treatment.

Heat treatments are highly specialized processes for softening, stress relieving, and hardening. They involve heating the metal to a critical temperature well below its melting point and controlling the rate of cooling. Heat treatments for steel are given on page 217 and for aluminium on page 229.

Ageing is an increase in strength and hardness which occurs in certain aluminium alloys after heat treatment and in steels as a continuation of work hardening.

Equivalent thicknesses of sheet metals

Table 78 relates the thickness gauges formerly used for *sheet* metals to millimetre thicknesses and code numbers for lead.

Strictly, thicknesses above 3 mm are classed as *plate*.

Properties of common metals

As a class of materials metals are dense, strong and often elastic at room temperatures, and have high electrical and thermal conductivity.

Tables 80 and 81 enable comparisons to be made between some of these properties in common metals, concretes, timbers and plastics.

The effects of fire on metals have been considered in chapter 1 and the main cause of deterioration, by corrosion, is considered here:

Corrosion

References include:
Metal Corrosion, T K Ross, Engineering Design
  Guides, OUP

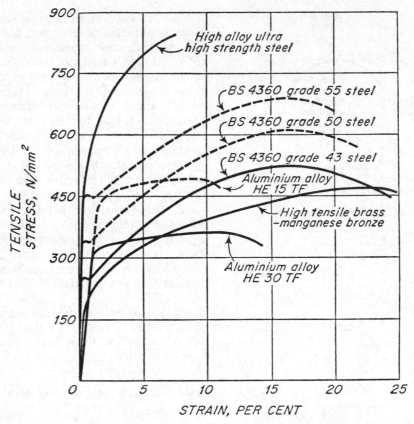

34 *Stress-strain curves for ferrous and non-ferrous metals*

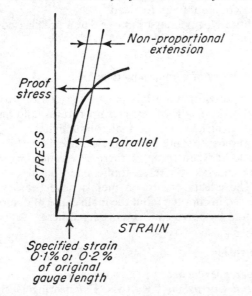

35 *Proof stress*

Corrosion – Attack and Defence
British Steel Corporation and NCST Trent Poly-
technic
DD24 *Recommendations for methods of Protec-
tion against corrosion of light section steel* BSI
BRE Digest: 59 *Protection against corrosion of
reinforcing steel in concrete*
BRE Digest: 70 *Painting: Iron and Steel*
BRE Digest: 71 *Painting: Non-ferrous metals and
coatings*
BS CP 1021: 1973 *Cathodic protection*

Metals tend to revert to stable compounds, the
nature and the rate of the process depending on
the characteristics of the particular metal, the
environment and temperature; corrosion is more
rapid at higher temperatures.

Generally, in reasonably clean atmospheres a
coating of corrosion products forms quickly, and

Former standard wire gauge SWG	Metals other than lead mm	Lead code no. formerly lb/ft²	Former English zinc gauge ZG
40	0.12		
33	0.25		
30	0.31		
27	0.42		
26	0.46		
25	0.51		10
24	0.56		
23	0.61		
	0.63		12
22	0.71		13
	0.79		14
21	0.81		
20	0.91		15
19	1.02		
	1.04		16
18	1.22		
	1.25	3 (1.32 mm)	
17	1.42		
16	1.63		
	1.80	4	
15	1.83		
14	2.03		
	2.24	5	
13	2.34		
	2.50		
12	2.64	6 (2.65 mm)	
11	2.95		
	3.15	7	
10	3.25		
	3.55	8	

Table 78   Equivalent thicknesses of sheet metals

in the case of non-ferrous metals and some ferrous metals it is often firmly adherent and stifles further corrosion. Nevertheless, in 1976 corrosion of metals in the UK was estimated to cost £4,000,000 per day.

Destructive corrosion usually occurs in moist conditions where different metals are in electrical contact, in the presence of atmospheric pollution and flue gases have damaged the copper roofing of Coventry Cathedral within 25 years.

Soil type	Effect
Light, sandy soils Chalk	not generally aggressive, non-saline sand, or chalk can be used to protect pipes in aggressive soils (provided care is taken to prevent the back-fill becoming a land drain)
Cinders Builders' rubble	very corrosive to steel, copper and aluminium
Heavy anaerobic clays	corrosive to ferrous metals
Saline soils	cause severe corrosion of aluminium and galvanized steel, and of lead if it is connected to copper
Sulphate bearing soils	in anaerobic conditions sulphate-reducing bacteria convert cast iron into ferrous sulphide leaving soft spongy graphite

Table 79   Corrosion by soils

Clearly, wherever practicable, metal structures and components should be designed so as to discourage condensation and to allow water to drain away rapidly. Where corrosive conditions are unavoidable dissimilar metals should be separated and the more corrosion-resistant metals, or possibly non-metallic materials, may show a favourable 'cost-in-use'. Discussions of typical environments and of corrosion mechanisms follow:

Environments

Water   BRE Digest: 98 *Durability of metals in natural waters* and CP 310: 1965 *Water Supply* are valuable references.

Natural waters contain mineral impurities, dissolved oxygen and carbon dioxide, the latter derived from rainfall. Water containing the equivalent of 50 parts per million of calcium carbonate, derived from chalk, may be classed as soft and over 350 ppm as very hard. Fortunately deposits of calcium carbonate combined with corrosion products can form a protective layer, thin

Material	Density kg/m³	Proof stress 0.1% N/mm²	Tensile strength[1] N/mm²	Elongation on 50 mm per cent	Modulus of elasticity E N/mm²	Hardness Brinell no.	Thermal conductivity W/m deg C	Linear thermal expansion mm/mm ×10⁻⁶ per deg C	Electrical conductivity at 20°C per cent of tough pitch high conductivity copper	Melting point °C
Gold (Au)	19 300								67	1063
Silver (Ag)	10 500								105	960
Lead (Pb) 99.9% pure	11 340		15 (rolled) 18 (extruded)	50 (on 153 mm)	13 800	4	35	29.5	8	327
Copper (Cu) 99.2–99.9% pure	8940	46–371	216–355	8–65	96 600–132 000	42–96	400	17	101.5[2] 102[3] 54–90[4] 45[5]	1083
Cupro-nickels 70–935 Cu:Ni	8900	114–525	386–509	7–46	119 000–152 000	69–162	21–69	16	4–8	1120–1240
Nickel (Ni) 99.99% pure	8880		316	28	200 000	85	62	13	16	1453
Phosphor bronzes 99.99% Cu:Sn:P	8887–8929	108–679	323–739	6–70	100 000	69–234	47–120	18–19	10–27	1032–1065
Monel Cu:Ni:Fe:Mn	8830		535–752	5–40	172 000–221 000	125–240	27	14	4	1350
'Nickel silvers' Cu:10–30% Ni:Zn	8750–8872	100–618	339–695	4–75	120 000–127 000	66–166	20–27	16–17	5–7	1100
Brass 60 Cu:40% Zn	8380	108[6] 386[7] 124[8]	371[6] 541[7] 386[8]	40[6] 5[7] 45[8]	103 000	75[6] 150[7] 75[8]	129	21	29	904
High tensile brass ('Manganese bronze') Cu:Zn:MN etc	8300–8400	247–463	530–725	13–35	103 000	140–200	90–112	21	20–25	990
'Alminium bronze' 5–105 Cu:Al etc	7570–8150	93–556	417–687	13–69	120 000	66–175	64–85	17–18	13–18	1041–1063

Material										
Stainless steels Fe:Cr:Ni:(Mo)		210	510[9] 540[10]	50		170	15	17	2.5	1440[9] 1430[10]
High strength steel	7850	350–430	495–617	19	207 000	150–180		12	12	1900
Mild (structural) steel			423–510	22		130	52–63	12	12	
Wrought iron			355	25–40		100		12	12	
Grey cast irons[11]	7150[12]	100–200	155–310	0.5–1.0	120 000[12]	140–250	45–50	11	2.0	1150–1350
Nodular and malleable cast irons[11]	7225[12]	193–440	310–800	20–2.0	172 000[12]	120–300	31–47	11	2.0	1150–1350
Zinc (Zn) 99.99% pure	7140		139[13] 216[14]	25[13] 10[14]	96 500	45–50	113	23[13] 40[14]	28	419
Aluminium alloy HE 30 TF	2700	270[15]	310[15]	18	69 950	60–100[16]	184–206[16]	23	32–52[16]	570–660
Aluminium (Al) 99.0% pure	2650	—	70–140	2–20	68 300–72 400	23 (extrusions) 22–42 (sheet)	214	24	60.5	660
99.99% pure		—	80–100	3–45		15 (extrusions) 15–30 (sheet)	244			

[1] The ranges given relate to methods of forming and condition. Strength increases in the following orders: Cast, rolled, extruded, drawn. Annealed, half-hard, hard with corresponding reductions in elongation.
[2] Oxygen-free high conductivity copper.
[3] Tough pitch high conductivity copper.
[4] Deoxidized non-arsenical copper.
[5] Deoxidized and tough pitch arsenical coppers.
[6] Hot-rolled.
[7] Cold-rolled.

[8] Cold-rolled and annealed.
[9] 18 Cr:10 Ni.
[10] 17 Cr:11 Ni:2.5 Mo.
[11] Ranges; except for melting point, are for low–high strength irons.
[12] Average values.
[13] Parallel to direction of rolling.
[14] Perpendicular to direction of rolling.
[15] BS 1474 minima for bars and sections.
[16] Various alloys

The principal sources of information are The Corporate Laboratories of the British Steel Corporation (BISRA); The British Cast Iron Research Association; The Development Associations for Copper, Zinc and Lead; The Aluminium Federation; Metals Reference Book, Vol. II, C. J. Smithells, Butterworths Ltd.

Table 80 *Properties of metals (approximate values) listed in order of decreasing density*

Material	Density kg/m³	Proof stress 0.1% N/mm²	Tensile strength N/mm²	Elongation on 50 mm %	Modulus of elasticity E N/mm²	Hardness Brinell no.	Thermal conductivity W/m °C	Linear thermal expansion mm/mm $\times 10^{-6}$ per °C	Electrical conductivity at 20° % of tough pitch high conductivity copper	Melting point per cent °C
Glass	2520		34–172		68 900		1.04	8.9		1500
Concrete:[17] dense aggregate	2240– 2400		4[18]		28 600[18]		1.0–1.04	10–14		—
lightweight aggregate	320– 2000		3[19]		8 000[19]		0.5[19]	6.5–8[19]		
Plastics	900– 2300		7[20]– 90[21]	5–800[20]	170[20]– 10 300[21]		0.2	7–210		80–295 (softening point)
	3.2–128[22]		0.14–0.55[22]	nil[21]				10–30		
	1900[24]		68–310[24]		20 600[24]					
Timbers[23]	380– 900		20–110		5 860– 18 600		0.14	4.5		

[17] Properties vary with density of concrete and type of aggregate.
[18] Approximate values for concrete having crushing strength of 28 N/mm².
[19] Approximate values for expanded clay aggregate concrete having crushing strength of 14 N/mm².

[20] Thermoplastics.
[21] Thermosetting plastics.
[22] Cellular plastics.
[23] The properties of timbers vary with species, density, moisture content and direction of loading – see chapter 2.
[24] Maximum for GRP

Table 81 Properties of some non-metals

egg-shell scales being far more effective than thick nodular scales.

Iron filings or rubbish often impede the formation of a protective scale of carbonate deposits on the surfaces of galvanized steel hot water tanks and acidic or alkaline waters of high chloride content caused *dezincification* of certain hot-pressed alpha-beta brasses – see pages 201 and 223.

Excessive amounts of detergents, bleaches and cleansers corrode copper and lead waste pipes, and zinc alloy waste traps.

Soils The corrosive effects of various soils is given in table 79.

Acids are present in rain, in some soils, water supplies, plasters and woods particularly Western red cedar, Douglas fir, oak and sweet chestnut, and they are produced by algae and mosses. Even copper and lead, which are otherwise very durable metals, may be perforated by washings from Western red cedar shingles or from mosses growing on roof tiles.

Salts Fortunately, most potable waters form a protective layer which inhibits corrosion. However, moorland waters which contain organic acids or inorganic salts may be plumbosolvent and lead pipes carrying drinking water can give rise to a serious health risk. Cuprosolvency is often indicated by green stains on ceramic fittings.

Sulphates in soils and in clay products, and calcium chloride can intensify corrosion. If used, calcium chloride additions to Portland cement must not exceed 2 per cent by weight as a solution (never flake) see page 155, if corrosion of steel reinforcement, or of aluminium, is to be avoided. Magnesium oxychloride which is used in magnesite flooring, is highly corrosive to all metals.

Alkalis Sodium and potassium hydroxides released by Portland cement are very harmful to zinc and aluminium and to lead in continuously damp conditions. Copper is not affected, and the protection afforded to ferrous metals is useful where steel is embedded in concrete, particularly in cement-rich mixes. However, the pH value increases as concrete carbonates and corrosion can result from variations in the rate of carbonation on the two sides of an embedded steel member.

Concrete cover to reinforcement must be dense and of adequate thickness – see page 179.

High calcium lime may corrode aluminium and is slightly corrosive to lead and zinc. It has no effect on copper and so long as it remains uncarbonated it protects ferrous metals.

Ammonia, eg from ammonia-stabilized latex adhesives for floorings, can contribute to *stress corrosion* – see page 202, of copper embedded in screeds.

Mechanisms of corrosion

Corrosion usually results from a complex electrochemical action. (Direct chemical action by dry gases at high temperatures rarely occurs in buildings.)

Where dissimilar metals which are electrically connected are immersed in a conducting liquid (an *electrolyte*), a short circuited galvanic cell is formed. Metal in the form of positively charged *ions* is removed at the *anode* but the *cathode* does not corrode. Thus, zinc protects steel *sacrificially* because it develops a lower potential and becomes the anode in a cell. Sometimes the anodic metal protects the cathode by *plating out*, eg silver plating on nickel (*EPNS*).

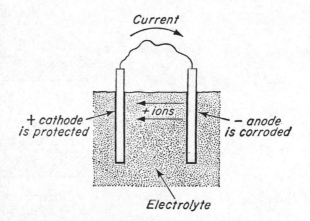

The rate at which electrolytic corrosion proceeds depends upon:

(i) the potential difference between metals in a particular environment. The following order for normal conditions varies due to many factors. Metals higher in the scale corrode preferentially, and the more remote two metals are in the scale, the greater is the reaction.

In particular, in practice the following pairs of

METALS

			Metal attacked
ANODIC	magnesium		
negative	zinc including		
'base metals'	galvanized		
	coatings		cast iron
	aluminium		mild steel
	cadmium		cadmium plated steel
	aluminium-magnesium		galvanized steel
	-silicon alloys	copper and	zinc
	copper-aluminium	copper alloys	aluminium
	alloys		
	iron and mild steel		
	chromium		
	lead	copper	
	tin	phosphor bronze	
	nickel	gunmetal	high tensile brass
	brasses	aluminium bronze	(manganese bronze)
	bronzes	bronze	
positive	copper	silicon	
'noble metals'	stainless steel (austenitic)	nickel	aluminium
CATHODIC	silver	mild steel	zinc, galvanized steel

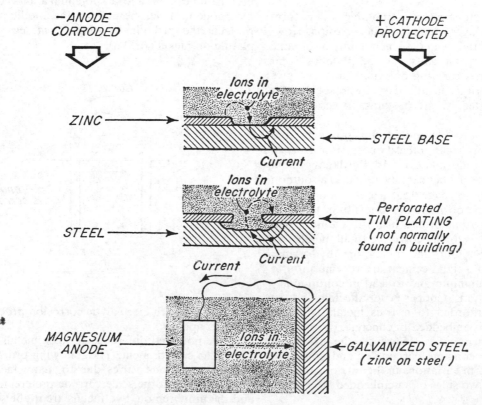

36   *Examples of bimetallic corrosion*

metals should not be used in contact in the presence of moisture:

(ii) the relative areas of metals. Corrosion is particularly serious where the area of the anode is small compared with that of the cathode.

(iii) characteristics of the electrolyte. Corrosion is more rapid where strong mineral acids or their salts, eg chlorides or sulphates, are present.

(iv) temperature. Often the rate of corrosion increases with temperature. In the case of combinations of zinc and steel a reversal of polarity occurs at about 70°C, so that if the temperature of water in galvanized steel cylinders exceeds this value, unless a protective scale has been allowed to develop, the normally protective zinc coating will tend to corrode the steel.

Electrolytic cells are formed in five ways:

1 By *different metals* (*bimetallic corrosion*) – see figure 36.

2 By *particles deposited from solution*:

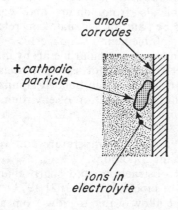

This form of corrosion occurs, for example, where hot water which passes through copper pipes picks up as little as 0.1 ppm of copper and flows into galvanized and aluminium vessels. (In well designed indirect hot water systems the free oxygen content of the water quickly diminishes and attack usually ceases.)

The action is more pronounced at high temperatures but rainwater which has passed over copper-clad roofs should never be taken into unprotected aluminium or cast iron gutters and pipes.

3 Less commonly by *different constituents in a metal*:

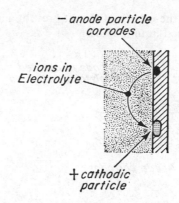

In certain waters selective dissolution of one constituent happens in certain high strength aluminium copper alloys and some brass pressings containing about 40 per cent zinc, may undergo *dezincification* in this way – see page 223.

4 *By differential aeration.* Corrosion takes place, even in pure metals, where there is a difference in oxygen concentration in an electrolyte: the parts in contact with an aerated electrolyte become cathodic and are protected, while the parts in contact with the non-aerated electrolyte become anodic, and corrode. For example, the open surface of a film of water which is held between two sheets of metal is more highly oxygenated than the enclosed water and the metal in contact with the latter corrodes. Similarly, under scale deposits and in depressions, the parts of metal having greater access to oxygen become cathodic relative to those where the electrolyte is confined, and the latter corrode preferentially. *Pitting corrosion* is further assisted by the fact that the cathodic area is much larger than the anodic area and because accumulations of corrosion

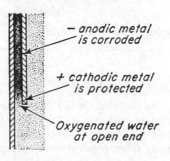

products at the mouth of the pit prevent aeration of the trapped liquid. Thus:

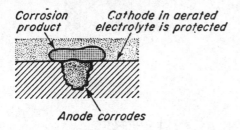

Corrosion product

Cathode in aerated electrolyte is protected

Anode corrodes

Corrosion also occurs where parts of metal components are in *different environments*, for example where a steel pipe passes through a damp plastered wall. Such pipes should be wrapped and pipes in chases in floors should be embedded in uniformly dense cement-sand mortar.

5 Electrolytic corrosion can also be caused by *stray current*, more particularly from a DC source.

Mechanically aided corrosion

(a) *Stress corrosion cracking* This is caused by the simultaneous action of a corrosive environment and a stress, whether induced in manufacture (eg by cold working) or applied in service. The critical environments for various metals are:

Metal	Environment
Certain copper-based alloys eg manganese bronze	Ammonia (eg arising from certain concrete foaming agents and adhesives)
Certain aluminium alloys Stainless steels	Chlorides (above 80°C only)
Lead	Sulphates
High-strength steels	Nitrates

(b) *Cavitation corrosion* A considerable drop in pressure in liquid induced by a change in pipe diameter, partial blockages etc, produces cavities. These collapse violently and remove the protective coating from pipes thereby exposing an ano-dic area which being relatively small leads to active electrolytic corrosion.

(c) *Impingement attack* Entrained air bubbles or solid particles impinging on metal surfaces, eg at a sharp bend in a pipe, remove the protective film and lead to corrosion as above. The velocity of water in copper pipes should be less than 1.5 m/s particularly in salt water systems.

Practical precautions against corrosion

Arising from the foregoing discussion, some advice is offered to prevent corrosion. However, in cases which are not straightforward, expert advice should be sought at the design stage.

1 Select metals to suit the environment.
2 Avoid damp conditions. In particular –
avoid crevices which could hold water or moist dirt by filling them with weld or mastic or by providing drain holes;
avoid contact with absorbent materials and protect metals from damp and corrosive agents, eg by paint or adequate concrete cover on steel reinforcement – see page 179.
3 Avoid contact of dissimilar metals by interposing an electrical insulator, eg bituminous paint, synthetic rubber or resin-bonded laminates.
4 Avoid differential aeration of environments in contact with a metal.
5 In water
(a) avoid excessive temperature and velocity, and stagnant conditions.
(b) avoid obstructions and sharp changes in direction of pipes – see CP 1021:1973.
(c) do not allow water to flow from a noble metal on to a base metal. If different metals must be used in the same water system cathodic metal must be placed downstream.
(d) provide cathodic protection by fixing a sacrificial anode electrically connected to the relatively noble metal, eg an immersed magnesium anode connected electrically to a galvanized steel cylinder or cistern protects the zinc coating while the protective scale is forming. This use of a *sacrificial* anode is particularly effective in soft waters and in many cases it is not necessary to replace them when they have all been deposited as scale.

The terms *anode* and *cathode* are used here in

relation to corrosion processes occurring within cells which are delivering current – as distinct from *electrolysis* where current derives from an independent source. Electrons flow in the opposite direction to the 'conventional current' shown on the diagrams.

(e) Specialized measures, mainly restricted to industrial applications, include treatments to reduce the electrical conductivity of water or its oxygen content, and the provision of an impressed current.

Some corrosive, and some protective electrolytic mechanisms are illustrated in figure 37. References have been given on page 193.

Forming metals

Complex components are often best made in more than one metal and employing many manufacturing processes. The choice of metals and processes require highly specialized knowledge. A reference is *Metals*, Vol 1, *Design Engineering Series*, Morgan-Grampian Ltd. Some of the more common methods of forming metals are outlined here:

Casting

The process of casting results in a coarse grain structure in which the regions of metal in contact with the mould can be particularly weak in ten-sion. Ordinary castings are generally more brittle and weaker than their 'hot worked' counterparts in which the grain size is smaller, more uniform and thus more suited to withstand impact loads and tensile stresses.

1 *Sand casting* is economical where a number of complex and identical objects such as rain-water goods and other components with internal passageways are required. If the cast metal is brittle, eg certain grades of cast iron, then walls of castings must be thick to prevent fracture. Other metals, eg cast steels, are more ductile and the reduced wall thickness leads to a weight reduction.

Most metals and alloys having melting points up to about 1 500°C can be cast. Lead is easy to cast and casting grades of iron, steel, bronzes and aluminium are available.

Although essentially simple, in that molten metal is poured into moulds and allowed to cool, casting requires considerable knowledge and skill. Patterns to provide the shape of moulds and cores are usually made in wood in sizes which allow for shrinkage of metal in cooling. Moulds and cores are usually formed in special foundry sand, the cores often being held in their correct positions by *core nails* which are in due course included in the casting. Pouring of metal must be continuous to ensure even cooling, otherwise the casting may warp or crack. When the metal has solidified unwanted projections are removed by the *fettling*

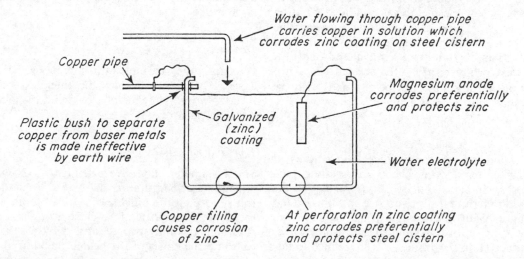

Water flowing through copper pipe carries copper in solution which corrodes zinc coating on steel cistern

Copper pipe

Plastic bush to separate copper from baser metals is made ineffective by earth wire

Galvanized (zinc) coating

Magnesium anode corrodes preferentially and protects zinc

Water electrolyte

Copper filing causes corrosion of zinc

At perforation in zinc coating zinc corrodes preferentially and protects steel cistern

37   Galvanic actions in a cistern

process. Patterns can be used many times but moulds and cores are broken up when the casting is removed. Sand castings are never completely smooth and may be distorted in cooling, so that surfaces which are required to have a precision fit to other surfaces must be cast oversize to allow for grinding or machining. Freedom from porosity in sand castings can never be guaranteed and continuously cast or wrought products should be used in preference for supporting members, eg in wall cladding.

2 *Continuous casting* is a process of casting small usually rectangular shapes called *slabs* or *billets* in a continuous process. With steel, the heavy and costly plant used in the ingot making and heavy rolling processes conventionally used to form semi-finished products, is eliminated. Continuous cast billets are often used in tube making.

3 *Shell moulding* Very accurate mouldings are made in some metals by using thin resin-bonded sand shells supported in sand.

4 *Die casting* Accurate and permanent moulds of steel are used for mass producing small castings of low melting point metals, eg aluminim copper and zinc alloys. The practice of gravity die casting is similar to ordinary foundry practice but in pressure die casting, the molten metal is injected under high pressure to make complex shapes.

5 *Centrifugal casting* Metal is poured into a cylindrical steel mould rotating about its horizontal axis. When the metal has solidified rotation stops and the pipe or tube is withdrawn. The product is stronger and is less likely to contain defects than its sand-cast equivalent. Centrifugally cast products are often termed *spun* products but these should not be confused with the spun sheet products on page 205.

Rolling

Hot rolling is used to produce long lengths of sections, eg bars, strips, sheets and sections such as angles, channels and I-sections. Heavy rollers are used to reduce the section size and thus extend its length. With steel a billet or slab is heated to above 700°C prior to rolling so that rolling forces are reduced and the grain structure refined to improve the ductility of the finished product. Steel I-

sections are produced up to 920 mm deep, and strip up to 1 830 mm wide.

Cold rolling, again using heavy rollers, improves the surface finish, dimensional accuracy, hardness and strength of those hot rolled products it is applied to, eg steel sheet, which can be cold rolled up to 1 830 mm wide.

Forging

Forging is a process of squeezing hot metal to shapes, the products being tougher than casting. Hammering by hand is rare today, most products being *drop forged* using mechanical hammers and suitable dies.

A useful reference is *Engineering Design Guide Part 1 Metal forming – Forging and related processes*

Extrusion

Heated metal is forced through a suitably shaped hole in a hardened steel die, to produce continuous solid or hollow sections including those which cannot be rolled such as finned sections. Metal can be disposed where it can be most effective and often sections can serve several functions. Dimensional accuracy depends largely upon the thickness of the 'arms' of sections. Generally a large range of standard dies is available and the cost of new dies can be justified for large orders.

Drawing

Wire and tubes are pulled through tapered dies or a series of dies to reduce the thickness of the metal. Normally the metal is cold and the process improves its strength.

Pipe and tube making

Small diameter tubes are extruded or drawn as already described. Larger diameters are made by rotary forging the outside of a tube bloom while the inside is supported by a close fitting mandrel. Alternatively, flat strip is formed by a series of rolls and the edges of the heated metal are fused together under pressure, or for higher quality work the edges of cold strip are welded electrically.

Forming from cold sheet

Suitable soft and ductile sheets can be shaped by cold forming. Prefinished products such as plastics coated steel sheets can be cold formed since most coatings withstand more deformation than the base metal.

Roll forming Corrugated and other profiled sheets, simple sections (beams and purlins) and complex sections are formed in any length by passing flat strip through a series of rollers. With steel the maximum strip thickness is normally 5 mm and the minimum economic run is about 3 000 m, although for simple sections smaller quantities may compete with press-baked sections.

Pressing The *rubber press* can produce lightly dished or patterned products and building panels. The sheet is pressed over a wooden former by rubber pads which are placed on the sheet and pressurized by an oil filled rubber bag contained in the upper part of an hydraulic press. Tool costs (wooden former) being very low, it is possible to produce small quantities economically.

The *press brake* employs a V shaped die and punch which is able to form a bend with each stroke of the punch. It can be used to make limited quantities of troughed sheets and other open sections. Low tool costs and great versatility facilitate small quantity production (even *one-offs*). Length, unlike in roll forming, is commonly limited to 4 m although some presses produce sections above 9 m long.

Deep pressings are made in hydraulic presses using highly accurate matched steel dies which prevent the metal from buckling during the extensive deformation which takes place. It is important to realise that large dies are very costly and require mass production runs such as those provided by baths and car body panels.

A reference is *Engineering Design Guide Part 2 Metal Forming – Pressing and related processes*.

Embossing and coining Thin sheet is embossed by a punch and die, or by passing between a pair of embossed rollers (*rigidized sheets*). In coining, a metal slug can be given different profiles on the two sides.

Stretch forming Sheet is held at the edges and stretched over a male former of simple shape.

Spinning Hemispheres, cones and similar regular shapes for lighting fittings and containers, up to about 1.2 m diameter, are shaped by forcing a rotating sheet against a rotating former. Copper, bronze, aluminium alloys and stainless steels are the metals most commonly spun while cold. Domed ends for pressure vessels are spun from heated steel plate up to 4.3 in diameter.

Explosive forming is suitable for making 'one-off' components incorporating a high degree of complex curvature for which alternative processes, eg matched die pressing, would be totally unecomonomic, or in the case of exotic metals used for example in aerospace work, technically difficult. The metal sheet is placed over a female mould which is often made of concrete suitably lined, the mould cavity is then evacuated and an explosive charge detonated in the water above the sheet forms it at high velocity into the mould, eg for nickel alloy dental plates.

Panel beating Irregular shapes are beaten from flat sheets, and damaged car bodies are repaired, by this highly skilled hand work.

Bossing Sheet lead, soft temper copper, aluminium and stainless steel can be formed to complex shapes by bossing by hand using boxwood shapes and a mallet.

Joining metals

Mechanical, soldering, brazing and welding methods are considered here. Adhesives, dealt with in chapter 14, may be preferable where stress concentrations in thin sheet would be caused by mechanical fixings and where large areas are to be joined, without impairing the appearance of either part. However, loss of strength in fires and creep may require to be taken into account in the selection and use of adhesives.

Mechanical joints

Nuts and bolts, and metal screws are useful where high strength joints may require to be separated later. Nuts and bolts require space on both sides to manipulate tools but single sided access is sufficient where a hole in one part receives a *self-tapping screw* which is hard enough to cut its own thread; where the hole is tapped to receive a *set screw*, where a *threaded stud* is cast or welded on one component, or where a *captive nut* is used. *Friction grip bolts* which indicate the correct degree of tightening are now commonly used in structural steelwork.

Ordinary *solid rivets* are inserted in prepared holes and a second head formed from the protruding shank with a pneumatic hammer. These require access from both sides and are of the type used in structural steelwork whereby the preheated rivet shrinks and compacts the joint upon cooling. Likewise, solution treated aluminium allow rivets age-harden after the joint has been formed. *Hollow rivets* are available for single sided access. The hollow shank contains a mandrel, which when pulled with a machine or lever tongs, expands the shank and forms a head.

Soldering and brazing Most metals can be joined with an alloy which melts at a lower temperature than the melting point of the parent metals and which, although different in composition, alloys with them. Surfaces must be clean, fluxes being used to prevent oxidization.

(a) *Soldering* This term usually refers to *soft soldering* with tin-lead and sometimes lead-silver alloys which melt at temperatures below 300°C. The solder is added with a copper bit, which must first be cleaned and *tinned* with solder. Wide surfaces can be *sweated* by tinning both parts and holding them together while they are heated and then while the solder cools and hardens.

(b) *Brazing* Brazing, sometimes called *hard soldering*, gives stronger joints than soft soldering, but as it is done at a higher temperature (over 600°C) it is not suitable for joining metals such as lead which have low melting points. Steels containing chromium or aluminium are difficult to braze due to the oxide layer. The use of *silver solder*, an alloy of silver and brass, is increasing, due to the strong joints (460 N/mm^2) which can be made at low temperatures.

Welding Welding methods can be classified as: (a) cold, (b) plastic and (c) fusion welding; carried out at normal, moderate and high temperatures, respectively. In all cases surfaces must be clean.

(a) *Cold welding* Soft metals such as lead and gold can be welded by hammering. Stainless steel

and other alloys can be joined by ultrasonic vibrations when the parts are lightly clamped together.

(b) *Plastic welding* Metals such as wrought iron and, less readily mild steel, can be hammer welded at a temperature below their melting points, although the strength of the joint is somewhat less than that of the parent metals.

In *resistance welding* heat is provided locally by the resistance of metal to a heavy electric current at low voltage and the parts are pressed together.

Spot, stitch, seam, projection and *butt* welds are of this type. *Spot welds* which can be used instead of rivets, are formed by small diameter electrodes which heat the metal locally and then press the parts together. *Stitch welds* are formed by an intermittently operated spot welding machine.

Projection welds are formed by passing a current through multiple points of contact provided, for example, by small surface projections on metal plates or by overlapping wires as in wire mesh manufacture. *Seam welds* are formed either by two rollers or by one roller and a plate acting as electrodes. Wire and rods are *butt welded* by pressing their ends together and passing a current between them. Threaded studs and similar items are electrically *stud welded* with special guns onto surfaces. In *flash welding* heat is applied by striking an arc before applying pressure. *Friction welding* is a specialized process in which heat is generated by high speed rotation of one part relative to the other with an axially applied load.

(c) *Fusion welding* Fusion welding involves melting of the metals to be joined, either by a gas flame or an electric arc. In some cases a filler rod, often of similar composition to that of the parent metals, is used. Properly formed fusion welds are as strong as the parent metals. The main methods are:

(i) *Gas welding* Gas flames are of oxyacetylene or propane for mild steel, atomic hydrogen for high temperatures up to about 4 000°C, and oxy-hydrogen for low temperature lead welding, known as *lead burning*. Highly skilled operators are required but the equipment is cheap and portable, and the method is useful for joining thin steel sheet, plate and sections and for non-ferrous metals.

(ii) *Arc welding* An arc is struck between two carbon electrodes or between a carbon electrode, or more commonly a welding rod, and the work, in each case with or without a filler rod. Arc welding is faster and gives deeper penetration than gas welding and is suitable for both thin and heavy sections and for both site work and automatic processes.

In both gas and arc welding, molten metal combining with oxygen and nitrogen from the air makes the joint brittle and less resistant to corrosion. To prevent this, joints must be protected by a suitable flux, by a blanket of inert gas provided from a cylinder, or by special coatings on electrodes.

In very heavy structures such as bridges it may be necessary to build up a weld in a number of separate passes. Alternatively, in vertical joints, joints can be formed in one pass by the electro-slag or electro-gas processes.

(iii) *Induction welding* Small metal parts are heated by being placed inside insulated coils through which an alternating current is passed. The method has the advantages that it is quick and little oxidization or discolouration occurs; but the high cost of equipment limits its use to high quality production lines.

(iv) *Thermit welding* This method, which uses a metallurgical reduction process to produce molten iron, is useful where a large quantity of molten metal is required.

BS 499:Part 1, 1983 is a *Glossary for welding, brazing and thermal cutting*.

Ferrous metals

Cl/SfB Yh 1/3

Iron is the fourth most common element in the earth. Iron objects were cast in the Middle East some 1 200 years BC, and some steel was made soon after.

By the eighteenth century iron was used only for small objects, charcoal used to smelt the iron ore consuming considerable quantities of timber. Coal is unsuitable as fuel but in 1709 Abraham Darby heated it to drive off impurities, and the resulting coke, together with developments in furnaces, made the mass production of iron possible.

This metal which was strong in compression could be cast, but not forged. In 1799 *cast iron* was used instead of timber, stone or brickwork, for the first iron bridge at Coalbrookdale, Shropshire, England. Cast iron was used extensively for so-called fire-proof columns and beams in the mills and warehouses of the industrial revolution, and in 1851 for the columns and arches of the Crystal Palace.

The cast iron used for the first railway rails was too brittle. A strong and ductile metal was available in the form of *wrought iron*, but production being extremely laborious, was very small. However, in 1856 the Bessemer Converter made the mass production of ductile steel possible.

Today, the tonnage of ferrous metals, mainly mild steel and cast iron, produced annually exceeds 90 per cent of the total for metals as a whole.

Although relatively small the output of special steels, eg stainless steel, is very important. In general, ordinary ferrous metals can be worked easily and are less costly than the non-ferrous metals used in building. On the other hand, except for *stainless steels* and *weathering steels*, eg *Cor-Ten*, ferrous metals which are not properly and continuously protected corrode readily (see page 193) and sections may suffer serious loss of strength. The products of corrosion occupy many times the volume of the original metal and expand with considerable force. Thus, granite curbs are split by rusting iron railings, brickwork is lifted by steel cramps and steel reinforcement splits concrete.

All metals are *non-combustible* (BS 476, Part 4) but ferrous metals lose strength rapidly in fires – see chapter 1.

A reference is *Metals*, Volume 1 edited by J Dancy, Morgan-Grampian (Publishers) Ltd which is mainly concerned with ferrous metals. BS 6562 is a *Glossary of terms used in the iron and steel industry*.

Ferrous metals are discussed here under the following headings:

Influence of carbon
Production of pig iron
Cast irons
Wrought iron
Steels

Influence of carbon

All ferrous metals contain carbon, derived originally from charcoal and, since 1709, from the coke used in smelting the ore. Variations in carbon contents of ferrous metals have important influences on the properties of ferrous metals.

In general, ductility and ease of welding reduces, and hardness increases with an increase in carbon content and, in the case of steels, tensile strength increases with an increase in carbon content up to about 1.5 per cent – see figure 38.

In cast irons, the form in which carbon as graphite occurs, as well as its quantity, largely determines their properties. In *grey cast irons* – by far the most widely used of the cast irons – carbon in the flake form results in low ductility, and both tensile strength and hardness generally decrease

as the carbon content increases. On the other hand, the strength and hardness of *nodular* and *malleable cast irons* increases as the carbon content of the matrix increases. The high carbon steels and all cast irons have relatively low melting points and are suitable for casting.

Production of pig iron

All ferrous metals are made from *pig iron*. This is produced in a *blast furnace*, typically 30 m high by 9 m diameter, by heating a mixture of iron ore and coke, with limestone and other materials designed to separate iron from the earthy material at about 1 100°C. A blast of hot air injected at the base of the furnace reacts with the coke to melt the iron. Some of the carbon in the coke combines

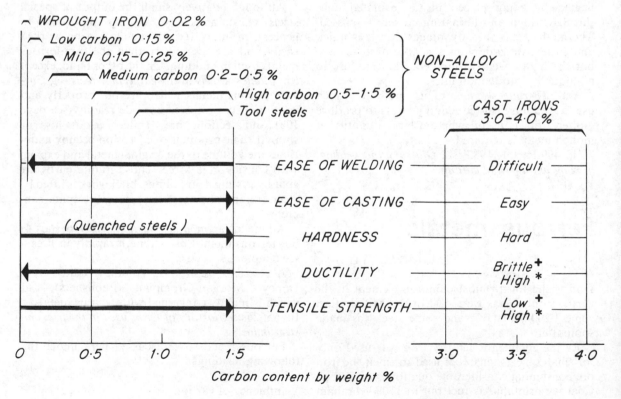

38   *Influence of carbon on properties of the main ferrous metals*

with oxygen and is given off as gas. Some of it combines with the molten pig iron, which acquires a high carbon content, some 3 to 4 per cent.

Blast furnace slag is an important by-product. It is used as concrete aggregate, see page 164; *road metal* and in cement manufacture, see page 149. By blowing water into the molten slag it is 'foamed' and *foamed slag* is used as aggregate for lightweight concrete, see page 188. *Slag wool* for thermal insulation is a further by-product. *Slag-ceram*, made from molten blast furnace slag using a precisely controlled heat treatment process can be cast in its final shape in solid or foamed form. It is likely to find use in building where strength, abrasion resistance or fire and chemical resistance are required.

Types of ferrous metals

We now consider cast irons, wrought iron and steels:

CAST IRONS

The British Cast Iron Research Association is at Alvechurch, Birmingham.

Cast iron is produced by re-melting pig iron together with steel and cast iron scrap, and also by recarbonizing steel charges. Its high carbon content makes it free-running and very suitable for intricate castings, but not for hot working. It can be machined and brazed. Cast irons do not possess the good weldability of mild steels but welds adequate for many purposes can be obtained by using a suitable welding process and technique. This applies particularly to the more ductile grades of nodular and malleable irons.

Ordinary grey irons break with dark grey crystalline fractures. They emit a 'dull' note when struck with a hard instrument. Resistance to corrosion is decidely better than that of mild steel and superior to that of wrought iron. Cast irons form an adherent coat of rust and because their strength properties necessitate thick sections they rarely suffer serious corrosion. Casting minimises the need for expensive machining operations, but as surfaces are not very smooth or true, and thicknesses tend to vary, machining is necessary where close contacting surfaces are required.

The tensile strengths of grey irons are from 155 to 400 N/mm^2 and typically about 300 N/mm^2, a realistic design stress being about 0.25 of the tensile strength. Elongation at failure is only 0.5 to 1 per cent.

Grey irons are economical where a number of identical, rigid, heat and corrosion resisting components are required and where weight due to their necessarily thick walls is not a disadvantage and may be desirable, eg bollards, gully gratings, telephone boxes and letter boxes.

Cast iron is an excellent base for vitreous enamel.[1]

British Standards include:

BS 416:1973 *Soil, waste and ventilating pipes*
BS 460:1974 *Rainwater pipes*
BS 497:Part 1:1976 *Manhole covers, road gully gratings and frames for drainage purposes*
BS 779:1976 *Cast iron boilers for central heating and indirect hot water supply (44 kW rating and above)*
BS 41:1973 *Cast iron spigot and socket flue or smokepipes and fittings*
BS 493:1970 *Air bricks*
BS 1189:1986 *Baths (vitreous enamelled)*
BS 1329:1974 *Metal hand rinse basins*
BS 1452:1977 *Grey iron castings (7 grades)*
BS 2035:1966 *Cast iron flanged pipes and fittings*
BS 4622:1970 (and to 1983) *Grey iron pipes and fittings*
BS 2789:1985 *Spheroidal or nodular graphite cast iron.*

Many components which were formerly made, almost exclusively, in cast iron are now often made in more recently developed materials. For example, plastics materials, which are lighter and do not require painting, are being used for drainage above and below ground, and for baths, which although less heat, scratch, and chemical resistant, are lighter than vitreous enamelled cast iron baths, do not absorb heat. Pitch-fibre pipes are flexible, and competitive in cost with cast iron for drains which do not carry solvents and which are buried sufficiently deep to require no protection against mechanical damage. Polyester 'concrete' (see page 172) is now used instead of cast iron for drainage channels and gratings.

Malleable cast irons, obtained by annealing

[1] See page 216 and *MBS Finishes* chapter 6.

white cast iron, are used extensively for small components and fittings, while the more recently developed *nodular or spheroidal cast irons* are used for a wide range of castings varying from small fittings to relatively large structural components. Tensile strengths are normally from 310 to 900 N/mm^2 and strengths over 950 N/mm^2 are achieved by special heat treatment. Unlike grey irons, nodular irons are ductile, with elongation from 20 to 2 per cent – reducing with increasing strength.

British Standards are:

BS 309:1972 *Whiteheart malleable iron castings*
BS 310:1972 *Blackheart malleable iron castings*
BS 4772:1980 *Ductile iron pipes and fittings.*

Highly alloyed cast irons are available for special resistance to corrosion, abrasion and to high and low temperatures.

Centrifugal casting

Pipes and other cylindrical products are cast by pouring molten iron into permanent moulds which are spun at high speed. The liquid is flung against the lining producing a casting having uniform wall thickness. The resulting iron is denser and stronger than gravity cast iron, and pipes can be made in longer lengths.

BS 1211:1958 describes *Centrifugally cast (spun) iron pressure pipes for water, gas and sewage.* Pipes with spigots and sockets are in three classes: *B, C* and *D* to resist 122, 183 and 244 m head of water, respectively. The Standard gives sizes and test procedures.

WROUGHT IRON

Wrought iron, a low carbon (about 0.02 to 0.03 per cent) ferrous metal which contains some slag, was made in the UK until 1976.

Pig iron, together with millscale (iron oxide) was melted in a small *puddling furnace.* Carbon and impurities react with the millscale to form slag. Balls of more or less pure iron are hammered, rolled and rehammered to expel slag, and to distribute more evenly the 1 per cent, or so, slag which remains. The latter, elongated in the direction of rolling, is seen in a typical fibrous fracture.

Wrought iron is moderately strong in tension,

failing at about 355 N/mm^2, and extremely ductile with 25 to 30 per cent elongation at failure. From early times wrought iron was used for members in tension, such as the original chain encircling the dome of St Paul's Cathedral, London and the suspension chains for Telford's Menai Straits bridge built in 1819. In 1889 the Eiffel Tower in Paris was one of the last large wrought iron structures.

Wrought iron is very tough and resistant to impact – hence its use for ships' chains. The development of railways would not have been possible but for Stephenson's replacement of the brittle cast iron rails with wrought iron. The metal can be forged, even when cold, and two pieces heated to white heat can be joined by hammering them together. Wrought iron is the best ferrous metal for hand wrought work, such as the elaborate screens by Tijou in St Paul's Cathedral, London. Resistance to corrosion is notably better than that of mild steel. However, the metal cannot be cast, tempered, or gas or arc welded.

STEELS

Steels can be described as finely crystalline ferrous metals which are produced by removing impurities, principally sulphur and phosphorous, from pig iron and then accurately adjusting the proportions of all the ingredients including manganese and silicon.

Means of producing steel on a large scale date from 1856 when Henry Bessemer devised his converter. In 1889 the Forth railway bridge was erected in steel and from then onwards the use of steel for large structures was established.

Steels are notable both for their high strength: cost relationship, and poor performance in fire. Ordinary steels have poor resistance to corrosion but a wide range of special qualities and forms for specific applications is now produced, including those with high resistance to corrosion and with tensile strengths more than 1 500 N/mm^2.

Organizations of The British Steel Corporation (BSC) include:

Technical advisory service
Teesside Laboratories, Ludgate Lane, Middlesborough, Cleveland.

Use of steel in construction:
Constructional Research and Development Organisation (CONSTRADO)
12 Addiscombe Road, Croydon, Surrey.

Selection and design of sheet steel products:
Coated products advisory service and Corrosion advisory service
Shotwick Research Laboratories, Clwyd.

Hot-dip galvanizing of fabricated products:
Redpath Dorman Long Ltd
Orchard Place, London E 14.

We now consider:
  Primary manufacture of steels
  Reshaping of steel ingots
  Type of steels:
    Structural steels including
      slow rusting weathering steels
  Sheet steels
  Alloy steels including
      stainless steels
  Heat treatments for steel
  Finishes

Primary manufacture

In making steel, the carbon, silicon, phosphorus and other elements in pig iron are reduced in quantity by oxidation.

The main processes are:

1 *Converter processes* The *Bessemer converter* provided the first method of making steel on a large scale, and steel then became both more economical and more reliable for structural members than wrought iron.

Converters (not to be confused with furnaces), oxidized the high carbon, manganese silicon and phosphorus content of molten pig iron by blowing air, or more recently oxygen, into the melt. In about 20 minutes phosphorus, the last undesirable element, reached an acceptable level. A controlled amount of carbon was added back to the melt and residual oxides and gases were removed by the addition of ferro-manganese.

Ores with a high phosphorous content could not be converted until 1878 when Sidney Gilchrist used basic furnace linings with a lime flux which reacts with acids.

Bessemer converters operated in the UK until 1975.

Today most steel is made in LD converters which use low cost *tonnage oxygen*. They combine good scrap consuming capacity with a high production rate of say, 400 tonnes in 15 minutes.

2 *The open hearth furnace* The open hearth process developed in 1861 is now obsolete. Gas and preheated air, together with oxygen and oil, later heated a charge of pig iron with up to 50 per cent of steel scrap. Some carbon and most of the silicon were oxidized by the flame, and the remaining impurities were removed by the addition of iron ore, mill scale (the oxidized skin of a steel ingot) and lime.

3 *Electric furnaces* Electric arc and high frequency induction furnaces are used to produce alloy steels from molten steel and selected scrap. Shorter refining periods are possible by injecting oxygen. The high temperatures attainable allow both the melting of alloy additions and the removal of impurities.

4 *Spray steelmaking* A development of this rapid process in which molten pig iron is finely atomized by a blast of oxygen and converted to steel as it falls through a reaction chamber may eventually permit the continuous production of steel sections.

Reshaping of steel ingots

Ingots produced by methods 1, 2 and 3 are reshaped into *sections* by:

Rolling Hot rolled sections have rough surfaces which must be mechanically smoothed if a smooth finish is required.
Sheet can be hot rolled down to 1.5 mm thickness.
Cold-rolling following hot-rolling gives clean surfaces and arrises and also increases the strength of the metal.

Extrusion Sections can be more complex, including hollow sections, than is possible by rolling into *forms* by:

Forging Hammering into shapes.
The above methods generally improve the mechanical properties of steel.

Casting Only high carbon steels are suitable and metals poured into moulds tend to be more brittle than those shaped by pressure.

Carbon content of steels

All steels contain carbon but the description *plain carbon steels* is used to distinguish those steels which do not contain substantial proportions of alloying elements. These are subdivided into:

Low carbon steels (up to 0.15 per cent carbon) are soft and suitable for 'iron' wire and thin sheet for tin plate.

Mild steels (0.15 to 0.25 per cent carbon) are strong, ductile and suitable for rolling into sections, strip and sheet but not usually for casting. They are easily worked and welded. The group includes normal and high strength 'low alloy' weldable structural steels – see *Structural steels* below.

Medium carbon steels (0.20 to 0.50 per cent carbon) are suitable for forgings and for general engineering purposes.

High carbon steels (0.50 to 1.50 per cent carbon) Tensile strength increases to about 90 N/mm^2 as the carbon content increases to about 1 per cent and this strength can be further increased by heat treatment. Hardness increases up to about 1.5 per cent carbon content, but ductility decreases and high carbon steels are too brittle for structural work. They are also difficult to weld.

High carbon steels can be hardened for use as files and cutting tools, and they can be treated to the springy condition without loss of hardness. Like high carbon iron, high carbon steels are suitable for casting, eg heavy machine frames.

Types of steels

Having distinguished the types of steel by their carbon contents we look at steels for use in building under three headings:

    1 Structural steels
    2 Sheet steel
    3 Alloy steels.

1 Structural steels

Structural steel continues to be cost-effective for members which do not require to be protected from fire, eg single storey and roof structures. The frames of multi-storey buildings are rapidly erected, and in the UK they are again less costly than reinforced concrete.

BS 4360:1979 describes *Weldable structural steels*. It includes dimensional tolerances for the sections described in BS 4 *Structural steel sections*, testing procedures and mechanical properties. Table 82 contains a few selected requirements, as a general guide only, for the four tensile strength ranges, each having subgrades with differing yield stress and notch ductility requirements. Rutile welding electrodes evolve hydrogen and if site conditions are not dry, cracking of welds and adjacent metal can result. BS 4360 includes weather-resistant steels with 0.25–0.55 copper and 0.70–1.25 chromium contents.

Figure 33 page 193 compares the tensile strength curves of the main structural steels with those of a typical non-ferrous metal.

The strengths quoted apply only at normal temperatures. Above 550°C there is loss of strength and considerable expansion. Most structural members must therefore be protected if they are to survive building fires – see page 41.

BS 4 *Structural steel sections* Part 1:1980 Hot rolled sections describes: *'Universal' beams, columns, bearing piles, etc*. It gives dimensions of sections and weights per metre.

BS 4848: *Hot-rolled structural steel sections* include: Part 2:1975 *Hollow sections* and Part 4:1972 *Equal and unequal angles*.

BS 2994:1976 *Cold rolled steel sections* specifies dimensions and properties for angles, tees, channels and compound sections made from channels.

BS 1449:Part 1:1983 *Steel plate, sheet and strip* describes plates from 3 mm thick upwards which can be welded to form a wide range of structures, eg heavy box columns and I beams larger than the standard beam sections.

BS 6363:1983 *Welded cold formed steel structural hollow sections* gives properties, and outside dimensions, ie circular 21.3–193.7 mm, square 20 × 20–150 × 150 mm, rectangular 50 × 25–20 × 100 mm.

Although more costly than 'open' sections, the resistance to bending and tension of hollow sections is excellent, and welded joints are neat in appearance. Corrosion protection is not needed to inner surfaces if sections are fully sealed.

BS 5950 *The structural use of steelwork in building*, Parts 1 and 2:1985 deal with *design, materials, fabrication and erection of hot rolled sections*.

Grade	Maximum carbon per cent	Maximum sulphur and phosphorus –each– per cent	Minimum tensile strength N/mm²	Minimum yield stress N/mm² up to 16 mm thick[2]			Minimum elongation $5.65\sqrt{S_0}$	Previous related BS	Uses
				Plates	Flat bars Sections	Hollows sections			
40A	0.27	—		—	—	—		3706	general
40B	0.25	—		230	240	—		—	engineering
40C	0.22	0.060	400/480	230	240	—	25	—	
40D	0.19	—		260	240	—		2762	notch-
40E	0.19	0.050		260	255	—		—	ductile[5]
43A1[1]	0.30[4]	—		—	—	—		3706	structural
43A[1]	0.30[4]	0.060		245	255	—		15	
43B[1]	0.26	—		245	255	—		—	
43C	0.22	—	430/510	245	255	255	22	2762	notch-
43D	0.19	0.050	(430/540[3])	280	255	255		2762	ductile[5]
43E	0.19	—		280	270	270			
50A[1]	0.27	—		—	—	—		—	high
50B	0.24	0.060		355	355	355		968	strength
50C[1]	0.24	—	490/620	355	355	355	20	—	steels
50D	0.22	0.050		355	355	355		—	notch-ductile[5]
55C[1]	0.26	0.050		450	450	450	—		
55E[1]	0.26		550/700	450	450	450	19		

[1] Including either 0.20/0.35 or 0.35/0.50 copper if specified in order.
[2] Yield stresses are lower for thicker plates and sections specified.
[3] Hollow sections.
[4] The average is lower.
[5] Have superior resistance to impact, eg for low temperature work.

Table 82   Extracts from BS 4360:1979 Weldable Structural Steels

Parts 5 and 7 will deal with design materials and workmanship using cold formed sections and Part 8 will be a CP for design of fire protection of structural steelwork.

British standards for the reinforcement of concrete are listed on page 178.

BS 405:1945 describes: *Expanded metal* (steel) for general purposes.

'Weathering steel' Plain carbon steels containing 0.2 per cent copper have better resistance to corrosion than mild steel but they must be protected from the weather. *Cor-Ten* (registered name), however, containing additions including 0.25 to 0.55 per cent copper, when exposed to alternate wetting and drying externally, develop a tenacious oxide coating, russet-copper darkening to purplish brown in colour.

The small number of buildings and structures constructed in *Cor-Ten* steel in this country have developed the oxide coating in 18 months to 2 years and in ten years the loss of thickness expected is $\frac{1}{2}$ to $\frac{1}{16}$ that of ordinary mild steel. *Cor-Ten* is shot blasted at works to ensure uniform weathering. The usual painted identification marks are not possible.

During the first few years it is important to drain away corrosion products formed so they do not stain adjacent walling and paving.

Incidentally, although protective treatments are not necessary, *weathering steel* provides a better base for paint than ordinary steels. If a painted surface is damaged, oxide will not creep under adjacent painted areas.

In addition to possessing slow rusting properties *Cor-Ten* is a *high-strength* steel approximating in this respect to grades 50B and 50C.

Cor-Ten is used in the manufacture of sections, plates, sheet and coil. The current extra cost of

Cor-Ten over that of ordinary mild steel of about 20 per cent can be set against savings in weight, protective treatments and maintenance.

2 Sheet steel

Sheet steel, described as not more than 3 mm thick, is used for wall and roof cladding, curtain wall panels, floor and roof trough decking, demountable partitions and furniture, ducting and rainwater goods. Sheet, and exceptionally mild steel sections up to 5 mm or even 6.3 mm thick, can be cold formed.

Sheet is available in mild steel, low alloy high strength steels including slow rusting *Cor-Ten* and in stainless steels in maximum widths of 1 830, 1 520 and 1 320 mm respectively.

Sheet is available uncoated, and with factory applied corrosion-resistant and decorative prefinishes which dispense with the need for costly and less effective shop and site treatments. Sheets prefinished with paint, PVC coating and laminates are produced in a wide range of colours and textures in widths up to 1 320 mm. With suitable precautions the usual forming methods can be used without damaging surfaces.

For a long maintenance-free life externally, sheets must be galvanized unless they are vitreous enamelled or similarly protected.

The more important British Standards for steel sheet products are:

- BS 1449 *Steel plate sheet and strip*. Part 1:1983 – up to 16 mm thick.
- BS 3083:1980 *Hot-dip zinc coated corrugated steel sheets for general purposes*.
- BS 1390:1972 (amd 1983) *Sheet steel baths for domestic purposes*.
- BS 1091:1963 (1980) *Pressed steel gutters, rainwater pipes, fittings and accessories*.
- BS 31:1940 (amd 1970) *Steel conduit and fittings for electrical wiring*.
- BS 1245:1975 *Metal door frames*.
- BS 5977: *Lintels*.

3 Alloy steels

Alloy steels contain more than 5 per cent of alloying elements to provide special properties such as ultra high strength, corrosion or heat resistance. In particular, alloy steels can be heat treated, see page 216, more effectively to provide required degrees of hardness and strength properties, throughout their thickness rather than at the surface only. British standards for alloy steels for concrete prestressing are listed on page 179. Further discussion here is confined to stainless steels.

Stainless steels A stainless steel containing chromium was developed in 1913. Although a relatively costly ferrous metal it is being used increasingly in building. It develops an invisible corrosion resistant film in air and has high resistance to organic and weak mineral acids. Stainless steel is compatible with non-metals and the metal, or washings from it, will not stain adjacent materials. Stainless steel may accelerate corrosion of mild steel, but galvanic attack on aluminium and zinc only in the most aggressive environments, and where the ratio of size of stainless steel to that of the other metal is large.

Thermal movement is low and stainless steel is also resistant to high temperatures. It is very hard and strong and has good appearance. See BRE Digest 121, *Stainless steel as a buidling material*.

BS 1449:Part 2:1983 *Stainless and heat resisting plate, sheet and strip* gives requirements for dimensions, chemical composition and mechanical properties. Stainless steels are available in rolled, extruded and drawn forms. They can be forged and cast and fabricated by normal methods including soldering, brazing and welding.

There are five standard *mill finishes* and four *polished finishes*. Welds may need to be polished. Large areas of polished sheet require a rigid and continuous backing to avoid a wavy surface appearance. Polishes include a lustrous undirectional dull polish obtained by grinding with fine abrasives and a mirror polish. Complex shapes which cannot be polished mechanically can be given bright to dull matt finishes by electrochemical means. The oxide coating which forms on stainless steel can be modified to give colours ranging from gold to dark brown. Strippable protective coatings should be used to prevent damage to surfaces during fabrication and erection.

In service, the hard, smooth corrosion-free surface of stainless steel does not hold dirt readily, and light accumulations of non-greasy dirt may often be washed off with plain water. Stubborn deposits should be removed with soap impregnated non-scratching nylon or other plastic pot scourers. In each case the surface should be rinsed with clean water.

Stainless steels are classified as:

martensitic, about 13 per cent chromium,
ferritic, usually about 17 per cent chromium and
austenitic, usually 16 to 19 per cent chromium together with 6 to 14 per cent nickel. Austenitic stainless steels which comprise about 70 per cent of production are non-magnetic, have high tensile strength and ductility, and welding and soldering properties are good. They cannot be hardened by heat treatment, but working improves proof stress and ultimate strength, with some loss of ductility.

Some properties of the two principal stainless steels used in building, both of which are austenitic, are given in table 83.

Properties	18 per cent[1] chromium 10 per cent nickel type 304	17 per cent chromium 11 per cent nickel $2\frac{1}{2}$ per cent molybdenum type 316
Tensile strength N/mm²	520–645	540–645
0.2 per cent proof stress N/mm²	205–430	205–430
Elongation on 50.65 mm (average per cent)	30–25	30–25
Modulus of elasticity (e) N/mm²	207 000	207 000

[1] Approximate average compositions

Table 83   Stainless steel plates BS 1449 minima for softened/work hardened conditions

Type 304 alloys are suitable for normal internal uses, and externally in the country. In industrial atmospheres where good appearance is important *Type 316 alloys* are essential.

Certain stainless steels containing at least 13 per cent iron can be coloured by immersion in hot solutions of chromic and sulphuric acids. Bronze, blue, gold, red, purple and green result as the film thickens. After rinsing in cold water the product is hardened in another acid solution.

Uses

Tubes BS 4127:Part 2 1972 describes *Light gauge stainless steel tubes 5 mm to 42 mm diameter* – suitable for plumbing and hot water heating for a working pressure up to 13 bar and for connection by capillary or compression fittings.

BS 6362:1984 describes *Stainless steel tubes suitable for screwing.*

Stainless steel tubes are cheaper and stronger, requiring less frequent support than copper tubes and they are more resistant to corrosion. Accelerated tests at the BRE have shown no adverse galvanic effects at copper/stainless steel junctions and they are accepted by the British Waterworks Association, and by virtually all water authorities in this country – see BRE Digest 83 *Plumbing with Stainless Steel.*

Sheet – uses include: *Fully supported roofing* – see BS CP 143 *Sheet roof and wall coverings.* The sheet is now produced with an integral lead coating.

Flashings are recommended to be of fully softened sheet 0.46 mm thick where exposed, and 0.38 mm thick where protected from rain.

Urinals BS 4880:Part 1:1973 describes *Stainless steel slab urinals. Bowl type urinals* and *wcs* – Stainless steel has considerable resistance to damage by vandals.

Hospital equipment
Kitchen and restaurant equipment
Sinks BS 1244 Part 2:1982 *Metal sinks* for domestic purposes describes *Sink tops*

Flexible flue liners for defective flues

Windows   A typical product comprises frames fabricated from 0.9 mm stainless steel and sashes in 0.7 mm sheet. Large sections are produced with stainless steel used as a sheathing on wood or aluminium cores.

Sandwich panels   Stainless steel sheet on both sides of plywood or on light-weight cores of foamed plastics or honeycombed paper or foil.

Foil up to 800 mm wide and 0.1–0.3 mm thick is rolled in various tempers and colours and in embossed patterns to increase rigidity. Uses include cladding to insulation, chipboards and other cores.

Structural uses Cold worked stainless steel may have tensile strength approaching 1 500 N/mm². Stainless steel was used for the 'new' chain encircling the dome of St Paul's Cathedral in 1926 and as reinforcement for concrete in the restoration of York Minister, where freedom from corrosion was crucial.

Fixings Stainless steels are very suitable for cramps, dowels and other fixings for securing precast concrete and stone wall claddings. Also for wall ties and frame and sole plate anchors.

BS 1243:1978 specifies *18:8 Stainless steel cavity wall ties.*

Heat treatments for steel

The mechanical properties of steels can be modified by subjecting them to one or more temperature cycles which alter the shape and size of the grains and the micro-constituents of the metal. As the carbon content of steels increases they become more amenable to heat treatment and a wider range of properties can be obtained. Common forms of heat treatment outlined in figure 39 are:

Hardening is obtained by heating steel above a critical temperature and then cooling it rapidly. Within limits the higher the temperature and the quicker the cooling the harder but less ductile is the result.

Tempering gives hardened steel increased ductility with only a slight loss of strength, by reheating to a temperature below the hardening temperature followed by cooling at any rate. The higher the reheating temperature the greater is the ductility, and also the loss in strength. For example, high strength friction bolts are tempered after being hardened in oil or water.

Annealing Steel is heated to a critical temperature above 700°C, held at this temperature for a period related to the thickness of the section, and then cooled slowly and at a controlled rate, usually in the furnace. Annealing softens the steel and removes internal stresses caused, for example, by welding or cold working.

Normalizing Like annealing, the steel is heated above a critical temperature, but more rapid cooling in air refines the grain size and higher strength results.

Case hardening The normal hardening and tempering processes become possible with low carbon steels if the surfaces are first carbonized by heating the steel while it is surrounded with a carbonaceous material.

FINISHES ON FERROUS METALS

Stainless and weathering steels can provide acceptable appearance, free from destructive corrosion. Other steels often require applied finishes for appearance or for protection.

Painting

This subject is dealt with in *MBS: Finishes* chapter 6. It is treated fully in BS 5493:1977 *Code of Practice for protective coating of iron and steel structures against corrosion.* Briefly, rust and mill scale must be removed. Protection is prolonged five-fold where this is done by shot-blasting or pickling rather than by weathering and wire-brushing. *Bonderizing* and *Parkerizing* are trade names for treatments in which the steel is treated in a solution of acid phosphates to produce a grey-black matt coating of insoluble phosphates which is a valuable rust inhibiting pretreatment for paint and which is particularly effective in minimizing the lateral spread of rust under paint coatings.

Factory blast-cleaned, primed and painted steel plates, sections, and sheet prefinished with plastics are likely to be much more durable than products having site-applied finishes.

Vitreous enamel

Molten glass applied on steel or cast iron forms a tough, highly corrosion resistant, colourfast coating. In London, the Underground signs and cladding to the Tower Hotel and Daily Mirror buildings are vitreous enamelled steel. The finish, which is easily cleaned, can be glossy or matt, eg for blackboards.

BS 4900:1976 specifies: *Vitreous enamel colours for building purposes.*

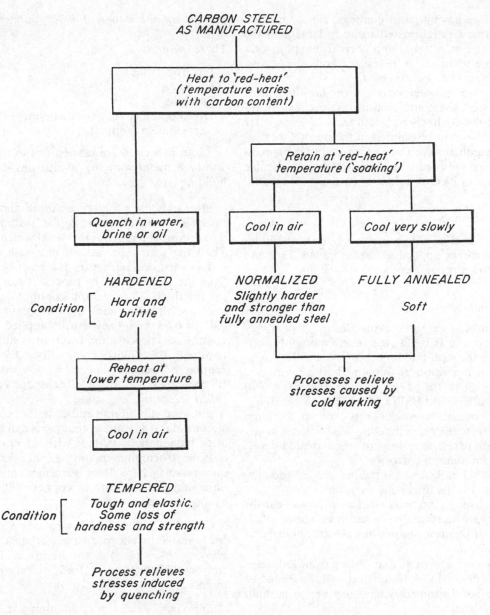

39   Heat treatments for steels

BS 6340 describes *Prefrabricated shower trays made from vitreous enamelled:* Part 6:1983 *cast iron*, and Part 7:1983 *Sheet steel*.

The edges and corners of components should be rounded to avoid spalling, see also *MBS: Finishes*, chapter 6.

Plastics coatings

Steel sheet or strip up to 1 320 mm wide and any length is available surfaced on one or both sides with PVC, acrylic, epoxy or phenolic coatings applied as liquids or laminates. These products,

217

like those having paint coatings, can be pressed and otherwise fabricated without damage to the coating. For external and corrosive exposures a zinc pretreatment is necessary and lives up to thirty years can be expected.

Polyester powder coatings on hot-dip galvanized steel sheet and sections as specified by BS 6497:1984 are likely to last at least 15 years, after which they can be painted in the normal way.

Components such as handrails, gutter brackets and wire mesh can be given tough and durable coatings of PVC or nylon – see pages 279 and 282.

Metal coatings

Many metal coatings including lead, cadmium and tin can be applied to ferrous metals. Zinc and aluminium provide cathodic protection.

Zinc coatings

Zinc affords cathodic protection at a relatively low cost and is by far the most commonly used coating for steel building components. The resistance to corrosion of unpainted zinc coatings is dealt with in BS 5493:1977 *The expected life of zinc coatings* and BSPD 24:1973 *Methods of protection against corrosion* on steel up to 5 mm thick, in temperate climates. While zinc is many times more resistant to corrosion than mild steel it is not immune to corrosion.

BS 1243 indicates that the life of a zinc coating is related to its thickness.

The lives of zinc coated steel products can be prolonged indefinitely – even in corrosive situations – if paint or plastics finishes are applied and maintained.

It is now apparent that light galvanized items such as pressed steel lintels and wall ties should be coated with bitumen where they are to be built into external walls.

Wherever possible, zinc coatings should be applied after all cutting and forming has been done. However, exposed cut edges of thin galvanized steel sheet are protected by the sacrificial action of the zinc coating (see page 200) and coatings are not damaged by normal forming operations.

Methods of application of metal coatings

These include:

Electroplating
Cladding
Spraying
Hot-dipping (usually zinc coatings)
Sherardizing (with zinc).

These processes are carried out in the factory although metal spraying is also possible on the building site.

Electroplating A very accurate and uniform thickness of metal can be applied without distortion or loss of temper in the metal which is plated. The latter forms the cathode in a bath and metal is deposited on it either from a reactive anode or from the electrolyte. The process is not economical for thick coatings, and coatings for zinc on steel are thinner than 'hot-dip coatings', from 0.01 to 0.03 mm. They can be applied also by a continuous process, and on one side only if required. BS 1706:1960 describes *Electroplated coatings of cadmium and zinc on iron and steel* and BS 6687:1986 *Electrolytically zinc coated steel flat rolled products*.

Tin, zinc, aluminium and cadmium coatings on steel have the advantage that parts can be formed after having been plated, without cracking the coating. Chromium plating is very hard. It requires a surface free from blemishes and is applied on a thin *flash* coating of copper followed by a coating of nickel.

Cladding Metals are hot-rolled on a stronger, less durable or cheaper base so that a degree of alloying takes place at the interface. Thus steel can be clad with brass, lead, nickel and aluminium.

Spraying Zinc or aluminium in wire or powder form is heated by an electric arc or gas flame, and the atomized metal is sprayed from a gun by compressed air. Adhesion is mainly mechanical and it is desirable to first roughen the surface by shotblasting. A skilled operator can apply an even matt finish, of a thickness suitable for painting or a greater thickness for grinding and polishing. In situ applications include structural steelwork such as bridges.

BS 2569:Part 1:1964 describes aluminium and zinc sprayed on iron against atmospheric corrosion.

Hot-dipping is the most commonly used process on steel products such as cisterns, corrugated and flat sheets, light structural sections and windows.

Essentially, after pickling in acid, drying and preheating, the object is dipped in molten zinc which forms an alloy layer with it. Baths exist in lengths up to 15 m long. Products sometimes have a characteristic *spelter* appearance, resembling frost on windows. Slight variations in thickness occur, particularly with shaped products due to the flow as the object is removed from the bath. Distortion resulting from the hot process can usually be avoided by correct design and galvanizing techniques, and it can be corrected mechanically.

The Building Regulations 1976 required that for the weather-resisting parts of walls and roofs, sheet steel compiled with either:

BS 2989:1982 *Continuously hot-dip zinc coated and iron-zinc coated steel wide strip, sheet/plate and slit wide strip* or BS 3083:1980 *Hot-dip zinc coated and iron-zinc coated steel sheets for general purposes*. This BS stipulates coating 'weights' of 350, 450 and 600 g/m^2 including both sides (assessed by the triple spot test).

Because galvanized butterfly cavity wall ties had corroded in 13 years, even in normal conditions, in 1981 the BS 1243:1978 *Specification for metal ties for cavity wall construction* requirement of 260 g/m^2, was increased to 980 g/m^2. However, the Building Regulations 1985, Approved Document A, states that in severe exposures austenitic stainless steel or non-ferrous cavity wall ties should be used.

Other *British Standards* are:

BS 417:Part 2:1973 *Galvanized mild steel cisterns and covers, tanks and cylinders*. (The lives of galvanized steel cisterns can be extended by the application of non-toxic bituminous paint coatings.)

BS 1565:1973 *Galvanized mild steel indirect cylinders, annular or saddle-back types*.

BS 6178 *Joist hangers* Part 1:1982 – *for building into masonry walls of domestic buildings*.

BS 1485:1983 *Zinc coated hexagonal steel wire netting* – dimensions only.

Sherardizing Relatively small objects such as nuts, bolts and door furniture are put into a cylinder containing zinc dust which is rotated and heated. An alloy of zinc and iron forms uniformly on the surfaces to a thickness which can be very accurately controlled. Screw threads and moving parts therefore do not have to be machined again after treatment. The grey, matt surface is a very suitable base for paint. It can be buffed and polished and if desired lacquered to prevent finger marks from showing.

BS 4921:1973 *Sherardized coatings on iron and steel* minimum thicknesses of 30 µm for Class 1 and 15 µm for Class 2 articles.

Non-ferrous metals
Cl/SfB Yh4/9

References: Metals Volume 2 Morgan Grampian (Publishers) Ltd.

The first cost of non-ferrous metals is usually much greater than that of ordinary ferrous metals, but the difference is often offset by their superior working properties and resistance to corrosion.

Tables 80 and 81 (pages 197 and 198) compare some of the more important properties of ferrous and non-ferrous metals with those of some non-metals.

The more common non-ferrous metals and their alloys are shown in figure 32 page 192 and are briefly described here, ie

Copper
Nickel, tin, cadmium and chromium
Zinc
Lead
Aluminium

Surface finishes on non-ferrous metals are described on page 234.

COPPER (Cu)

The Copper Development Association, Mutton

Lane, Potters Bar, Hertfordshire, provides technical advice and many excellent publications dealing with all aspects of the properties and uses of copper and its alloys, eg C 106, 102 and 101 below.

The three grades of copper used in building are:

Deoxidized copper This is used for domestic plumbing tubes where welding is to be carried out and for general engineering purposes (C 106).

Fire refined tough pitch copper This contains oxygen and is stronger, and has higher thermal and electrical conductivity and higher resistance to atmospheric corrosion than deoxidized copper. It is used as sheet for fully supported roof coverings (C 102).

Electrolytic tough pitch high conductivity copper This metal is similar to fire refined tough pitch copper but contains less impurities. It is largely used for electrical conductors (C 101).

Properties

The salmon-red colour of clean copper is well known. Its alloys vary from red, gold and pale yellow of soft silver in colour. In ordinary atmospheres and waters copper develops a protective skin. In certain environments a green patina slowly develops, an effect which can be obtained more rapidly by chemical methods, although in the case of copper roofing success depends very much on the climatic conditions prevailing at the time of treatment. Washings from copper may stain adjacent materials and inhibit the growth of lichen and they may give rise to the corrosion of other metals – see page 199.

Copper is, in general, very resistant to corrosive agents, particularly to sea water, but it is attacked by strong mineral acids and ammonia. Water containing a high proportion of free carbon dioxide is cuprosolvent. Up to 1.5 mg/l copper can be tolerated in drinking water, but low concentrations of copper greatly accelerate pitting corrosion of galvanized steel. Pitting corrosion of copper tubes has been caused by the cathodic scale deposited by soft moorland waters containing manganese salts, occasionally by certain hard well waters, and by the scale of carbon produced in the process of extrusion. The latter scale is removed at works by the method described in BS 2871. However, the formation of a sound protective film in copper cylinders is assured by the installation of aluminium protector rods, approved by the BNF Metals Technology Centre, which control the electro-chemical potential of the copper.

Copper is supplied in the fully annealed *dead/ soft, half hard* and *full hard* conditions.

Copper in the annealed or hot worked conditions is relatively strong, and it is extremely ductile. Its strength characteristics and hardness are increased by cold working, as shown below;

Condition	Tensile strength N/mm²	Hardness-diamond pyramid test	Elongation per cent 50 mm
As cast	155–170	45–55	25–30
After cold working	310–385	80–115	5–20
Annealed after cold working	215–245	40–50	50–60

Table 84   Copper conditions

Available forms

Rod is rolled from continuously cast sections.

Wire   0.025 to 5 mm diameter is drawn from rod.

Tube from about 1.5 to 610 mm diameter drawn from cylindrical billets or cold drawn from hollow extrusions.

Plate, ie flat material larger than 305 × 10 mm up to about 3.7 m long.

Sheet and strip from 0.5 to 10 mm thick sheet wider than 450 mm.

Foil   0.15 mm maximum thickness, any width.

Plate, sheet, strip and foil are either hot rolled from slabs or cakes or formed by electro-deposition.

Working copper

Unalloyed copper can be hot rolled, forged and extruded. In annealed condition it is eminently suitable for site working although work hardening may necessitate further annealing by heating to a dull red heat.

Copper and also alloys can be joined by welding, brazing and soldering.

Uses

British Standard Specifications include:

BS 3198:1981 *Copper hot water storage combination units for domestic purposes.*

BS 2870:1980 *Rolled copper and copper alloys – Sheet strip and foil.* A section deals with sheet for fully-supported roofing, flashings, etc. 0.60 mm sheet fixed on roofs in London in accordance with the recommendations of CP 143:Part 12: 1979 is less liable to perforation than lead and normally has a life of many centuries.

0.315 mm sheet is bonded to boards as preformed roofing units. Foil is bonded to bituminous felt which is supplied in rolls and is fixed similarly to traditional copper sheets roofing.

BS 743:1970 (1983) *Materials for damp-proof courses* requires them to be annealed copper (Grade O) 0.25 mm thick or 0.46 mm thick if part projects as a drip or flashing. Separate drips or flashings must be 0.5 mm thick.

BS 2871:Part 1. 1971 *Copper tubes of water, gas and sanitation* and Part 2:1972 *Copper tubes for general purposes*

BS 864, Part 2:1983 *Capillary and compression fittings for copper tubes.*

BS 699:1984 *Copper direct cylinders for domestic purposes.*

BS 1566: *Copper indirect cylinders for domestic purposes* Part 1: 1984 *Double feed* and Part 2:1972 *Single feed.*

BS 1431:1960 (1980) *Wrought copper and wrought zinc rainwater goods.*

BS 1878:1973 (1979) *Corrugated copper jointing strip for expansion joints in general building construction.*

BSs 1432–1434 and 4608 *Copper for electrical purposes.*

Copper is the main ingredient in brasses, bronzes and gunmetals – see figure 40. Small proportions of copper are added to other metals for various reasons: for example to improve the resistance to corrosion of structural steel, to facilitate the manufacture of cast iron and to increase the strength of aluminium alloys.

Thin films of copper or copper alloys can be electro-deposited on other metals and on non-metallic materials such as plastics the surfaces of which may have been made electrically conductive and facsimiles in relief (electro-types), are made in this way. Small bearings and bushes can be made from powdered alloys impregnated with lubricants, compressed and sintered. Copper powder is used in some paints and copper oxide provides colours in glass and ceramic glazes. Copper salts are used in timber preservatives – see page 82.

Copper-based alloys

These alloys have high resistance to corrosion, and high electrical and thermal conductivities. They have good mechanical properties and can be forged, pressed and easily machined. Certain alloys are suitable for casting in sand moulds, or for continuous casting which gives greater density and strength. They can be joined by welding, brazing and soldering. The darkening of copper alloys with exposure, can be preserved by washing, and coating with wax polish.

Approximate compositions of the common alloys are shown in figure 40.

We consider them here under the broad headings; (a) Brasses, (b) Brass-based alloys, (c) Bronzes, (d) Bronze-based alloys, (e) Copper-nickel alloys, (f) Copper-silicon alloys and (g) Copper-aluminium alloys.

(a) Brasses

Brasses are used for hinges in good quality joinery, and for screws and other fixings where a non-rusting metal is required. In recent years polished brass has again become fashionable for lighting fittings, door furniture, decorative balustrades and the like. Traditionally, brasses were kept bright by metal polish, which necessitated protecting adjacent materials, such as wood and marble, with masks. Now decorative brass is usually lacquered, but externally this treatment requires frequent renewal. Brasses are classified according to their zinc content, resulting metallographic structures and mechanical properties as *alpha, alpha-beta* and *beta brasses.*

Zinc content ALPHA BRASSES
per cent

Less than 3 — *Cap copper* is very suitable for cold working although like other Alpha brasses it then usually requires to be low-temperature annealed to prevent the possibility of stress corrosion (or 'season cracking').

10 to 20 — *Gilding brass* can be heavily worked, and being rich golden in colour it is particularly suitable for decorative work.

30 — *Cartridge brass* is very ductile and suitable for deep pressing, spinning and for drawing.

36 to 38 — *Common brass* is used for most ordinary cold pressings.

ALPHA-BETA BRASSES

37 to 45 — These brasses are best shaped while hot, cold working being reserved for finishing to size and to effect work hardening.

BETA BRASSES

45 to 50 — These brasses are too brittle for general use. However in *brazing brass* the melting point is low and the high zinc content is reduced by volatilization, and by diffusion into the metal being joined.

(b) Brass-based alloys

Copper-zinc alloys with other additions include:

Admiralty brass Here the addition of tin to cartridge brass improves resistance to many forms of corrosion.

Aluminium brass This is more resistant to high velocity water than Admiralty brass.

Leaded brasses Lead increases ductility, but reduces strength, it improves machining properties and provides *engraving brass*.

DZR (CZ 132), which contains 2 per cent lead, is widely used for plumbing and electrical fittings in buildings. Precise control of composition, including 1 per cent arsenic, and of heat treatment keeps this hot-working brass single phase, thereby avoiding *dezincification*. This form of electrolytic corrosion occurred in the old 60:40 *Muntz* or *yellow metal* where water had a high chloride content, leaving porous copper with a bulky zinc corrosion product.

The DZR alloy is accepted by most water authorities, but in some areas it is necessary, either to increase the temporary hardness of the water, or to use copper or gunmetal.

High-tensile brasses, commonly called *manganese bronzes*, are based on 60:40 brasses with additions of Mn, Fe, Al, Sn and/or Ni up to about 7 per cent in all. Tensile strength in the chill cast or forged condition is as high as 695 N/mm^2. High-tensile brass in the long term, becomes dark bronze in colour, or it can be artificially *toned*. The alloy has extremely high resistance to atmospheric corrosion, including salt and acid laden atmospheres, and does not stain adjacent materials. Although costly, the metal can be used as reinforcement in concrete to eliminate the risk of corrosion. However, in a moist atmosphere, galvanic corrosion can occur in contact with copper, phosphor bronze or aluminium bronze and its use is not recommended for load bearing fixings due to a possible risk of *stress corrosion*. The alloy is available in cast, and thick-walled tube form and as extrusions for windows, handrails and shopfronts. It is very suitable for forgings including cramps for restraining (but not supporting) wall claddings.

Naval brass The addition of tin improves resistance to corrosion, and to some extent, strength. It is suitable for hot rolling, forging and for sand and die casting.

(c) True bronzes

True bronzes are alloys of copper with tin and are often called *tin bronzes* to distinguish them from other copper alloys. They are harder and more resistant to corrosion than either constituent. Although costly, molten bronze is extremely fluid and ideal for large and intricate castings; the rich brown colour of which has been much favoured for sculpture. Bronze is still often employed for 'prestige' work in the form of nameplates often engraved and filled with enamel, door furniture, extrusions and sheet drawn on wood or metal cores for shop front and similar work.

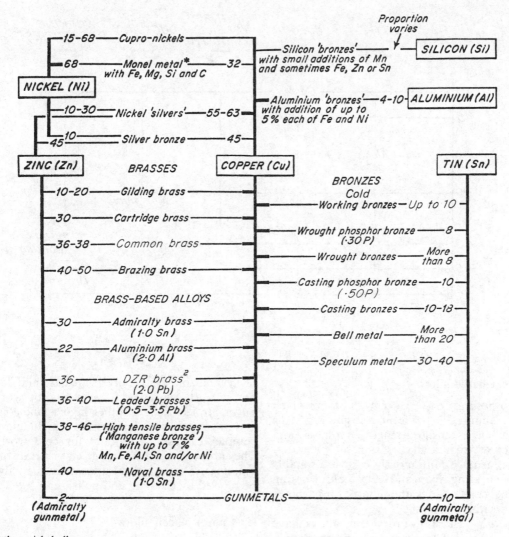

[1] Strictly a nickel alloy
[2] Dezincification resistant brass (CZ132)

Small amounts of lead are often added to brasses and gunmetals to improve their machinability. Additions of nickel, aluminium and/or manganese can improve casting properties, resistance to wear and or corrosion.

40   Copper-based alloys – with approximate percentages of main constituents

Bronzes with up to 6 per cent tin content are very similar to alpha brasses and like them they can be cold rolled or drawn, with consequent increase in the strength of the metal. *Wrought bronzes* with more than 8 per cent tin content can be cold worked after they have been annealed and have very high resistance to corrosion. *Casting bronzes* contain 10 to 18 per cent tin; ductility is reduced but in *bell metal* musical qualities result. *Speculum metal* with an even higher content takes a mirror polish and can be used for electroplating.

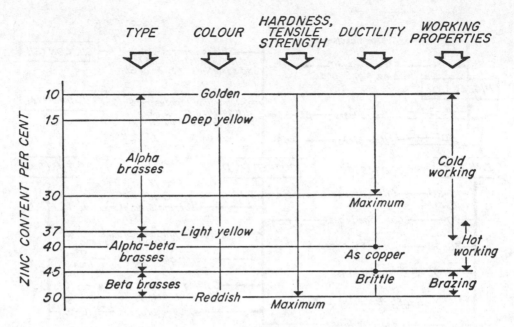

41 Brasses

(d) Bronze-based alloys

Phosphor bronze An 8 per cent tin bronze which contains about 0.3 per cent phosphorus. This alloy has high tensile strength, hardness, and resistance to wear and corrosion by sea water. Castings, usually continuously cast, are suitable for load bearing fixings. Heavy cold working makes the metal springy and suitable for purposes such as weather stripping

Gunmetals These tin/bronze/zinc alloys, have excellent resistance to corrosion, and strength is moderately high. They are very suitable for casting, more so with the addition of lead which also improves machining and anti-friction properties.

(e) Copper-nickel alloys

Tensile strength, ductility and hardness increase with nickel content.

Cupro-nickels are suitable for both hot and cold working. British 'silver' coins contain 75 per cent copper and 25 per cent nickel.

Nickel 'silvers' and *'silver bronzes'*, which are neither silvers or true bronzes, contain copper, nickel and zinc. Nickel silvers are hard and ductile but only the purer alloys can be hot worked. Silver bronzes have a rather higher zinc content and are suitable for hot working, including very complex extrusions, but not for cold working. They are very suitable for high class metalwork in buildings. Their excellent appearance can always be revived by cleaning and polishing.

(f) Copper-silicon alloys

Silicon 'bronze' alloys have good corrosion resistance – particularly in sulphurous atmospheres and are used as access fittings for chimneys. Having high strength also, they are widely used as masonry fixings.

(g) Copper-aluminium alloys

Aluminium 'bronzes' are bright golden-yellow in colour, very strong and resistant to corrosion. They can be annealed to make them soft and ductile.

Wrought alloys such as *Delta bronze no. VII* containing about 10 per cent aluminium, usually iron and sometimes also nickel and manganese, are available as rod and other extrusions including those suitable for load-bearing fixings for wall claddings.

Casting grades are available but the need for specialized foundry techniques restricts their use.

NICKEL (Ni)

Nickel is a white metal which is resistant to many acids, is hard and takes a high polish. It is used for electroplating food vessels, equipment for the chemical industry and on steel as a base for chromium plating.

Nickel is included in stainless steels and some copper alloys. *Monel metal* is a proprietary nickel containing 30 per cent copper and small proportions of iron, manganese and silicon. It has both high strength, about 510 N/mm^2, and elongation, about 45 per cent. Monel metal has extremely high resistance to corrosion and retains its properties at very high temperatures. It can be both cast hot and worked cold.

TIN (Sn)

Tin is a very costly, soft, weak metal with a low melting point, about 232°C. Extremely resistant to corrosion it is used as a coating on steel sheet (*tin-plate*), and as a constituent in true bronzes and with lead in many alloys for bearings and in tinmen's and plumbers' solders.

CADMIUM (Cd)

Cadmium is malleable and ductile at room temperatures but brittle at 80°C. It is used for electroplating steel components such as screws and as an alloying element in metals for bearings

CHROMIUM (Cr)

Chromium is well known for its good resistance to corrosion as a plating and in stainless steels and other corrosion resistant alloys. It is extremely hard and scratch resistant.

ZINC (Zn)

Metallic zinc was first used in England in 1760. It is now obtained either by electrolytic or thermal smelting of ores which contain the sulphide (zinc blende).

Information concerning the properties and uses of zinc can be obtained from the Zinc Development Association, 34 Berkeley Square, London, W1X 6AJ.

Properties

While the strength in tension of commercial rolled zinc (minimum 98.5 per cent zinc) is moderate, its ductility permits a moderate degree of bending and forming at ordinary temperatures but at very low temperatures the metal is brittle. Strength properties vary according to the direction of the grain.

The zinc/copper/titanium alloy although stronger is more ductile and has better creep strength than unalloyed zinc.

On exposure to ordinary atmospheres for some three to six months the initially bright surface of zinc tarnishes to a matt grey colour with the formation of an adherent protective layer consisting principally of basic zinc carbonate. The metal has good resistance to both inland and marine atmospheres, but is liable to slow and uniform attack by industrial atmospheres polluted with sulphur acids. Under average urban conditions, the maintenance-free life of zinc roofing conforming to CP 143:Part 5:1964[1] is forty years for a roof laid to the minimum recommended fall of 1 in 80 (approx 1°) and in rural atmospheres or with steeper pitches the life will be longer. The metal is unaffected by most potable waters unless they are of an acid (eg peaty) nature or contain large amounts of carbon dioxide in solution.

Zinc is unaffected by Portland cement or lime in mortars once they have set. However, coating with a hard-drying bitumen paint is recommended where zinc is embedded in materials containing soluble salts, notably chlorides and sulphates, and gypsum plasters may attack unprotected zinc under damp conditions. Contact

[1] For external work the thicknesses of zinc should not be less than 0.64 mm and for roofing the minimum thickness recommended is 0.78 mm (CP 143 Part 5).

corrosion by lead, tinman's solder, iron or aluminium is unlikely, but contact with bare copper or copper rich alloys should be avoided. Thus drainage from copper should not discharge on to zinc surfaces and copper lightning conductors on zinc roofs should be suitably coated – see CP 143, Part 5.

In areas where water has a high temporary hardness, lime deposits often stifle electro-chemical corrosion but copper tubes should preferably not be used in conjunction with galvanized cisterns and never with hot water cylinders.

Damp timbers may attack zinc and in particular contact with oak and Western red cedar must be avoided by interposing underlay felt. Water must not be allowed to drain from Western red cedar shingles on to zinc.

The coefficient of thermal expansion is relatively high (a 1 m length of unalloyed zinc expands nearly 26 mm with 56°C rise in temperature). This movement must be allowed for in roof coverings and particularly in components such as integral gutters.

Forms

Zinc is available mainly as sheet and strip but tubes, wire, rods and extrusions are also made. Traditionally, the standard sizes of zinc sheet have been $2\,400 \times 920$ mm and $2\,100 \times 920$ mm, but the material is now available as continuous strip in widths up to 1 m.

In the past the thickness of rolled zinc sheet was specified by the English Zinc Gauge (ZG), but now all sheet metals with the exception of lead are measured in millimetres – see table 78, page 195.

Uses

The main uses for zinc in building construction are: protective coatings on steel (ie galvanizing, Sherardizing, zinc spraying, see page 219); in brass and other copper/zinc alloys (see page 221 and in the form of sheet or strip for roof coverings, wall cladding, gutters and flashings either in unalloyed metal (BS 849:1939), or in zinc/copper/titanium alloy which can be laid in continuous lengths up to 9 m.

CP 143 Part 5:1964 recommends 14 ZG, and never less than 12 ZG (approximately 0.79 and

0.63 mm respectively), for permanent fully-supported roofing.

Zinc alloy pressure castings provide very accurately formed components. Their walls can be as thin as 0.5 mm. Die casting allows great freedom to the designer, and the smooth surfaces of casting can be plated or stove enamelled. Two zinc alloys containing about 4 per cent aluminium are described in BS 1004:1972 (amended to 1979), BS 5338:1976 *Code of practice for zinc alloy pressure die casting for engineering* includes recommendations for materials and design of products.

The method is economical for large quantities eg of door and window furniture.

BS 1431:1960 (1980) describes: *Wrought copper and zinc rainwater goods.*

High purity zinc and special alloys are used for cathodic protection of steel structures and of the hulls of ships in sea water.

Superplastic alloys containing aluminium, are suitable for vacuum forming.

Zinc oxide, zinc dust and lithopone are used as pigments in various types of paints (see *MBS: Finishes,* chapter 6).

LEAD (Pb)

Information can be obtained from the Lead Development Association, 34 Berkeley Square, London W1X 6AJ.

Lead, normally 99.9 per cent pure, is the densest, softest, weakest and one of the most durable metals commonly used in building.

Ductility is very high and lead can be cold worked into complex shapes without work hardening. Joining can be done by soldering or by *lead-burning.* See page 206. Continuous machine-cast lead sheet is now available. Although cast lead is claimed to have superior durability and creep resistance the price is now comparable to that of milled sheet. Creep of lead which is subject to continual stress within the elastic limit, is not serious at normal temperatures if thicknesses of metal and the frequency of supports are adequate. BS 602 and 1085: 1970 for lead pipes, stipulates minimum wall thicknesses which ensure that stresses will not exceed a safe level if horizontal pipes are supported at not more than 500 mm centres.

The coefficient of linear thermal expansion is the highest of the common metals (see table 80, page 196) and fatigue resistance is relatively low, so that if lead which is liable to thermal changes is not free to expand and contract cracking may occur.

Lead has a low melting point thus:

	°C
Lead	327
Zinc	419
Aluminium	660
Copper	1 083
Mild steel	1 900

Toxicity

Lead soluble compounds are poisonous. Most paints less than twenty years old contain little, or no lead, but some old paints, particularly gloss paint, found mainly on metalwork in pre-war buildings contain a great deal. Old paint containing lead should not be burnt off or rubbed down 'dry'. Paints containing more than 0.5 per cent soluble lead are required to be labelled: *'Must not be used on surfaces liable to be chewed or sucked by children'*. Paints containing more than 1 per cent soluble lead must also be labelled: *'Harmful if swallowed. Danger of cumulative effects. Keep away from food, drink and animal feeding stuffs. When using do not eat, drink or smoke'*. Thus, paint containing lead should not be burnt off or rubbed down 'dry'. Paint debris should be removed by vacuuming and it should not be burnt.

Resistance to corrosion

Freshly cut lead has a bright metallic lustre but when exposed to the atmosphere it forms an oxide film followed by a protective grey-blue coating of lead carbonate, which being insoluble does not stain other materials. Lead is not normally attacked electrolytically by other metals and it resists inorganic acids in varying degrees according to their concentration and the temperature. However, it can be attacked severely by certain organic acids, eg acetic acid and acids produced by timbers such as oak and Western red cedar and by lichens. Some waters which contain organic acids or free carbon dioxide are plumbo-solvent.

More than 0.1 mg/l lead in drinking water is a danger to health and in the UK the use of lead is not permitted where such concentrations occur.

Lead is not attacked by most soils but it is liable to corrosion by organic acids in wet peat soils and by incompletely burnt ashes in man-made ground. It is not appreciably affected by lime mortar but attack by Portland cement mortar or concrete can be serious if they remain wet.

Where necessary lead should be isolated from other materials by felt or building paper or by painting with bitumen; which should be on a hessian wrapping where pipes are below ground.

Uses

Lead is usually manufactured as pig lead, milled sheet, extruded tubes or as *lead wool*. Lead compounds are also used in paints and alloys.

Sheet

BS 1178:1982 *Milled lead sheet for building purposes* includes requirements for quality.

Lead is more conveniently measured by mass than by thickness, and code numbers approximate to the old 'weight in pounds per square foot' descriptions – see table 78, page 195. Sheet is used for:

Weathering and flashing   Lead is suitable particularly for complex shapes, eg for flashings dressed over pantiles.

Roof coverings and wall cladding, see *MBS: Finishes*. The cost of lead has discouraged its use for roof coverings on new buildings although 0.75 mm lead-clad flat galvanized steel sheet (*Almex*) cladding is now being used.

Damp-proof courses   BS 743 1970 (amd 1983) *Materials for dpcs* provides for lead not less than 19.5 kg/m^2 (Code no 4), 1.8 mm thick. Lead becomes economical in lead-cored bituminous felt damp-proof courses (lead not less than 1.20 kg/m^2 to BS 743).

Pipes   Lead pipes are extruded. Soil, waste and rainwater pipes, heads and gutters used to be made in cast lead and some survive on old buildings. Until recently, extruded lead pipes were used for water services, waste pipes and traps. BS 602, 1085: 1970 deals with *Lead and lead alloy pipes for other than chemical purposes*.

Other uses of lead

In suitable thicknesses, lead, being dense and limp, is effective for sound insulation and for the absorption of vibration, eg beneath machines. An important use is for radiation shielding from X-rays, reactors and radioactive appliances; lead is included in X-ray protective glass.

BS 4513:1969 (1981) describes *Lead bricks for radiation shielding*.

Lead is occasionally cast for sculpture, usually on a bronze armature, and for ornamental rainwater pipes and heads.

Lead is still the most effective material for sealing joints in iron gas and water mains and around iron railing standards. The metal can either be 'run' in the molten condition or *lead wool* or *yarn* can be caulked.

Lead compounds were important pigments for paints, more particularly for external use.

Lead alloys

Solders are lead-tin alloys, sometimes with a proportion of antimony to BS 219:1977, *Soft solders*. Plumbers' solder is plastic at 180°–240°C, so joints can be *wiped*. Tinmen's solder is plastic in a smaller range of temperatures. Two copper bearing leads, BS 334: 1982 *Chemical lead, type A, silver copper lead* and *type B, tellurium lead* which are used in the chemical industry, have superior strength properties. In the form of regulus metal antimonial lead is used for cast fittings in laboratories and chemical plants. In engineering, lead alloy bearing metals permit deformation and require less lubrication.

ALUMINIUM (Al)

Information concerning aluminium can be obtained from the Aluminium Federation, 60 Calthorpe Road, Five Ways, Birmingham 15.

Aluminium is the third most common element, so its price tends to be stable. Economical use is aided by ease of forming high strength/weight ratio and good resistance to corrosion and possible avoidance of the costs of painting.

However, it is not readily recovered as a metal. When first isolated in 1825 it cost £20 per pound.

Heroult (France) developed an electrolytic process of production in 1889, and the aluminium statue of Eros in Piccadilly Circus, London, was cast and erected in 1893.

About four tonnes of bauxite clay and 17 000 kilowatt hours of electricity are needed to produce one tonne of aluminium.

The metal can be classified as being either *pure* – containing 99 per cent or more aluminium, or as *alloys*. Properties which are common to both types are discussed here.

Density is low, 2 700 kg/m^3, about the same as granite and about one third that of steel, 7 850 kg/m^3. Aluminium is second only to copper in thermal conductivity. Electrical conductivity is about sixty per cent that of copper. Aluminium has long been used for electrical conductors on the grid system and its use for building installations is being developed.

The coefficient of linear thermal expansion is 24×10^{-6}/deg C at 20–100°C, about twice that for steel.

In ordinary atmospheres a thin, but dense, whitish film of oxide forms almost instantaneously and under damp conditions of external exposure roughening of the surface may follow if it is not kept clean. It is not recommended for exposure to marine atmospheres. The products of corrosion do not stain adjacent materials and are not toxic to animals or plants. Aluminium may suffer electrolytic corrosion in damp or wet conditions and contact must be avoided with copper, copper alloys such as brass and to a less extent with bare mild steel. Contact with zinc, stainless steel and lead is normally safe, but, BS 5516: *Code of practice for patent glazing* states that glazing bars should not have lead wings where there may be salt spray.

Metallic salts such as those originating from water which has passed through copper pipes or over copper roof coverings must not be allowed to come into contact with aluminium.

Acids, such as those which may be used to clean building materials or which arise from decaying vegetable growths on tiled roofs, attack aluminium.

Free alkalis such as those in wet Portland cement also attack aluminium. The metal should be protected from fresh concrete or mortar by bitumen of the solution type.

Appearance Aluminium, either with original mill or cast, or later anodized (see page 233) finish, is very liable to disfigurement by scratches, knocks or stains. Cement and lime adhere strongly and leave marks where they are removed. Components should be carefully handled and protected during construction by removable wrappings. Preferably they are fixed after all wet constructiom and finishes are completed.

Uses

Aluminium is being used for rainwater gutters and pipes. They are light, strong and do not need to be painted. They are made in sheet or are extruded in long lengths. Aluminium is not recommended for pipes conveying drinking water or for waste pipes and traps which would be attacked by bleach, strong detergents or soda.

BS 3660:1976 is a *Glossary of terms used in the wrought aluminium industry*.

British Standards employ the following symbols:

(a) *Form of material* – wrought aluminium and aluminium alloys for general engineering purposes.

S/NS Plate, sheet and strip }
C Clad plate, sheet and strip }
T/HT Drawn tube   BS 1471:1972
FIB/NF and HF
Forging stock and forgings   BS 1472:1972

RIB Rivet stock (pure)
NB and HB Bolt and screw } BS 1473:1972
   stock (alloy)   }
EI/NE and HE Bar, extruded round
   tube and sections   BS 1474:1972
G/NG and HG Wire   BS 1975:1972

Designations are for pure/alloy aluminiums respectively.

(b) *Heat treatment*
Prefixes are:
H Strain hardened (wrought material) designated *H1* to *H8* in ascending order of tensile strength. Material subjected to cold work after annealing, to hot forming or to a combination of cold work and partial annealing/stabilizing.
N Non-heat-treatable alloys
TB, TB7, TD, TE, alloys which have been
TF, TF7,TH, TS   subjected to various heat
                treatment processes

(c) *Conditions*
Suffix letters indicate the *temper* or condition of heat treatment:
M As manufactured. Material which has been subjected to shaping processes in which there is no special control over thermal treatment or strain hardening.
O Annealed. (Wrought material). Material which is fully annealed.

'Pure' aluminium

In the annealed condition pure aluminium is weak, but work such as cold rolling or hammering increases its strength with some loss of ductility. It cannot be hardened by heat-treatment.

In normal atmospheres pure aluminiums have very good resistance to corrosion. There are four grades:

		Per cent	Suitable uses
S1	*super-purity (softest-grade)*	99.99	*flashing* } roofing $\frac{1}{4}$ *H* condition
S1A	*high-purity*	99.80	
S1B	*medium-purity*	99.50	
S1C	*commercial-purity*	99	most aluminium foils

Pure aluminiums are rarely suitable for casting. Owing to their high ductility they are particularly suitable in sheet form for weatherings and flashings which may require to be hand formed on site, and for fully supported roofing. Pure aluminium extrusions are suitable for non-load-bearing components such as edge trims, joint covers and mouldings.

Foil, defined as before, less than 0.15 mm thick can be used as thermal insulation (it has high reflectivity and low emissivity), eg adhered to *insulating plasterboards*. It is also an effective vapour barrier.

Aluminium alloys

For building purposes, the alloys most commonly used may contain magnesium, manganese, and silicon, together with a number of minor additions which increase the strength of pure aluminium. Strength is further improved by cold working or by heat treatments which are described later.

Alloying reduces thermal and electrical conductivities, but they remain high. Thus, aluminium alloys are used for corrugated and troughed roof sheeting. A roofing system of interlocking roll-formed sections, can be used for roof pitches down to 1°C. 'Secret' sliding fixings accommodate thermal movement so that transverse laps are needed only at 30 m centres. Window, door and shop-fitting sections, venetian blind slats and for structural members.

Alloys, particularly those containing copper, are less resistant to corrosion than pure aluminium, and may become unsightly and cause sliding windows to 'stick'. However, in 'breathable' air, corrosion is unlikely to affect strength. Sheet and strip are sometimes clad with pure aluminium to improve their corrosion resistance.

Wrought aluminium alloys

These alloys which are suitable for rolling, pressing and extrusion are of two kinds:
(a) heat-treatable and (b) non-heat-treatable.
(a) *Heat-treatable alloys (prefix H)*
The tensile strength of these alloys is increased by heat treatment.

The heat-treatable alloys used in building include *H9* which contains magnesium and silicon which is largely used for extrusion for window frames, and the principal structural alloy *H30* which contains magnesium, silicon and manganese, but no copper.

These are three processes of heat treatment:
1 *Solution treatment (TB)* consists of heating in the range 500 to 530°C (the exact temperature will depend on the alloy) followed by rapid cooling and *natural ageing* which is accompanied by an increase in strength.
2 *Precipitation treatment (TE)* (or *artificial ageing*) consists of heating in the range 100 to 200°C for a suitable period according to the composition of the alloy.
3 *Full heat-treatment (TF)* combines solution and precipitation treatments. For example solution heat-treatment raises the tensile strength of the H30 structural alloy to a minimum of 190 N/mm^2, and full heat-treatment raises it to 310 N/mm^2 minimum with some loss of ductility.

(b) *Non-heat-treatable alloys* (prefix N)
Like pure aluminium these alloys gain strength by cold working, but not by heat treatment.

Table 85 shows examples of common alloys in the *H4* condition, all of which listed can be welded with little loss of strength. The suitability of certain grades of aluminium for extrusion leads to profiles in which material can be disposed so they fulfil several functions. Thus, glazing bars which support glass over a long span, keep out the weather and also provide a channel on the inside for the collection of condensate, are cheaper in aluminium than in other materials.

Alloy	Alloying addition per cent	Tensile strength (guaranteed minimum) N/mm^2	Properties and uses
N3	1.25 manganese	140	The simplest alloy. Corrugated and trough sheeting
N4	2.0 magnesium	225	
N5	3.5 magnesium	275	
N8	4.5 magnesium 0.75 manganese	345	High resistance to corrosion. Marine work

Table 85   *Aluminium non-heat-treatable alloys*

Casting aluminium alloys *(Suffix LM)*

These are used in building mainly for door and window furniture, rainwater goods, and simulations of nineteenth century decorative cast iron statuary. The choice of alloy depends on whether the metal will be sand cast, gravity or pressure die cast, on the finish required and the strength that is necessary, some alloys being heat-treatable.

BS 2997:1958 (1980) describes: *Aluminium rainwater goods.*

Structural aluminium alloys

Modulus of elasticity, coefficient of expansion, density and melting point are, for all practical purposes, the same for all alloys but tensile strength, ductility and durability vary with composition.

Alloys are available having a tensile strength equal to that of mild steel, but the modulus of elasticity of aluminium ($68\,900\,N/mm^2$) is only about a third. Thus, for the same sectional dimensions, spans and loadings, aluminium sections deflect nearly three times as much as steel sections. Where this is not acceptable an aluminium section of greater effective depth or second moment of inertia is required. Nevertheless, in order to support the same load with the same deflection, an aluminium structure will rarely be more than half the weight of its steel equivalent.

Special consideration must be given to the resistance to torsional buckling, and local buckling of aluminium struts with thin-walls. These factors are discussed in CP 118:1969 *Structural use of aluminium*. A draft for public comment based on limited state design principles will replace the 1969 code.

Table 86 compares the mechanical properties of a typical aluminium structural alloy with those of structural mild steel.

Aluminium increases in strength at low temperatures, by as much as 33.3 per cent at $-196°C$, hence its use for the storage of liquid methane. However it loses strength at high temperatures more rapidly than steel. The *HE30-TF* alloy behaves as follows:

°C	Tensile strength N/mm^2
0	367
20	352
100	301
150	261
200	212
250	113
400	30.9

BS 1161:1977 describes *Aluminium alloy sections for structural purposes*.

Properties	Structural aluminium HE 30-TF	Mild steel BS 4360 grade 43
Density kg/m³	2700	7850
Ultimate tensile stress N/mm²	280–310	430–510
Modulus of elasticity (E) N/mm²	69 950	207 000
Elongation (min) per cent	8 (on 50 mm)	22 (on 50 mm)
0.20 per cent proof stress (min) N/mm²	239–270 N/mm²	—
Permissible bending stress (max) N/mm²	162	162
Permissible stress/weight ratio (approx)	6	2

Table 86 Properties of an aluminium structural alloy and mild steel

Superplastic aluminium alloys

Superplasticity, the ability of a material to extend many times its length before breaking, occurs in certain metals at critical temperatures. This property enables sheets to be thermoformed into complex shapes in a single-sided mould by air pressure with savings in costs of tooling, assembly and finishing operations.

Superal produced by Superform Metals Ltd, of Worcester, England, is an aluminium alloy containing 6 per cent copper which reverts to non-super-plasticity in forming.

Sheets 0.8 to 4.0 mm thick can be anodised as a pretreatment for painting, coated with pure aluminium, or chromium plated.

More than 3000 1.2 × 1.8 m cladding panels are being used at the Visual Arts Centre of the University of East Anglia, England.

Other aluminium alloy products

British Standards include:

BS 5286:1978 *Aluminium framed sliding* glass doors

BS 4873:1972 *Aluminium alloy windows*

BS 4868:1972 *Profiled aluminium sheet for building*

BS 2997:1959 (1980) *Aluminium rainwater goods (cast, extruded and wrought)*.

Jointing

Aluminium members can be joined with bolts or rivets, by soldering, welding or by adhesives. In long members such as gutters, sliding joints are needed to accommodate thermal movement.

Bolting and riveting

Bolting and cold riveting avoid heat which may upset the properties of alloys, and the corners of window frames are often jointed in this way.

Soldering

Soldering is performed either with a flux (which may be corrosive and must be removed) or with a friction solder without flux. Care must be taken in the choice of materials and in design to avoid bi-metallic corrosion between the solder and the aluminium. Generally, where moist conditions are expected joints should be wrapped or painted.

Solder joints are sometimes brittle and advice should always be obtained from the manufacturers.

Welding

Most aluminium alloys can be welded. Heat-treatable materials lose strength in the heat affected zone but non-heat treatable alloys give welds of up to 90 per cent efficient. Difficulties arise in welding on the building site.

For example, the older oxy-acetylene process requires the use of an active flux which is corrosive and it is rarely possible to completely remove it on building sites.

Reliable welds can be produced without flux by the inert gas process using an electric arc with a protective envelope of argon gas. However, the equipment is heavy, the gas shield is disturbed by draughts and like oxy-acetylene welding its use is nearly always confined to the factory.

Adhesive bonding

Adhesive bonds, which are becoming increasingly reliable, are used, for example, for the corner joints of some windows.

FINISHES ON NON-FERROUS METALS

Non-ferrous metals are often chosen because they do not require protection from destructive corrosion and because their appearance is often agreeable without special treatments. Surface finishes which preserve or enhance the natural appearance of non-ferrous metals are described below, followed by applied finishes.

The condition of a surface largely determines the appearance of finishes. CP 3012:1972 deals with *Cleaning and preparation of metal surfaces* and BS 4479:1969 gives *Recommendations for the design of metal articles that are to be coated*.

Mechanical treatments

Rough castings, forgings and welds may need *grinding*, or sometimes *hand filing*.

Shot blasting produces a rough texture.

Sand blasting gives a matt finish, the nature of which can be varied according to the sand, the air pressure used and the distance of the nozzle from the work.

Scratch brush working is usually done with rotating stainless steel brushes, the coarseness, size and speed of which determine the result. A 'satin' texture is very sensitive to finger marks and a lacquer should be applied, or in the case of aluminium the metal can be *anodized*.

Polishing, as an integral finish or as a base for applied finishes, is done with progressively softer mops and finer abrasives.

Chemical treatments

These include bronze metal antique — 'BMA' – a

heat and chemical treatment on brass, which provides a simulated bronze appearance.

Anodizing

The oxide film, which forms naturally on aluminium, is thickened by making aluminium the anode in an electrical cell in which the electrolyte is usually sulphuric acid, and the surface is then sealed, in water.

Although there are a few tanks up to 2.4 × 1.0 × 12.0, components are better anodized before assembly, because welds, especially with heat-treatable alloys, can cause local darkening, and process liquors trapped in mechanical joints may leak out later and cause stains.

Anodizing enhances the appearance of smooth aluminium surfaces – for example finger marks become less visible. However, appearance is very dependent upon the composition, condition and texture of the base. Commonly anodized alloys are:

HE9   (BS 1474:1972)

SIC ⎤
NS4 ⎦   (BS 1470:1972)

Before anodizing, surfaces can be mechanically abraded, chemically etched or polished to a mirror-like finish. Descriptions include: matt, gloss, light satin, bright and smooth. The clear uncoloured finish is often called 'natural' or 'silver' anodized.

Surfaces are porous before they are sealed, and can be dyed. Some colours are very intense, but few are fade proof externally where gold, black, brown, dark green and blue are most suitable. Alloys with inherent colour, which do not require dyeing after anodizing have been developed.

Matching of colours in adjacent components may prove difficult.

Anodizing of sufficient thickness improves the durability of aluminium.

Recommended minimum average thickness for anodic finishes are shown in table 87.

Thicknesses greater than 0.003 mm for purposes such as heat reflectors, may suffer thermal cracking as can coatings more than 0.025 mm thick which may be desired where they are exposed to severe pollution. The latter thickness may also be difficult to seal.

In 1985 a harder tin-based electrolytic finish

		mm	
External uses	Typical applications in UK	0.025	BS3987 :1974
	In mild or rural atmospheres away from industrial pollution or marine influences, or where frequent washing will be done	0.020 or 0.015	
	Externally, where maintenance will be frequent, the installation is temporary or some deterioration is acceptable	0.010 or 0.005	
Internal uses	Excessively aggressive atmospheres, especially where serious wear and condensation on surfaces is likely	0.025	BS1615 :1972
	Serious wear	0.020	
	Moderate wear	0.015	
	Much general usage	0.010	
	Little wear or corrosive hazard	0.005	
	Some interior fittings and reflectors, etc	0.003	
	Base for paint or lacquer (unsealed)	0.001	

Table 87   Recommended thickness of anodized coatings

gives shades of bronze 8 microns thick and black in 12 microns. Externally 25 microns thickness is advised.

BS 1615 designates coatings as AA25, AA20, etc, according to their thickness in mm.

BS 5599:1978 is a *Specification for hard anodic oxide coatings on aluminium for engineering purposes* (minimum average thickness 25 microns).

BS 3987:1974 (amd. 1982) *Anodic coatings on wrought aluminium for external architectural applications* states that with regular maintenance suitable anodic finishes can give satisfactory service for many years. However, it is particularly important to remove dirt from surfaces, especially in damp conditions. In rural areas an annual wash is usually sufficient, but in industrial areas a monthly wash with water containing detergent, which may be done when windows are cleaned, may be required.

Applied finishes

With suitable preparation, non-ferrous metals can be painted or lacquered for protection or for decoration – see *MBS: Finishes*, chapter 6.

Other applied finishes are:

Electroplating

This process has received attention under *Ferrous Metals*, page 218. A wide variety of metals can be applied, including tin and copper. Nickel-chromium plating, which gives a very hard finish on steel, is commonly used on brass and zinc alloys. *Speculum metal*, a silver-like copper-tin alloy (see pages 223 and 224), has good resistance to corrosion and wear. The resistance to corrosion of copper and its alloys makes them suitable bases for plating with costly metals such as gold and silver.

Vitreous enamel

Essentially glass, can be applied on copper and aluminium – see *MBS: Finishes*, chapter 6.

Plastics

The durability, electrical properties, warmth and colour provided by plastics may justify their use as coatings on non-ferrous metals – see chapter 13.

Durable white coloured polyester finishes are electrostatically deposited and heat-cured on aluminium windows and rainwater goods. BS 4842, revised 1984 specifies *Liquid organic coatings for aluminium for external architectural purposes.*

Maintenance of finishes

Frequent cleaning may be needed to preserve a bright appearance, although a natural patina may be preferred. Finishes such as anodizing should be kept clean with soapy water or detergent. When dry, they can be polished with a light application of furniture cream. Metal polishes and more drastic abrasives prevent the formation of a natural patina. They should never be used on chemically and/or heat-modified surfaces, anodized aluminium, or on plated or other applied finishes.

10   Fibres and fibre-reinforced products CI/SfB Y

Fibres, comprising natural and man-made hair-like structures are widely used in building, either loose or in formed products.

BS 5803 deals with *Thermal insulation for pitched roof spaces in dwellings*. Part 1:1979 *Specification for man-made mineral fibre thermal insulation mats*.

BS 6232, Parts 1 and 2:1982 is entitled: *Thermal insulation of cavity walls by filling with blown man-made mineral fibre* – for walls with masonry and/or concrete leaves up to 12 m high.

Ceramic fibres are used for thermal insulation. See page 123.

Loose fibres of combustible materials are a fire hazard and loose asbestos fibres are extremely dangerous to health – see page 238.

Wool, cotton, linen, coir, glass and polymer fibres twisted together form *yarn* which is woven, felted, bonded or knitted into textiles carpets, etc.

Fibre reinforced composites

The uniform dispersal of fibres in a cementitious matrix distributes stresses and improves reistance to microcracking. Vegetable and animal fibres have been used to reinforce bricks and mud and lime plasters from very early times.

Carbon fibres have great potential as reinforcement, but at present their cost is prohibitive for building work.

Today, fibre reinforced products include:

	Fibres	*Matrices*	
Glass			
	– ordinary grade		
		Plastics (GRP)	Chapter 12
		Bitumen	Chapter 13
		Gypsum (GRG) plaster	Chapter 11
	– alkali-resistant (Cem-FIL)	Portland cement (GRC)	
	Steel (*Wirand*)	Portland cement	This chapter
	Polymer	Portland cement	
	Mineral	Starch	
	Asbestos	Portland cement Lime/silica	
		Bitumen	Chapter 11

Animal and Vegetable	Polymers Bitumen Pitch	Chapter 13
Rock wool	–	This chapter
Slag wool	–	
Wood		
– compact	–	
	Oil or resin	

1 Glass fibre reinforced gypsum (GRG)

This tough, non-combustible composite, with low moisture movement does not require costly alkali-resistant glass fibres. It is not suitable in wet conditions.

References are:
Gypsum – MBS: Finishes, chapter 2.
Mechanical properties of glass-fibre reinforced gypsum, M A Ali and F Grimer.
BRS Current Paper 33/69 and BRS Current Paper 12/71 *Glass fibre reinforced cement and gypsum products* A J Majumdar.

The last named paper states that gypsum with a low water:plaster ratio, although weak in tension, may have a compressive strength of 70 N/mm², but GRG with 10 per cent glass fibres by weight is 2.5 to 3 and 3 to 4 times stronger, for flexural and tensile strengths respectively, and impact strength is 20 to 30 times greater than that of unreinforced plaster.

Impact strengths of GRG and timber are comparable, but the tensile strength of GRG although only a quarter of that of timber along the grain, is about four times that of timber across the grain. The modulus of elasticity of GRG is about 1.5 times that of timber.

A 7 mm GRG panel with 6 to 8 per cent glass by weight can withstand a temperature up to 1000°C on one face for one hour without flame penetration. Lengths of glass fibre for effective mixing, and glass fibre:cement ratios are critical.

It is better to consider GRG as a fire-resistant alternative for timber than as a substitute for reinforced concrete.

The BRE developed pressing and spray/suction techniques for:

school partitions
floor units
diamond shaped units for stressed skin arches
double-skinned fire-check doors with exfoliated vermiculite filled cores.

A typical GRG product is a 600 mm^2 tile for suspended ceilings. Acoustic properties are modified by varying the density of tiles, and by perforations backed by sound absorbents.

2 Glass fibre reinforced cement (GRC)

BRE Digest 216 describes: *GRC*
This fibre-reinforced composite consists of cement (usually Portland cement), with or without aggregate, reinforced with *Cem-FIL* (Pilkington Flat Glass Ltd) glass fibres containing zirconium, which are resistant to the alkaline cement environment in which ordinary glass would soon become brittle. Cem-FIL[1] Fibres were developed at the BRS in 1967, and then by Pilkington Flat Glass Ltd who licence their use.

BS 6432:1984 describes *Methods for determining properties of glass reinforced cement material.* The *Glass fibre reinforced cement Association (GRCA)*, Gerrards Cross, Buckinghamshire. SL9 7LD. has a Code of Practice for cladding and a code for formwork is being prepared.

Important design decisions include optimum glass fibre content and orientation and the inclusion of workability admixtures such as polyethylene oxide.

White cement can be used, but as with other cement products, added pigments do not always give satisfactory colours or tonal uniformity. Exposure of aggregates, and textures obtained from moulds or by spraying, may make dirtying less objectionable.

No moisture appears inside when rain is blown on a 10 mm sheet at 33 m/s and most GRC formulations make a vapour barrier unnecessary.

The maximum moisture movement in external cladding in this country is about 0.07 per cent and thermal movement from 25–45°C, is 13–20 × 10^{-6} per deg C.

Moisture movement is similar to that of autoclaved aerated concrete and thermal movement to that of dense concretes.

[1]Trade name.

With suitable formulations, a single skin of GRC can provide two hours fire resistance (integrity and stability criteria BS 476, Part 8). However, higher density forms for GRC build up internal steam pressure and may explode in fires.

The modulus of rupture of GRC increases quite rapidly up to about 6 per cent fibre content (by volume).

Glass fibres, most effectively in a two-dimensional array, improve the resistance to cracking of both plain and reinforced concretes.

A composite with a 5 per cent (by weight) planar distribution of *Cem-FIL* fibres with a density of 2000 kg/m^3 give a strength of 16 N/mm^2 at 28 days and a modulus of elasticity of 20k N/mm^2. Another composite with 6 per cent (by volume) fibre give 3–4 times the tensile strength and 15–20 times the impact strength of plain concrete.

Although the use of glass fibres instead of steel reinforcement would remove the need to protect reinforcement from corrosion. BRE News 30 stated that because the initially excellent tensile and impact strengths and ductility of GRC reduced considerably under normal weathering conditions, structural uses must await sufficient evidence to enable safety limits to be established.

Manufacture

Chopped fibre, in about 80 per cent of the mixing water containing an admixture, is poured into a mixer and dry fine sand and cement are added gradually, followed by the rest of the water. Pressing then reduces with water:cement ratio.

GRC products can also be made by injection moulding, by extrusion, and in the spray/suction method the cement slurry and finely chopped fibres are sprayed on an horizontal moving base through which water is extracted. On the site, GRC can be sprayed, using a rather drier mix.

Uses of dense GRC

Although *Cem-FIL* fibres are costly, some glass fibre-cement moulded window frames, valley gutters, posts, bollards, planters and wall panels are competitive, and being thin walled, show a substantial saving in weight compared with their precast concrete equivalents. The elegant and finely finished 3.3 m × 1.5 m wall units to the Crédit Lyonnais building in the City of London have two 10 mm skins with a 40 mm air space. GRC permanent formwork can have superior resistance to abrasion, frost and chemicals and a better finish than the in situ concrete it encases.

Other GRC products include lids for cable ducts and coal bunkers and roofing 'slates'.

BS DD 76: *Pre-cast concrete pipes of composite construction*, Part 1:1981 describes products strengthened by *continuous alkali-resistant glass rovings*.

Lightweight GRC

The BRE has developed insulating boards 8–12 mm thick in which cenospheres (hollow spheres of silicate materials obtained from pulverized fuel ash), are substituted for a proportion of the cement approximately halving the density of the boards.

LWGRC is, however, denser and thermal conductivity is greater than that of asbestos insulating boards. Bending strength is similar, tensile strength is better and impact strength substantially better.

Suggested uses include: asbestos-free fire protection, permanent formwork for concrete and cores for veneered panels.

3 Glass fibre reinforced polymer/concrete

(GRPC) is a glass fibre reinforced polyester resin bound concrete – not to be confused with polymer impregnated concrete – see page 172.

GRPC panels exposed in this country since 1976 are stated not to show deterioration. The 'concrete' can resemble ordinary concrete in appearance, but is less dense. It has higher tensile, compressive and flexural strength and better resistance to moisture and aggresive atmospheres. Cover to protect reinforcement can be less than with ordinary concrete.

GRPC is not easily ignited (*Class P*, BS 476: Part 5) and *Surface spread of flame* is *Class 0*.

Panels made by Panelcraft Ltd, Maidenhead, Berkshire, have exposed aggregate surfaces and with foamed polyurethane cores.

4 Steel fibre reinforced concrete

BRE Current Paper 69/74 stated that steel fibres in concrete impart increased ductility, energy absorption and resistance to rapidly induced stresses. Flexural strength and resistance to cracking is improved, but tensional torsional and compressive strengths only marginally. Abrasion and spall resistances are improved. As with other fibre reinforced concretes, members can be thinner than those in conventional reinforced concrete, because protective 'cover' to corrodible steel is not required.

BSPD76:Part 2:1983 discusses *Precast concrete pipes strengthened by chopped zinc-coated steel fibres*.

Manufacture In mix design, the fibre geometry (*Wirand* mild steel fibres are 25 mm × 0.25 mm diameter mild steel) and the fibre:aggregate ratio are critical.

Uses include: road and floor toppings, thin section structural and semi-structural building components, pipes and duct covers.

5 Polymer fibre reinforced concretes

John Laing Concrete Ltd have developed *Faircrete* concrete, which with either dense or lightweight aggregate, contains about 20 per cent entrained air and a proportion of polypropyline fibres. Malleability and cohesion is such that the wet mix can be moulded or impressed to give intricate shapes and textures which it then retains without formwork.

Wests' Piling use another type of polypropylene fibrillated fibre to improve the impact strength of concrete tubes which are driven into the ground on a steel mandrel. The *Caricrete* tubes are filled with concrete.

6 Mineral fibre reinforced composites

BS 6676 describes *Thermal insulation of cavity walls using man-made mineral* fibre batts (slabs), Parts 1 and 2:1986

A composite containing rockwool fibres has a starch/cellulose binder. In consequence, the material is *combustible* (BS 476, Part 4), but surfaces may be *Class 0*, and have *Class 1 Spread of flame* (BS 476, Part 7).

The products should not be subjected to a relative humidity exceeding 70 per cent at 18°C.

Monolux 40 (Cape Boards Ltd) comprises an inorganic fibre reinforced calcium silicate matrix. It is non-combustible and Spread of flame is Class 1 (BS 476) and Class 0. It is used for fire protection to steelwork, and as a core for fire doors and veneered panels. Density is 640 kg/m^3.

Monolux 40 is made in 12–32 mm sheets and in battens. It is also used for TRADA firecheck glazing channels to retain the edges of wired glass. These are coated with an intumescent paint which expands in fire producing an insulating foam. The coating should be protected from rain. The GLC granted waivers, subject to conditions, for channels 44 mm wide to give $\frac{1}{2}$ hour, and 50 mm wide to give 1 hour fire resistance, with panes up to 1.2 m^2.

7 Vegetable and animal fibre reinforced composites

Reinforcing fibres include wood, hessian, sisal, jute cloth and animal hair. Straw in mud and hair in lime plaster were used from early times.

A 6 mm thick general purpose board consists of autoclaved calcium silicate reinforced with wood fibres. Density is 875 kg/m^3 and conductivity is 0.17 W/m deg C. It has low alkalinity and is easy to decorate. It is absorbent, loses some strength when wet and requires an impermeable coating if it is exposed to rain. The board is easy to work and can be nailed without pre-drilling.

Uses include ceiling and partition linings, eaves soffits and fire insulation.

Other products under this heading include certain bituminous roofing and dpc felts (BS 747:1977 and BS 743:1970) and pitch fibre pipes and conduits – see page 245. Sisal fibre reinforced plaster wall panels are used in Australia.

8 Asbestos fibre reinforced composites

Asbestos, a silicate of magnesium, occurs as a glassy rock which can be split into extremely thin fibres. These are strong in tension and flexible and have good resistance to alkalis, neutral salts and organic solvents, and the varieties used for building products have good resistance to acids.

Asbestos is *non-combustible* and able to withstand temperatures up to 900°C without change. Asbestos dust causes asbestosis and mesotheloma – a cancer of the lungs, and products which may release asbestos fibres into the air should be removed or encapsulated by a flexible membrane. This work must be done by a licensed operator listed by the *Asbestos removals contractors' association, 45 Sheen Lane, London, SW14*.

The vast areas of existing asbestos cement roofing and cladding present no hazard while the fibres are retained in the cement matrix. However, *Health and Safety Guidance Note EH36 Work with asbestos cement* (1984) forbids the use of cutting or grinding discs, circular and jig saws and cleaning of asbestos cement by abrasion and high power water jets. Guidance note EH37 is *Work with asbestos insulation board*.

All work involving asbestos must be done in accordance with the requirements of the *Health and Safety Commission* and advice is given by the *Asbestos information committee*, 2 Old Burlington Street, London, W1.

The fibres have been used as reinforcement with Portland cement, lime, plastics and bitumen binders, eg in asbestos-cement and asbestos-silica-lime products, in vinyl floor tiles and in bitumen felts. Other uses are fabrics and ropes.

The following products are dealt with here:
(a) Asbestos-cement/silica, lime, cellulose and bitumen bonded products.
(b) Asbestos/inorganic agent bonded products.
(c) Resin bonded asbestos products.
(d) Sprayed asbestos.
(e) Bitumen bonded asbestos.

(a) Asbestos-cement/silica, lime and cellulose-bonded products

The materials used in these mixtures are asbestos fibres bonded with Portland cement, fine silica and hydrated high calcium lime, cellulose and/or bitumen. Pigments are sometimes included.

Products are formed in open moulds by hand, more usually by the *Hatschek* process, and in some cases by extrusion. In the Hatshek process the slurry mixture is picked up by wire-mesh covered cylinders, transferred to a felt and then wound round a roller until the required thickness is obtained. The length of the *wet flat* corresponds to the circumference of the roller. Both moulded and cylinder-formed products are pressed to provide the required density and final shape, ie flat and profiled sheets, pipes, etc.

Asbestos-cement products are air cured for approximately four weeks. Asbestos-silica-lime and asbestos-silica-cement products are steamed under pressure in an autoclave. Steam curing induces thorough hydration, reduces efflorescence, and in asbestos-silica-lime products is essential to promote chemical bond between lime and silica. 2 to 2.5 per cent of Portland cement may be used in the latter case as a catalyst.

Products are unaffected by alkalis but the cement binder may be attacked by acids, depending upon their concentration and the density of the product. Products can be painted with alkali-resistant paint (see *MBS: Finishes*). Some products are available with factory applied finishes, and some with extremely durable PVF film surfaces in fade-resistant colours.

Table 88 shows that strength and thermal conductivity increase with density. The converse is true in respect of fire resistance.

BS 690 deals with *Asbestos-cement slates and sheets.*

Part 1:1963 is withdrawn.
Part 2:1981 is a *Specification for asbestos-cement and cellulose-asbestos cement flat sheets.*
Part 3:1973 *Corrugated sheets.*
Part 4:1974 *Slates.*
Part 5:1975 *Lining sheets and panels.*
Part 6:1976 *Fittings for use with corrugated sheets.*

Asbestos-cement and cellulose-asbestos-cement flat sheets

BS 690: Part 2;1981 describes these sheets.

1 Asbestos-cement flat sheets

Products are classified according to bending strengths and densities and designated either *semi-* or *fully-compressed* – see table 87.

British Standard	Class	Dry density kg/m³	Thermal conductivity (k) W/m deg C (max)	Bending strength (min) N/mm²
690: Part 2: 1981	Fully-compressed asbestos-cement	1600 (min)	0.65[1] (approx)	20/28[2]
	Semi-compressed asbestos-cement	1200 (min)	0.431 (approx)	13/16[2]
	Fully-compressed cellulose-asbestos-cement	1300 (min)		16/20[2]
	Semi-compressed cellulose-asbestos-cement	(min)		12/16[6]
3536:1974*	Asbestos wallboards	(a) 900–1200 (b) 1200–1450	0.300 0.360	10 15
	Asbestos insulating boards	(a) 500–720 (b) 720–900	0.145 0.175	5 8

[1] Not BS requirements [2] Parallel to-/At right angles to- asbestos fibres in sheet * High fibre content

Table 88 *Asbestos reinforced flat sheets and boards*

Preferred sizes are:

```
                              1830 ⎫
      1800 ⎫                  2400 ⎪
      2400 ⎬ × 900 mm         2440 ⎬ × 1220 mm
      3000 ⎭                  3000 ⎪
                              3050 ⎭
      1800    × 1200 mm
                              3000    × 1500 mm
```

Thicknesses At present, 3–12 mm thicknesses are made in the UK and thicknesses up to 25 mm are imported.

Semi-compressed flat sheets contain about 70–80 per cent Portland cement by weight and it is important to realise that they are not fire resisting and tend to burst explosively in fires.

They are used as wall cladding, eaves soffits, bath panels and for similar purposes.

Sheets have one natural fairly smooth finish, and are available ready-painted.

Fully-compressed asbestos-cement flat sheets are stronger, more flexible and have better resistance to impact and abrasion than semi-compressed sheets.

The thinner sheets can be bent to radii.

Surfaces can be natural integral white or grey cements, blue, polished, or matt finished for painting. White and coloured mineral enamel and bonded natural stone aggregate coatings are available.

Boards with an epoxide resin finish have good resistance to acids and alkalis and to temperatures up to 130°C and are suitable for laboratory bench tops and fume cupboards. Other applications for thicker boards in this density are wall cladding, fascias, shelves, stair treads and hearths. Window boards and sills and copings are extruded, with voids.

Other fully-compressed asbestos-cement products include:

BS 3497:1979 *Unimpregnated asbestos-cement boards (incombustible) for electrical purposes 3–50 mm thick.*

2 Cellulose-asbestos-cement flat sheets

Sheets of this composition are also classified as either *semi-* or *fully compressed*, according to density and strength – see table 88.

The sheets are more flexible than asbestos-cement, but they are not suitable for use externally.

Sizes are as for asbestos-cement flat sheets, but thicknesses are 3, 4.5 and 6 mm for semi-compressed, and 3 mm only, for fully-compressed sheets. Table 88 shows dry densities, thermal conductivities and bending strengths, enabling comparisons to be made between: semi- and fully compressed asbestos-cement and cellulose-asbestos-cement flat sheets and asbestos wallboards and insulating boards.

Corrugated asbestos-cement sheets

BS 690:Part 3:1973 describes these sheets which remain an economical and efficient, intermittently supported, roof and wall covering. They are widely used, mainly for industrial buildings.

The specification has five profile classes, each with a stated minimum load-bearing capacity. It includes an impermeability test, and a frost resistance test requiring twenty-five cycles of freezing and thawing.

Corrugated asbestos-cement sheets are moderately strong, and resistance to continuously applied loads increases with age. Resistance to impact, however, decreases in a short period of years, and for roofing, adequately supported corrugated, trough and other inherently strong profiles should be employed, and cat walks must always be provided for and used by, maintenance personnel. Expansion, shrinkage or curling may cause sheets to crack if they are fixed too rigidly to supports.

The slightly rough and open texture of some products increases with weathering and the sheets tend to darken with age. With appropriate precautions, products can be cut on site with a hand saw or shaped with a coarse file.

Although non-combustible and classified 'Ext SAA' (BS 476:Part 3:1975), if restrained the material cracks in building fires, which although sometimes beneficial in venting smoke, the product does not merit any *fire resistance* grading when tested in accordance with BS 476:Part 8: 1972).

In industrial atmospheres the expected life of forty years can be extended by alkali-resistant protective finishes. If impervious paint is used externally, back painting is necessary.

Aluminium, lead and zinc should be protected from alkalis which are released by new asbestos-cement.

Most profiles can be obtained curved to a minimum radius of 1400 mm. A range of fittings is available, including those such as roof venti-lators with soaker flanges which when used in conjunction with asbestos-cement roofing dis-pense with the need for separate flashings.

Relevant products are dealt with by BS 690: Part 5:1975 *Lining sheets and panels* and Part 6:1976 *Fittings for use with corrugated sheets*.

References

BS 5247: *Code of practice for sheet roof and wall coverings*, Part 14:1975 *Corrugated asbestos-cement*.
MBS Finishes, chapter 7.

Asbestos-cement slates

BS 690:Part 4:1974 *Slates* describes these man-made roofing slates – see *MBS:Finishes*, chapter 7.

The specification describes three sizes, all 4 mm thick and two degrees of compression ie:

Fully compressed slates These must have a smooth finish on one side, minimum density 1800 kg/m³ and minimum bending strength 22.5 N/mm².

Semi-compressed slates have a smooth finish on one side only, minimum density 1450 kg/m³ and minimum bending strength 16.0 N/mm².

The impermeability test allows traces of mois-ture, but not drips on the undersides of speci-mens. Tests are required to be in accordance with BS 4624.

Other asbestos-cement products

BS 567:1973 *Asbestos-cement flue pipes and fit-tings (light quality)* 150 mm maximum dia-meter. These are intended for gas appliances not exceeding 45KW.
BS 835:1973 *Asbestos-cement flue pipes and fit-tings (heavy quality)*. These have thicker walls than the BS 567 pipes.
BS 569:1973 *Asbestos-cement rainwater goods*.

BS 2777:1974 *Asbestos cement cisterns*, in sizes 17–701 litres.
BS 4656:1981 *Asbestos-cement pipes, joints and fittings for sewerage and drainage*. These are manufactured in three classes, with differing crushing strengths, in diameters from 100 mm to 2500 mm. The pipes are suitable for gravity flow at atmospheric pressure.
BS 3717:1972 *Asbestos-cement decking for roofs*
BS 486:1981 *Asbestos-cement pressure pipes and joints*. These are used for water, sewage pump-ing mains and similar purposes. Diameters are 50–2500 mm. Pipes are supplied with or with-out protective coatings.

Asbestos-cement products not covered by BSs include 300² mm injection moulded tiles for foot traffic on mastic asphalt and built-up felt roof-ings. Specially moulded products have included: large window louvres, corrugated profiles to sup-port and protect river banks, and fish tanks. Bitumen-impregnated asbestos-cement boards with matt and high gloss enamel finishes are suitable for electrical switchboards.

(b) Asbestos/inorganic agent bonded products

Products of this type are described by BS 3536:Part 2:1974 *Asbestos insulating boards and asbestos wallboards*.

These boards contain a greater proportion of asbestos fibre than asbestos-cement sheets (BS 690), together with mineral agents and inorganic fillers. They may contain cellular material and colouring matter (BS 1014). They must be non-combustible (BS 476).

One side must be smooth or decorated. The boards are intended mainly for internal use, exter-nally they should be protected from moisture.

Insulating boards

These boards (*Type 1*), are used primarily for thermal insulation and fire protection. Moisture absorption is high with associated loss of thermal insulation, but water causes no deterioration and the material regains its original properties in drying out. They are stable in extreme tempera-tures and moisture conditions. Boards can be nailed without pre-boring, and with due precautions, sawn or *Surform* planed. The 6 mm sheet can be bent to a radius of about 2.5 m.

Wallboards

These boards (*Type 2*), are denser than insulating boards and are intended for use as wall and partition linings, which are not likely to be abraded in use.

The 5 mm sheet can be best to a radius of about 2.5 m.

Sizes and thicknesses

Insulating boards 6, 9 and 12 mm thicknesses are made up to 2099 × 1199 mm and 16, 19, 22, 25, 28 and 30 mm thicknesses up to 2999 × 1199.

Wallboards 3.0, 4.5, 6, 9 and 12 mm thicknesses are made up to 2999 × 1199 mm.
Minimum sizes are 5 mm less in each case.

Densities, thermal conductivities and bending strengths are stated in table 88.

(c) Resin bonded asbestos sheets

These consist of synthetic resin impregnated asbestos paper, fabric or felt, bonded under heat and pressure. Dependent upon the resin employed laminates resist temperatures up to 220°C and absorb 0.2 to 1.0 per cent water after twenty four hours immersion.

The laminates can be machined if appropriate health precautions are taken.

(d) Sprayed asbestos

BS 3590 which required at least 55 per cent asbestos content was withdrawn in 1978. Existing sprayed asbestos should now either be sealed and kept sealed, or it should be removed, in both cases by licensed contractors – see page 238.

Asbestos in sprayed coatings has been replaced by other fibres or coatings containing lightweight aggregate such as exfoliated vermiculite are being used instead.

(e) Bitumen bonded asbestos

Bituminous roofing and dpc felts are described in BS 747:1977 and BS 743:1970 respectively.

9 Rock wool

This fibrous material produced by heating and spinning rock is used as thermal insulation – see page 42.

10 Slag wool

This product which is spun from molten blast furnace slag is similar to rock wool – see page 42.

11 Bituminous products

The bituminous products comprise; bitumen, natural or derivative, coal tar and pitch. Their general properties are discussed first followed by description of each material.

Properties

(a) Bituminous products are usually dark brown or black and composite products rely for any colour upon aggregate or pigments.

By the addition of pigments to 'albino' bitumens, which are light grey, colours usually confined to reds and brown, as possible.

(b) They resist the passage of water and water vapour. Materials applied hot or constituted with a solvent are more likely to be impervious than bituminous emulsions. To avoid flow, bituminous tanking materials must be supported by structural walls or floors in positions such as basements where water is under pressure.

(c) Although bituminous materials are combustible, composite products such as mastic asphalt and pitch mastic are not readily ignited and do not support combustion.

(d) In general, bituminous products are durable materials resistant to acids, alkalis, sugar, milk and brewing liquids and natural bitumens are more durable than derivate ones. Although pitch is more resistant than bitumen to fats and oils contact with these should be avoided. Oil in floor polishes can be damaging and oil in ordinary paints causes crazing. Conversely, bituminous paints 'bleed through' ordinary paints.

(e) Bituminous products, particularly pitch, are damaged by heat and sunlight. Thermal insulation below roof coverings prevent heat being absorbed by the structure and the covering becomes hotter and is liable to deteriorate in consequence. To reduce this risk, exposed surfaces should be finished with a light-reflecting material, eg lime wash and tallow, suitable aluminium paint or with mineral particles or stone chippings.

(f) Bituminous products flow under mechanical stress, particularly at high temperatures. The appropriate grade must be used for industrial floors, damp-proof courses and wherever resistance to flow, spreading and indentation is required.

(g) On the other hand, bituminous products embrittle at low temperatures. Tanking to basements is normally free from damaging stresses, but roofing should be isolated from structural movements, eg by laying mastic asphalt on sheathing felt.

Thermal stresses in roof coverings are further reduced by placing insulation over, rather than under, the impervious layer – see the *inverted roof MBS: Finishes*, chapter 7.

(h) Bituminous products are damaged by over-heating before use but to avoid cooling during laying, they should be heated as near the work as possible.

(i) Natural bitumen and coal tar products may be poisonous and only specialized products are suitable for contact with drinking water. Coal tar contains phenol which is toxic to plants and moulds. This is of value in coal tar creosote (BS 144:1973) see page 82, and some phenol is replaced in pitches which are required to inhibit plant growth.

(j) Pitch mastic evolves fumes when it is hot and in confined spaces care must be taken to avoid inhaling them.

(k) Although the bituminous materials are a 'family' it is important that they should not be intermixed or applied over one another.

Bituminous products comprise:

BITUMEN

Bitumen is a non-crystalline solid or viscous material comprising complex hydrocarbons which is soluble in carbon disulphide, softens when it is heated, is waterproof and has good powers of adhesion. It was used in building work in 3000 BC. Bitumen lacks 'body' and stability and is rarely used by itself but as an ingredient in mastic asphalt, adhesives and paints and as a saturant. *Bitumastic* is a registered trade name and should be used only for the proprietary product to which it refers. Confusingly, in the USA bitumen is called asphalt and because the cheapest thermoplastic tile contained bitumen, it is known there as an *asphalt tile*, whereas in this country an asphalt tile is a compressed mastic asphalt product containing aggregate.

Natural bitumen occurs in Lake Trinidad, West Indies, where it is associated with finely divided mineral matter as *Lake asphalt* and from which it is extracted and refined. It also occurs in limestones known as *rock asphalts* in France, Sicily, Switzerland and Germany.

Derivative bitumens are distilled from petroleum.

BS 3690:Part 3:1983 specifies *Mixtures. Bitumens for building and civil engineering*.

Uses

Mastic asphalt

The ingredients in mastic asphalts are:
Suitably graded aggregate:
Cushed limestone (or silica rock for acid resistance) and/or,
Natural asphaltic rock, a limestone which contains bitumen (*rock asphalt*),
Asphaltic cement – natural, or derivative bitumen,
Fillers, to reduce temperature movement and
Fluxing oils are sometimes added to adjust hardness.

Natural rock mastic asphalt, as it is called, carries more bitumen without loss of stability and is more durable and also more costly, than a mastic based on ordinary limestone.

Mastic asphalts are used in building for:
Dpcs – see *MBS: S and F, Part 1*, tanking of basements – see *MBS: S and F, Part 2*.

Floorings – see *MBS: Finishes*, chapter 1.
Roof coverings – see *MBS; Finishes*, chapter 7.
British Standards for mastic asphalt include:

BS 988, 1076, 1097, 1451:1973 *Mastic asphalt for building (limestone aggregate)*.
BS 6577:1985 *Mastic asphalt for building (natural rock asphalt aggregate)* specifies requirements for roofing, tanking and flooring grades, including coarse aggregate where applicable.
BS 1446:1973 *Mastic asphalt (natural rock asphalt fine aggregate) for roads and footways*.
BS 1447:1973 *Mastic asphalt (limestone fine aggregate) for roads and footways*.
BS 594:Parts 1 and 2:1985 *Hot rolled asphalt for roads and other paved areas* (known in many parts of the USA as bituminous concrete).
BS 4987:1973 *Coated macadam for roads and other paved areas*.

Other uses of bitumen include:
1 Paints – see *MBS: Finishes*, chapter 6.
2 Damp-proof membranes,
3 Adhesives for wood-block flooring, cork slabs insulating linings and felts,
4 A saturant for roofing felts (BS 747:1970), for use in the British Isles – including bitumen-polyester felts.
5 A saturant for damp-proof courses (BS 743:1970).
BS 6398:1983 *Bituminous damp proof courses for masonry* includes classification of base materials and refers to *high bond strength dpcs*.
6 A saturant for sheathing and sarking felts.
7 Waterproof building papers (BS 1521:1972),
8 Building papers, breather type (BS 4016:1972),
9 Bonded glass-fibre products – see page 235,
10 Impregnated foamed polyurethane joint fillers,
11 Impregnated cork joint fillers.

COAL TAR

Coal tar runs at a lower temperature than bitumen and oxidizes much more easily. Ordinary coal tar is suitable for many of the applications of bitumen other than the very heavy duties.

Tar-latex compositions, eg *Synthapruf*, have very good adhesion and slight permeability.

Other uses include:

BS 4987:1973 *Coated macadam for roads and other paved areas*
BS 76:1974 *Tars for road purposes.*

PITCH

Pitch is the residue after distilling tar from coal *fluxed back* with some of the by-products. Coal tar and pitch soften at lower temperature, are less plastic, less able to resist temperature variations and are less costly than bitumen. Pitch is not suitable for damp-proof courses and as roofing it should be used only for temporary work.

BS 1310:1974 describes *Coal tar pitches for building purposes.*

Uses include:

An ingredient in pitch mastic flooring – see BS 5902:1980 and *MBS: Finishes*, chapter 1.
Paints, see *MBS: Finishes*, chapter 6.
Fluxed, eg for damp-proof membranes.
A saturant for felts (BS 743:1970 *Materials for dpcs*).
BS 2760:1973 *Pitch-impregnated fibre pipes for below and above ground drainage.*
BS 4108:1973 *Pitch-impregnated fibre conduit, 50–150 mm bore.*

Glass is used in building mainly as flat glass, and for products such as lenses, glass fibres and foamed (cellular) glass. Another use is vitreous enamel coatings on metals. Glass was common in Egypt five hundred years BC but, in this country, little glass for windows was used in secular buildings before the Elizabethan period. Up to the end of the eighteenth century *crown glass* was the most common form of glazing. Reasonably flat sheets were produced by rotating molten glass, firstly on the tube used for blowing and then on a solid rod. This resulted in a circular sheet with a *bullion* in the centre surrounded by concentric striations due to air bubbles and variations in thickness and consequently in colour. Thus, the process severely limited both clarity of vision and the sizes of the panes cut from the glass surrounding the bullions. The latter which was originally the poor man's glass is now 'collectable'. The machine-made simulations which are available today lack the visual interest of 'antique' glass. Later *blown sheet glass* was produced by blowing glass into cylindrical moulds, allowing the glass to cool, and then cutting, reheating and flattening it. No sheet was truly flat, causing the interesting and sparkling reflections seen in adjacent panes of old windows. Clearly, ordinary modern glass should not be used in restoring old buildings. By 1851 mechanical production of blown glass made possible the glazing of the vast Crystal Palace with panes as large as 1245 mm × 254 mm. The first continuous process of drawing glass was developed in 1913 and for sixty years most flat glass was *drawn sheet*. In 1923 Pilkington Flat Glass Ltd developed a continuous process of *grinding and polishing* rough cast glass on both sides to form plate glass and in 1959 they developed the *float glass* process by which most clear glass is produced today.

Glass and its use in modern building is dealt with in:

Window glass design guide edited by D Turner, Architectural Press Ltd

BS 952 *Glass for glazing*

Part 1: 1978 *Classification,*

Part 2: 1980 *Terminology for work on glass,*

Part 3: *Quality of glass for glazing* is being prepared.

BS 6262:1983 *CP for the glazing of buildings* is concerned with general glazing.

BS 5516:1977 *Code of Practice for Patent glazing* Publications of Pilkington Flat Glass Ltd, St Helen's, Lancashire, and glass merchants.

This chapter deals with:

Manufacture

Molten glass batches produced continuously in 'tanks' consist mainly of sand, soda ash, limestone and dolomite, a small amount of alumina, a few residual materials and broken glass (*cullet*). Iron oxide in ordinary glass made in this country gives it a slight green tint – although this can be removed by the inclusion of manganese with arsenic or selenium. Other additions impart special properties such as tints, opacity, sparkle, chemical and fire resistance, low thermal movement, heat absorption or rejection and improved working properties.

Products are produced by several basic processes:

Blowing by mouth, for *antique* glasses and by machine for bottles, etc.

Drawing ordinary *sheet glass* for windows.

Grinding and polishing rough cast glass.

Today, the main processes are:

Rolling for *rough cast* and *patterned* glasses (wire can be incorporated in the glass during rolling).

Floating to give parallel and flat surfaces so that vision is not distorted.

Pressing for lenses, hollow glass blocks, etc.

Extrusion, blowing and drawing of fibres for insulation and textiles.
Foaming – see page 266.

Annealing

After forming to the shape required, glass must be cooled slowly (*annealed*) to relieve the strains which would otherwise result. The description *annealed glass* is used to distinguish ordinary glass from the *toughened* product – see page 258.

Properties

Headings are:

Appearance	Ultra-violet trans-
Density	mission
Melting point	Durability
Visible light trans-	Strength properties
mission	Thermal movement
Solar heat trans-	Thermal insulation
mission	Sound insulation
	Behaviour in fire

Appearance

Ordinary glass is transparent and more or less colourless. Transparent, translucent and opaque glasses can be coloured in four ways:
1 *Pot* colour is uniform throughout its thickness.
2 *Flashed* glass has an extremely thin layer of transparent coloured glass applied to one surface during blowing.
3 Coloured ceramic pigment fused on one side.
4 *Painting* – see page 271.

Surfaces of glass which are formed directly from the kiln are bright and lustrous, ie *fire-finished*. This applies to *antique* blown glass, drawn sheet and float glass.

Density

2560 kg/m³ (for comparison: *Perspex* (ICI) 807, Aluminium 2771, Steel 7850 kg/m³).
Glass weighs 2.5 kg/m²/mm thickness.
Flat glass is often described by 'weight', ie mass per unit area.

Melting point

1500°C approx. (Aluminium 660, Steel 1900°C)

Solar heat transmission

Ordinary glass is relatively transparent but solar heat rejecting glasses are available – see page 261.

BRE Report *Hazards from the concentration of solar radiation by textured window glass* states that the lens effect of antique bullions (see above) has caused ignition of fabrics. This is less likely to occur if bullions are varnished or if mock bullion glass is used – see page 265.

Visible light transmission

Average percentage values for ordinary glass and *Perspex* (ICI) made in UK are:

	Directed light	Diffuse light
4 mm clear float	92	87
6 mm clear float	90	85
6 mm Georgian polished wired	80	85
Patterned glasses	–	80
6 mm Georgian wired cast	–	75
Perspex (for comparison)	92	85

The refractive index for glass is about 1.52.
Obscuration is provided by patterned glasses – see page 257.

Ultra-violet transmission

Ordinary glass transmits a very small proportion of the sun's ultra-violet rays, and virtually none in the so-called *health band*, ie about 35 per cent at 334 nonometres and less than 2 per cent at 313 nono-metres.

Durability

Glass is extremely durable in normal conditions. BS 952:Part 1:1978 *Classification of glass for glazing* states 'All window glass should be of such quality that surface deterioration will not develop after glazing under normal conditions of use, provided the glass is cleaned at reasonable intervals.' It is, however, attacked by hydrofluoric and phosphoric acids and by strong alkalis, eg caustic soda. Alkaline paint removers which are not properly cleaned off, and even water running onto glass from new concrete, may cause trouble.

Glass can be damaged by permanent condensation in unventilated cavities between sheets of glass and they should not be allowed to remain wet when stacked face to face.

Strength properties

Glass in building is required to resist loads including wind loads, impact by persons and animals and sometimes thermal and other stresses.

Ordinary glass is elastic right up to its breaking point, so that an ordinary single glazed pane will safely deflect up to 1/125th of its span. On the other hand, glass is completely brittle so there is no *permanent set* which in ductile materials gives warning of impending failure. Strength in tension is very dependent on minute flaws, particularly those on surfaces and even glass of uniform manufacture varies very widely in strength so that safe stresses have to be computed from the results of tests on a very large number of samples. Another characteristic is that resistance to shock loads is about twice that to static loads which can be withstood indefinitely. Patterns, and wire in glass reduce its strength and opaque glasses are appreciably weaker than clear glass. The support provided to glass also affects its strength performance. Solar control glasses are particularly liable to thermal stress cracking – see page 261. Allowable working stresses are usually calculated for a risk of not more than 1 per cent if glass is subjected to the most severe combination of loadings which could ever foreseeably occur, the likelihood of such a combination happening simultaneously being so remote that risk is virtually eliminated.

It is important to realize that maximum manufacturing sizes are generally larger than safe glazing sizes, which depend upon the stresses to which the glass will be subjected in service. These stresses vary with the uses of buildings, the number and behaviour of people likely to be near the glass, its area, length:breath ratio and the method of fixing.

CP3:Chapter V:Part 2:1972 gives *Basic wind speeds* in m/s (ie three second gusts 10 m above the ground which are likely to occur once every 50 years). Examples for towns are:

38 London
41 Bournemouth
44 Coventry, Leicester, Plymouth
45 Cardiff, Exeter, Newcastle on Tyne
46 Blackpool, Leeds, Liverpool, Sheffield, Swansea
47 Bradford
49 Aberdeen, Inverness
52 Londonderry

External glazing A procedure for determining the thickness of transparent glass for vertical windows glazed on four edges is given in CP 152 as follows:

Determine the *design wind speed* by multiplying the *basic wind speed* by three factors:

1 *Topography factor* – usually 1.0, but for steepsided enclosed valleys sheltered from all winds 0.9 and for very exposed hill slopes and crests where acceleration of the wind is known to occur 1.1.

2 *Ground roughness factor* – for actual height above ground and nature of terrain values range from 0.56 at 3 m or less above ground in city centres to 1.27 at 200 m above the ground in open country without obstructions.

3 *Life factor* – usually taken as 1.0.

Read the maximum wind loading which results from the calculated *design wind speed*, ie:

Maximum wind loading N/m^2	Design wind speed m/s
370	20
570	25
830	30
1 130	35
1 470	40
1 860	45
2 300	50
2 780	55
3 320	60
3 880	65
4 500	70

Calculate the *glass factor* – area of pane (mm^2) divided by its perimeter (mm).

Read from table 5 in BS 6262 the *thickness of glass* using the *maximum wind loading* and *glass factor*.

Figures 42–45 reproduced by kind permission

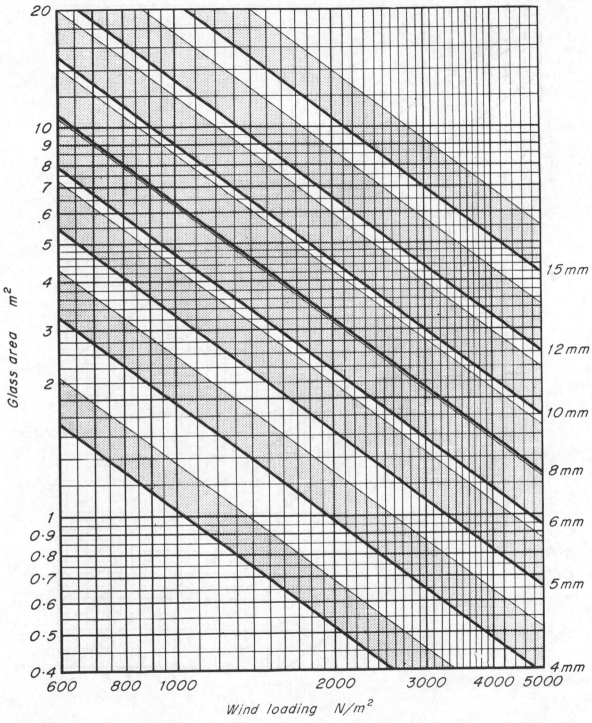

42   *Minimum thickness of Clear glasses annealed (except laminated) subject to 3 second mean wind loadings (Glass held vertically and at 4 edges)* (Pilkington Flat Glass Ltd)

249

GLASS

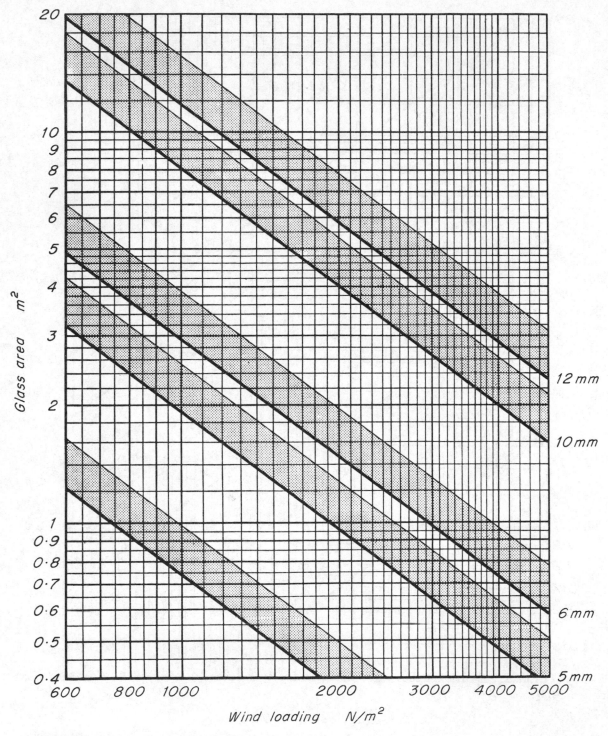

43 *Minimum thickness of wired, rough cast and patterned glasses subject to 3 second mean wind loadings (Glass held vertically and at 4 edges)* (Pilkington Flat Glass Ltd)

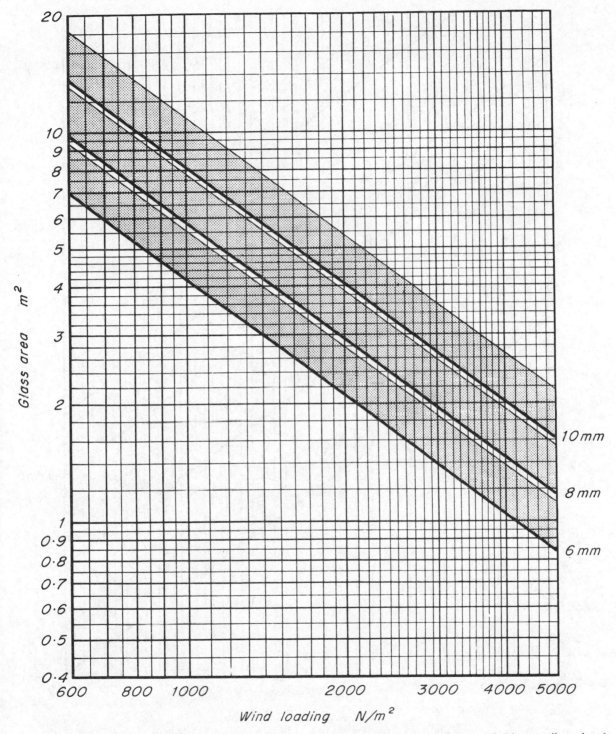

44   *Minimum thickness of Laminated glasses subject to 3 second mean wind loadings (Glass held vertically and at 4 edges)* (Pilkington Flat Glass Ltd)

251

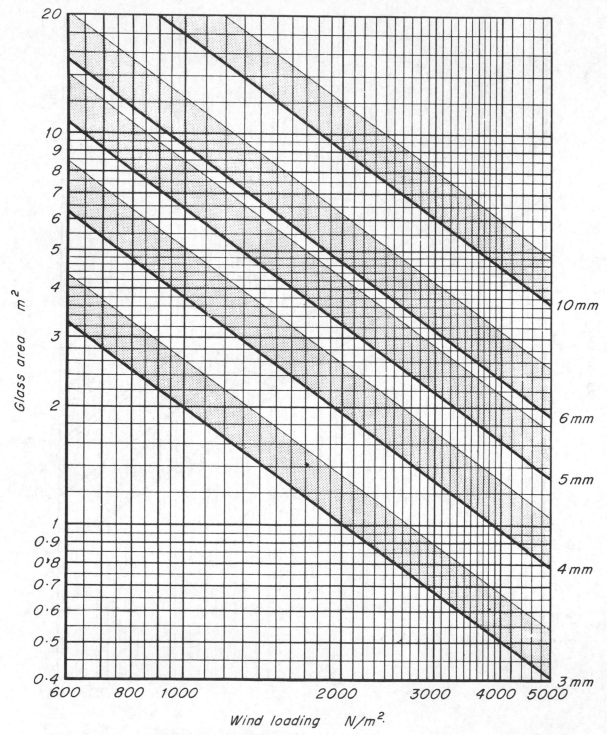

45   *Minimum thickness of Clear glasses in Double glazing subject to 3 second mean wind loadings (Glass held vertically and at 4 edges)* (Pilkington Flat Glass Ltd)

of Pilkington Flat Glass Ltd show minimum 'safe thickness' as recommended by BS 6262 for the more common glasses glazed on four edges in panes of stated area and length:width ratio subjected to various wind loadings. The thick black lines are for square panes and the extremities of the shaded bands are for length:width ratios of 3 to 1 and more. Proportional positions on the bands can be taken for ratios between 1 to 1 and 3 to 1.

BS 6262 contains nine graphs for various glasses used in single and double glazing.

Thicknesses for clear glass in sealed double glazing units having panes of equal thickness, can be obtained by doubling the wind loadings on figure 42. Thicknesses for patterned glass in sealed units can be obtained similarly from figure 42 provided the other glass in the unit is at least as thick as the patterned glass.

Thicker glass is needed where panes are fixed on less than four edges, at more than 15° from the vertical, where a life exceeding fifty years is expected or where the risk of accidental or wilful damage is high, eg doors in schools and glazing near floors. The technical service of Pilkington Flat Glass Ltd should be consulted in these and other cases which are not straightforward.

Safety from injury

Broken glass is a serious potential hazard. In 1975, 30,000 people were injured, some fatally, by glass in windows and doors in their homes in the UK. The recommendations of BS 6206:1981 *Impact performance requirements for flat safety glass and safety plastics for use in buildings* designed to minimise such accidents, should be carefully observed. Where the risk of breakage cannot be sufficiently reduced, eg by protective rails or by indicating its presence by visible motifs, suitably thick glass should be used or glass which is unlikely to cause serious injury if it is broken, ie toughened glass, laminated safety glass, or where it is required to be fire-resisting, glass such as copper-light or wired glass which do not cause injury so readily if they are broken. Alternatively, strong and durable scrim can be fixed with a durable adhesive to the back of ordinary glass, but the result is unlikely to be as safe as laminated glass.

BS 6206 requires *safety glass* and plastics to either 'not break' or 'break safe', ie into relatively harmless pieces or suffer insufficient penetration to cause serious injury. Products are classified as A, B or C relative to three 'dropheights' of a standard impactor. Safety glass should be permanently marked so it can be identified after it has been fixed.

The following thicknesses are generally considered to be reasonably safe for ordinary internal glazing except where otherwise recommended:

Thickness mm	Maximum area m²	
	Annealed glass	Toughened or laminated* glass
3	0.2	} not
4	0.3	} recommended
5	0.9	1.0
6	0.7	3.0
10	1.5	6.0
12	3.0	7.0

*Plastics interlayer not less than 0.76 mm thick

BS 6262:1982 requires that if annealed glass is used for large sliding doors, including patio doors for domestic buildings, it should be at least 6 mm thick, or two 5 mm thicknesses for double glazing. The BS points out that thicknesses determined by wind loadings may require to be increased where there is an appreciable risk of impact by humans, especially by boisterous children.

Where glass might not otherwise been seen, its presence should be indicated to avoid the risk of accidental human impact.

Common *risk areas* in buildings are identified as:

1 Doors in general use for access to rooms and passageways deemed to be *Major single risk areas.*

2 *Fixed side or subsidiary glass panels associated with doors which might be mistaken for a door or unimpeded access.*

Any glass in fully glazed doors and door side panels likely to be subjected to high pedestrian traffic should be toughened or laminated wired glass.

3 *Low level glazing* which is unprotected by barrier rails or by being recessed etc. in corridors,

landings, stairways and balconies subject to high pedestrian traffic up to 800 mm from a floor, or in balustrades where there is a difference in level up to 1100 mm is in *risk areas*. All balustrades must have sufficient resistance to impact to protect against falls and comply with BS 6180:1982 *Protective barriers in and about buildings*. All low level glazing should be laminated or toughened glass.

4 *Bathing and shower screens* in bathrooms and swimming pools, where people may slip on wet surfaces, are high risks and should be laminated or toughened glass.

5 *High level glazing* may be specially hazardous in circulation areas, large blocks of flats, gymnasia and similar situations, other than in single dwellings. In buildings used for specialist activities, where there is a risk of breakage to high level glazing, laminated, toughened or wired glass should be used.

Generally, roof glazing should be toughened, wired or laminated.

Security

Toughened glasses can be regarded as anti-vandal glasses. Laminated anti-vandal, anti-bandit and bullet resistant glasses are considered on page 263.

BS 5357:1976 is a *Code of Practice for installation of security glazing*.

Thermal movement

Because the coefficient of thermal movement for glass, of 76 to 80 $\times$ 10^{-7} per deg C, is lower than that of the materials in which it is normally fixed, allowance should be made for movement. Also, thermal stresses arising where one part of a glass pane is at a different temperature from other parts can lead to breakage. This is particularly likely where dark coloured glass, or clear glass with a dark background are exposed to the sun while the edges being shaded by beads are at a much lower temperature and are put into tension. Dark heat-absorbing frames are to be preferred to white or polished aluminium frames. A ventilated cavity behind glass helps in cooling it, and incidentally in removing condensation especially if it 'breathes' to the outside. If insulation material

must be in contact with the back of glass a toughened glass should be used. It is also important that the edges of such glass should be cut cleanly, nipped or shelled edges, particularly of heat-rejecting glass, create points of weakness in the critical shaded zone.

Thermal insulation

Although glass is dense and is a good conductor of heat (k = 1.05 W/m deg C) its surface resistances are high so that doubling the thickness of a 6 mm glass pane increases the overall thermal resistance by only 3 per cent. Double glazing almost halves the heat lost through a single pane, the optimum gap being about 20 mm. Typical double glazing provides thermal insulation approximately that of a 105 mm brick wall. Table 89 compares thermal transmittance through glazing and walls.

	'U' – W/m²deg C		
	S¹	N¹	Ex¹
Single glazing	5.0	5.6	6.7
Sealed double glazing with gaps:			
6 mm	3.2	3.4	3.8
12 mm	2.8	3.0	3.3
20 mm	2.8	2.9	3.2
80 mm glass blocks		2.5	
105 mm solid brick wall with 16 mm dense plaster		3.0	
102.5 mm brick outer leaf, 50 mm unventilated cavity, 100 mm lightweight concrete block inner leaf 16 mm dense plaster		0.96	

¹Exposures: S–sheltered N–normal Ex–severe

Table 89 Thermal transmittance of glazing compared with walls

The Building Regulations 1985, Approved Document L, limits the areas of glazing and relates them to heat losses through walls.

Because double glazing provides better thermal insulation from conducted heat than single glazing, condensation is less likely to form on the room side glass surface of the assembly. However, if the inner and outer panes are not hermetically sealed at their edges water vapour can condense on the inside of the outer pane as it does on single glazing. This tendency is reduced if the cavity is ventilated to the outside and the inner pane is sealed from the room atmosphere.

Effective U-values of glazing can be regarded as 'heat out' minus useful 'heat in'. The latter is dependent on the direction the glass faces. In the case of *Kappafloat* double glazed units, the normal U-value of 1.9 W/m² deg C is reduced to an effective U-value of about 0.1 W/m² deg C.

The performance of double glazing is improved 50 per cent by *Kappafloat* (Pilkington Flat Glass Ltd) having low emissivity reflective coating (see page 260) making it as efficient as triple glazing.

Detailed data taking into account percentage frame area and frame material are contained in Pilkington Flat Glass Ltd publication *Glass and Insulation*

Factory sealed units contain dry air so that condensation cannot form inside the cavity and dust cannot enter. Variations in the external air pressure and temperature subject the glass to stresses which tend to break down the less effective edge seals.

Multi-pane glazing units are discussed on page 263. Hollow blocks, page 264, and channel units in double thickness, page 265, are other forms of double glazing.

Sound insulation

Table 90 gives sound reduction values for various weights of glass and types of windows. It will be seen that the normal air leakages which occur around 'closed' opening lights are very important. Sealed double glazing is therefore necessary for a superior level of insulation. Cavities less than 100 mm wide are comparatively less effective for sound than they are for thermal insulation. The sound insulating advantages given by wider gaps can be further improved by the use of sound absorbents around the edges of windows. Also, one thicker glass pane eg 12 mm is effective particularly at the low frequencies characteristic of traffic noise.

Hollow glass blocks (see page 264), give 35 to 40 dB reduction.

Behaviour in fire

Although non-combustible, ordinary glass breaks and later melts in fires, and double glazing shows

Glass thickness mm	Single glazing		Double glazing with gaps less than 50, 100 and 200 mm										
			Opening lights – closed				Fixed lights and opening lights with seals[1]						
	Opening lights closed	Fixed lights and opening lights with seals[1]	No absorbent material to sides of gap		With absorbent material to sides of gap		No absorbent materials to sides of gap			With absorbent material to sides of gap			
			100 mm	200 mm	100 mm	200 mm	less than 50 mm	100 mm	200 mm	less than 50 mm	100 mm	200 mm	
3	18	23	22[2]	24[2]	28[2]	31	26[2]	34	38	28[2]	37	41	
4	18	25	22[2]	24[2]	28[2]	31	28	36	40	30[2]	39	43	
6	18	27	22[2]	24[2]	30[2]	33	30	38	42	32[2]	41	45	
10	20	30	25	27	33	36	33[2]	41	45[2]	35[2]	44	48[2]	
12	22[2]	31	—	—	—	—	—	—	46	—	—	49	
25	—	34	—	—	—	—	—	—	—	—	—	—	

[1]Air-tight weatherstrip or special cushion seals   [2]Estimated values

Table 90   Approximate (± 3 dB) average sound reduction in decibels (dB 100–3150 Hz)

Up to 30 min.	Up to 60 min.	Up to 90 min.
6 mm wired glass	Copperlight glazing	6 mm wired glass
(a) In panels up to 0.4 m² fixed with wood to metal beads, or with glazing compound and springs or clips, in timber frames not less than 44 mm wide or thick (including rebates) (b) In panels up to 1.2 m² with metal beads in metal frames (former GLC Bylaws 1979)	Annealed glass in squares up to 0.015 m² in direct contact with metal cames. Individual copperlights up to 0.4 m², but composite panels can be formed using metal dividing bars	(a) In panels up to 1.6 m² with metal beads and in metal frames having melting points not less than 980°C[1] (b) Laminated panels up to 1.6 m²
98 mm glass blocks Laid in cement-lime-sand mortar, with reinforcing mesh in every third horizontal joint, in panels up to 2.4 m wide and high. Panels recessed 12.5 mm at sides and head into surrounding non-combustible construction, bedded in glass fibres and with non-hardening mastic between faces of panels and recesses (former GLC Bylaws 1979)		*Fire resisting glass* – suitably fixed up to 1.6 m²

[1] ie not aluminium or lead

Table 91   Guide to Fire Resistance of 'fixed shut' glazing – based on BS 6262:1982 and former GLC Bylaws 1979. Note: Periods relate to flame and smoke penetration – not to heat transmission – see page 39.

no significant advantage over single glazing. Glass is a good conductor of heat and radiation from glass can ignite any combustible materials, eg wood beads if used, on the side remote from the fire. Also, radiation renders escape routes impassable.

The fire resistance of panes varies with the type and thickness of glass, size, height:width ratio and with the method of fixing, types of frame and surrounding construction.

Certain types of glass in sufficient thicknesses and suitably fixed, can provide a degree of *fire resistance* in terms of integrity and resistance to collapse, but not insulation. (See table 91 for fixed-shut glazing.)

BS 6262 deals with fire hazards associated with glazing in buildings.

Types of glass and products

We consider the types of glass under the following headings:

1   *Translucent glasses* (rolled) – page 257
Rough cast
Patterned
Wired rough cast
Wired patterned
2   *Transparent glasses* – page 258
Drawn sheet
Polished plate and float
Wired clear plate (ground and polished wired rough cast)
3   *Special glasses and products* – page 258
4   *Glass fibre products* – page 264
5   Foamed glass – page 264

1 Translucent glasses

Rough cast Rough cast glass, or just *cast glass* formerly cast on sand beds, is now made by passing molten glass between rollers which impart a characteristic pattern on one side. The other side is smooth, but the method does not permit optically true surfaces to be obtained. It is made in only one quality. Tints are available for solar control or for decorative purposes.

Table 91 shows the normal thicknesses of which 6 mm is the most common but up to 38 mm can be ordered.

| | | *mm* | | | | | | | | | |
	2	2.4	3	4	5	6	10	12	15	19	25
Translucent glasses											
Rough cast			•	•	•	•	•				
Patterned glasses			•			•					
Armour glass											
Patchwork					•						
Cotswold							•				
Wired rough cast						•[2]					
Wired patterned						•					
(*Linkon*)											
Diffuse reflection			•								
Armourclad–											
float or rolled						•	•				
Transparent glasses											
Drawn sheet	•[1]			•	•						
Clear float			•	•	•	•	•	•	•	•	•
Wired clear plate						•					
Spectrafloat						•	•	•				
Anti-sun grey											
green and bronze				•		•	•	•			
	2	2.4	3	4	5	6	10	12	15	19	25
						mm					

[1] Not recommended for general glazing
[2] Also available 7 mm thick

Table 92 Thicknesses of glasses

Wired rough cast glass Wire mesh, electrically welded at the intersections, is embedded centrally in the thickness of rough cast glass during rolling. The wire does not reinforce glass but holds the pieces together in the event of breakage. Wired glass cannot be toughened but to prevent injury from flying fragments it can be used in doors in schools and similar situations. Wired glass is normally used for roof lights, except small domes. Broken wired glass in roofing should be promptly replaced to avoid sudden collapse if the wires rust. Fire resistance of wired glass is given in table 90.

Messrs Pilkington's product has 13 mm sided square wire mesh known as *Georgian* pattern in 6 and 7 mm rough cast glass in sizes up to 3710 mm × 1830 mm.

Patterned glasses

The expression *patterned glass* is conventionally applied to products having patterns impressed by rollers on molten glass.

In the first type of glass there are fifteen patterns, but in particular cases effective choice is often limited by non-availability of a toughened form, desired thickness, degree of obscuration, tints or of wire in the glass.

Ease of cleaning must also be considered, remembering that patterns such as the *reeded* pattern with grooves which are shallow and parallel, are easy to clean, unlike those with small and deep depressions at frequent centres.

Thicknesses of 3 mm and 6 mm are standard. Eight patterns are available in amber tint, and ten toughened.

Pilkingtons' patterned glasses are classified from (a) to (c) below by increasing degrees of obscuration.

(a)	(b)	(c)
Flemish	Autumn	Patchwork
Reeded	Deep Flemish	Linkon
Hammered	Driftwood	Everglade
Pimpernel	Arctic	Cotswold
Sycamore		Stippolyte
		Mayflower

Obscuration depends on the depth and complexity of patterns, the relative light intensities on the two sides of the glass and the distance of the object from the glass. Where an obscured glass is required in thicknesses greater than those available in rolled patterned glasses, rough cast, acid etched, and sand-blasted surfaces are available.

It is advisable to see a sample of an obscured glass before specifying it.

2 Transparent glasses

Clear sheet glass (flat drawn sheet) Since 1913 glass has been drawn up vertically from a tank of molten glass into an annealing tower by wheels which grip the edges of the sheet, the rate of drawing determining the thickness of the glass. At the top of the tower the glass is cut into suitable lengths.

BS 952:1964 stated that: 'Sheet glass has natural fire-finished surfaces but because the two surfaces are never perfectly flat and parallel, there is always some distortion of vision and reflection'. The effects of this can be minimized by glazing the glass with the line of draw parallel to the ground. In the UK the drawn sheet process is being superseded by the float process and it is now used only for 3 mm and 4 mm thicknesses. 5 mm and 6 mm glass is imported.

A limited range of tints is available.

Clear plate glass BS 952 describes this glass as having: 'Flat and parallel surfaces providing clear undistorted vision and reflection, produced either by grinding and polishing thick rough cast glass or by the float process'.

Polished plate glass is produced by passing a ribbon of rough cast glass through machines which in turn, grind and smooth the two sides simultaneously.

The process is now used only for thicknesses over 25 mm and up to 38 mm and for 6 mm polished *wired clear plate* in 'Georgian' pattern which is normally available in sizes up to 3300 × 1830 mm.

There is only one quality and thickness of wired polished glass, but it can be ordered to be selected so that the wires are true and square, and it can be cut so the wires are aligned either vertically or horizontally in adjoining squares.

Float glass has been produced since 1959 by floating a continuous ribbon of molten glass across the surface of molten tin, in a controlled atmosphere, whereby the upper surface is fire-polished and the other surface is polished by contact with the metal. Both sides are flat and

parallel. Pilkingtons do not select for quality, but glass merchants may do so.

Today, most flat plate glass in thicknesses up to 25 mm, including solar control glasses is float glass.

3 Special glasses and products

(a) Toughened or tempered glass When suitable glass is heated and then suddenly cooled by jets of cold air the surfaces are put into compression and the interior into tension. Most toughening is now done with the glass horizontal. Where glass is suspended by tongs there is some distortion.

Toughened glass is much stronger, tougher and also more flexible than the original annealed glass.

Where it is fixed normally, resistance to applied loadings – depending partly on the nature of the particular glass – is 4 to 5 times that of annealed glass of the same thickness. The modulus of elasticity of glass is not altered by toughening, so that deflection like safe strength in tension, increases about four times, which may cause unnecessary concern, for example to those who use toughened glass stairs!

Uses which exploit the strength of toughened glass include *suspended all-glass assemblies* with glass supporting fins, but without opaque frames, mullions or glazing bars. Pilkingtons designed a bronze tinted *Armourplate* assembly at the Willis Faber and Dumas complex, Ipswich, 15 m high with a 310 m periphery.

Single assemblies can be designed up to 18 m in height on a 1.5 m module and up to 25 m on a 1.2 m with no limit to length. Toughened glass back walls for squash courts are an important development and other uses include frameless balustrades and machinery guards.

Strength characteristics are not affected by sub-zero temperatures, but toughened glass is 'detoughened' if it is exposed to temperatures above 300°C for too long or too frequently and it is not fire-resisting. However it has very superior resistance to thermal shock. For example, it can be heated above 200°C and have cold water poured over it. It withstands splashing with molten metal on one side, and can be heated evenly to 250°C while the other side is at ordinary atmospheric temperature without suffering damage.

Toughened glass can also be regarded as anti-vandal glass.

The remarkable properties of toughened glass depend upon the balance of stresses being maintained, if the outer skin is disturbed the energy locked up in the glass is expended in shattering the whole piece into small fragments. Unfortunately the surfaces of toughened glass are no more resistant to scratching than those of ordinary glass and the edges are equally vulnerable to damage. All cutting to size, grinding, drilling and any other work on the glass must therefore be done before toughening is carried out.

Unlike those of annealed glass, fragments of fully toughened glass are small, do not have sharp edges and are unlikely to cause injury to individuals. Fully toughened glass is, therefore, one type of *safety glass*.

Loss of vision is a serious disadvantage in vehicle windscreens, although these may have partially toughened *zones* in which the particles are larger.

It must be remembered, that fixed toughened glass tends to deny emergency exits achieved by breaking ordinary glass.

Toughened glass may be recognised by a brand mark, or by a pattern of slight irridescent discolourations when viewed in certain lighting conditions, such as polarized light. Glass which is toughened while suspended by tongs shows small indentations along one edge.

Toughened glasses manufactured by Pilkington Flat Glass Ltd are:

Armourplate clear float glass. Doors are usually 12 mm thick.

Armourcast rough cast glass. Internal doors are usually 10 mm thick.

Antisun and *Spectrafloat* glasses see below.

Armourclad clear plate and cast glasses have coloured ceramic fused on one side. There are ten standard colours and others can be ordered. When used as a cladding the coloured surface is placed inside.

Armourclad insulating panels, have a 25 mm backing of glass fibres giving an overall U value of 0.5 W/m² deg C.

Patterned glasses Certain white and coloured patterned glasses, eg *Cotswold* and *Driftwood* can be toughened.

Limitations upon the size of holes and their proximity to edges and to each other are given in figure 46. Many configurations of notches are practicable but the radii of re-entrant corners must not be less than the thickness of the glass.

There are further restrictions on the diameter of holes in glass which is from 50 to 150 mm wide.

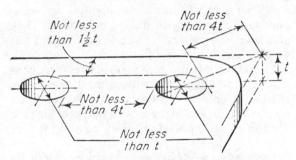

46 *Minimum dimensions appertaining to holes in toughened glass (t = thickness of glass)*

Inquiries should be submitted regarding holes, especially those more than 25 mm diameter.

In fixing toughened glass, edge clearance must avoid contact with hard materials.

(b) One-way glasses No glass gives one-way vision under all lighting conditions. *One-way* glasses depend upon the lighting on the 'viewing' side being less intense than on the reverse. They include:

Venetian mirror Broad strips of mirror which face the room alternating with narrow lines of clear glass. The usual proportions lie between 6 and 12:1.

Transparent mirror In this product a thin metallic film applied to float/plate glass reflects the major proportion of light which falls upon it. The metal film must be protected from abrasion.

Tinted glasses Glass with integral colour eg *Antisun* and laminated glass with a coloured interlayer.

(c) Diffuse reflection glass This glass is useful for glazing pictures and instrument dials. Both surfaces are very lightly textured so that when fixed within 25 mm of the object to be viewed it is completely transparent and there is no reflection.

(d) Solar control glasses Ordinary glass transmits a high proportion of short-wave solar radiation. Objects in buildings, which are heated by short-waves, re-radiate heat of a much longer wave-length to which glass is opaque. Thereby, heat is 'trapped' producing the well known 'hot house' effect.

Solar heat control glasses act mainly either by reflection[1] or by absorption. Pilkingtons' glass products reduce solar heat transmission from 16 to 65 per cent with improved comfort for occupants and reduced costs in air conditioning. They also give privacy and reduce sky glare and reflected glare.

Light transmission with these glasses is reduced from 20 to 78 per cent, with in most cases some change in colour. However, what is seen soon appears as normal to the human eye, provided the illusion is not spoiled by opening windows or doors!

Some of these glasses provide colour externally and the reflective types mirror surrounding buildings and landscape. See tables 93 and 94.

Pilkington Flat Glass Ltd products are:

Spectrafloat At a high temperature glass is electrically conductive enabling metal ions to be driven to a controlled depth and concentration into one surface of the flat glass ribbon. Because the metal ions are below the surface *Spectrafloat* can be used in single glazing, but the modified side should not be exposed to the weather.

At present normal production is limited to a bronze tint in 6, 10 and 12 mm thicknesses, which can be toughened if required.

Solarshield rejects heat mainly by reflection from a metallic film on the inner face of two panes which are laminated together.

The metal coating is usually put on the inside and the appearance from the outside is bronze, but if the coating is outside the external appearance is more gold than bronze. In both cases appearance is subject to subtle changes with variations in the weather and in the viewing angles. Pilkingtons strongly recommend that these characteristics should be judged by full-sized mock-ups erected on the building site.

Suncool is the name applied to double glazing units with 12 mm air spaces and a vacuum deposited coating in gold, coral gold, azure, bronze or silver on one of the airspace glass surfaces. Solar heat transmission is reduced to as little as 16 per cent with 20 per cent visible light transmission. The coating reduces the thermal transmittance (U) from 3.0 to 1.8 W/m^2 deg C.

Antisun softly body-tinted float glasses in green, bronze and grey give a measure of relief from solar radiation with a higher light transmission. As little as 5 per cent solar heat may be reflected on incidence, but as much as 51 per cent reradiated after being absorbed. *Antisun* glasses have been used in airport control buildings and the London British Telecom tower.

A pane of clear glass on the inside of heat absorbing glass reduces the amount of absorbed energy which is admitted into buildings and ventilators immediately above windows help to cool them. Fortunately, since there is usually more air movement outside most of the heat is dissipated outwards. All solar control glasses absorb more heat and expand more than clear glass, sometimes resulting in thermal stress failures. In particular, where edges are shielded from radiation by beads, and possibly lose heat by contact with a cold building having high thermal capacity, they remain cooler than the glass exposed to sunlight and the stresses resulting from the difference in temperature may exceed the breaking stress of the glass. Breakage may also follow a rapid rise in temperature on one side of the glass and from the insulating effect of multi-glazing.

It is essential that the edges of solar control glasses are cut cleanly to avoid weak spots from which thermal cracks would develop. A clean wheelcut edge is the strongest that can be consistently achieved in practice. Some 'feathering' of the edges of thicker glasses is acceptable, but not 'vented' edges and 'nipping' of edges of oversized panes can never be allowed. Special care is needed in glazing solar control glasses to accommodate thermal movement. For example, edge clearances must be at least 3 mm all round for panes where neither dimension exceeds 750 mm, and 5 mm where one dimension is greater. Similarly, at least 3 mm must be allowed between the faces of the glass and the upstands of rebates and glazing beads.

---

[1]Solar control reflective films are available which can be applied to existing glass. Great care must be taken to avoid the use of abrasives in cleaning.

Glass	Thickness	Visible light transmission / Total solar radiant heat transmission	Colour	Reflection	Absorption	Direct transmission	Total rejection
	mm	percentages		percentages			
Clear float – for comparison	6	87/84	clear	7	13	80	16
Anti-sun – body tinted	4	63/68	bronze	6	37	57	32
		54/66	grey	6	40	54	34
	5	78/65	green	6	42	52	35
Spectrafloat – surface modified	6	49/66	bronze	10	34	56	34
Anti-sun float – body tinted	6	75/60	green	6	49	54	40
		41/60	grey	5	51	44	40
		50/60	bronze				
Solarshield metal coated and laminated – sheet and float	6.4 *coolray*	38/38	gold	33	42	25	62
		17/35	grey	24	59	17	65
		14/33	silver blue	27	58	15	67
		20/24	gold and bronze	47	42	11	76
		20/18	deep gold	57	36	7	82
Spectrafloat – surface modified	10, 12	46/61	bronze	9	43	48	39
Anti-sun float – body tinted	10	24/48	grey	5	68	27	52
		32/48	bronze				
	12	18/44	grey	5	74	21	56
		26/44	bronze				
Solarshield metal coated and laminated float	8.4, 10.4	as 6.4 mm *Solarshield*					62-82

Note: The values are for radiation at normal or nearly normal incidence. *Shading coefficients* (the ratio of total heat gain to 0.87 the total heat gain of a notional glass 3 and 4 mm thick) enable designers to take into account different angles of incidence

Table 93 Performances of Pilkington Flat Glass Ltd single solar control glasses in order of thickness, type and solar rejection.

This subject is dealt with in *Glass windows Bulletin no. 10*, Pilkington Flat Glass Ltd.

(e) Fire resisting glass Wired glass is described on pages 257 and 258, glass blocks on page 264 and copperlight glazing on pages 265 and 266. Table 91 shows their performances in fire. These all obstruct vision in varying degrees but Pilkingtons' recently-introduced *Pyrostop* is fully transparent in normal conditions. The product

Unit		Designations			Solar radiant heat			
		Visible light trans-mission	Total solar radiant heat trans-mission	Colour	Refl-ection	Absorp-tion	Direct trans-mission	Total rejec-tion
		percentages			percentages			
6 mm *float* +	6 mm *float*	76/73		clear	12	24	64	27
4 mm *float* +	4 mm *Antisun float*	55/57		bronze	9	43	48	33
		48/55		grey	8	46	46	45
6 mm *float* +	6 mm *Spectrafloat*	43/56		bronze	12	43	45	44
	12 mm *Spectrafloat*	40/50		bronze	10	51	39	50
	6 mm *Antisun float*	44/48		bronze	6	59	35	52
		37/48		grey				
		66/48		green	7	57	36	52
	12 mm *Antisun float*	23/31		bronze	5	78	17	69
		16/31		grey				
	6 mm *float* with *Sun Cool* metallic layer on inner surface	47/33		azure	24	51	25	67
		33/29		silver	51	31	18	71
	Tinted glass usually outer pane	36/28		coral-gold	51	31	18	72
		18/26		silver	52	38	10	74
		33/25		bronze	35	47	18	75
		23/22		gold	38	47	15	78
		20/20		silver	59	31	10	80
		23/18		coral-gold	43	45	12	82
		18/17		coral-gold	42	48	10	83
		20/16		bronze	43	47	10	84

Table 94   Solar performances of typical Pilkington Glass Ltd double-glazing units in order of total heat rejection

incorporates multi-laminated panes of float glass and layers of intumescent material. It is 11 and 20 mm thick for internal use and 14 and 24 mm for external glazing.

Normally, the product is transparent but, when suitably fixed, in fire it prevents the transmission of radiant and conductive heat, flames and smoke for up to 1 hour.

(f) **Heat resisting glass** eg for doors to solid fuel space heaters, is usually a boro-silicate glass which, having a low coefficient of expansion is more resistant to high temperatures than ordinary and toughened glasses – although the latter is resistant to thermal shock (see page 258). Thickness varies and the glass may contain air-bubbles and other defects.

BS 952:1972 gave the following thicknesses and maximum sizes:

	Thickness mm	Normal maximum size mm
Blown	2.0–7.0	609.6 × 609.6
Pressed	5.0–10.0	508 mm diameter approx.
	12.0–15.0	254 mm diameter approx.

(g) X-ray resistant glass Usually a polished plate glass containing a large proportion of lead oxide, which has a high degree of opacity to X-rays.

BS 4031:1966 describes *X-ray protective lead glasses.*

(h) Opal glasses – now regarded as 'antique' glasses, may be white or coloured and vary from the faint milkiness of flashed opal to virtual opacity.

(i) Coloured opaque glass This is an easily cleaned finish for table tops, kitchen and bathroom walls but it cannot be toughened and is not made in the UK. Methods of glazing were described in CP 152:1972 by edge framing, clips or screws, or on mastic.

Fixing should be firm but not rigid and adequate allowance must be made in all cases for thermal movement. There should always be a clearance joint filled with mastic between the edge of an installation and surrounding structures.

(j) 'Antique glasses A wide range of clear and coloured glasses are hand made by traditional means and are therefore costly. Varying thicknesses cause variations in depth of colour, characteristic 'defects' contribute visual interest, and the surfaces are fire-finished and consequently lustrous. *Antique* glasses are used for decorative work, to diffuse vision and for restoration of old buildings in which modern glass would be out of place.

(k) Laminated glass Ordinary laminated glass products consist of two sheets of annealed glass cemented together with a polyvinyl butyral interlayer.

Unlike toughened glass, most laminated glasses can be cut to size and worked after manufacture. It is more resistant to breakage than wired glass, and because broken fragments remain attached to the interlayer, laminated glass can be described as a *safety glass.* (Wired glass is not a *safety glass.*) The safety aspect of laminated glass is further increased if toughened glass is used instead of annealed glass, although some loss of vision results. If it is not penetrated, laminated glass remains weather-resistant after being broken.

Special purpose laminated glasses include:

1 *Security* laminated glasses are designated: *anti-vandal*, *anti-bandit* and *bullet resistant*. It is important that the appropriate products are used in particular cases.

BS 5544: 1978 is a: *Specification for anti-bandit glazing*, intended to delay access for a short time, against manual attack.

BS 5051 deals with *Bullet resistant glazing*. Part 1: 1973 – *for interior use* and Part 2: 1979 – *for exterior use.*

BS 5327: 1976 is a: *Code of practice for installation of security glazing.*

2 *One-way vision* (see page 259) is provided by tinted interlayers or by tinted glasses.

3 *Solar radiation control* These products have either a thin metal deposit on an inner glass surface, or a reflective coating on a float glass. The glass or interlayer may also be tinted.

4 *Anti-fading* is provided by a special interlayer, 0.76 mm or more thick, which absorbs 99 per cent of ultra-violet radiation without loss of normal transparency.

5 *Sound control* A special interlayer gives better attenuation of sound at coincident and higher frequencies, although compared with 'solid' glass of the same thickness, the mean dB value is less.

6 *Obscuration* is provided by patterned or surface-treated glasses, or by diffusing interlayers. Designs can be included in interlayers.

7 *Fire resistance* A product which includes 6 mm wired glass has the advantages of safety glass and 90 minutes fire resistance is given by suitably fixed panes up to 1.6 m² in area.

8 *Manifestation* Parallel wires at 30 mm

spacings in the interlayer give visible warning of the presence of glass.

9 *Alarm* An electrical alarm can be activated if a wire element in the interlayer is broken.

(l) Double glazing BRE Digest 140 deals with: *Double glazing and double windows.*

The advantages of *double glazing* in reducing thermal loss and condensation and in providing sound insulation have been discussed on pages 254 and 255.

Where an outer pane has a heat reflective layer on its inner surface the reduced emissivity of the coated surface reduces transmission through the air space – see page 260.

Typical factory made units comprise two panes hermetically sealed at the edges with 3–20 mm gaps containing dry air.

Gaps filled with suitable gases can further reduce thermal transmittance and in future double glazing with a 6 mm gap may be equivalent to a brick cavity wall.

Variations in temperature and atmospheric pressure cause stresses which break down the less effective edge seals, while in extreme conditions differences between the atmospheric pressure and that of the enclosed air can cause explosion or implosion. As these possibilities increase with the width of the air space, thicker glass may be needed with gaps more than 5 mm. The manufacturers should be consulted if it is intended to take units to altitudes above 800 m.

Pilkington Flat Glass Ltd make hermetically polysulphide edge-sealed *Insulight* units, and *Suncool Insulight* solar control units with reflective interlayers in shades of azure, bronze, silver and gold. Other UK manufacturers also make insulating glass units. BS 5713:1979 describes *Hermetically sealed flat double glazing units.*

In new work, rebates with sufficient depth and upstand should be provided so that standard units can be used, but for glazing narrow rebates in existing sashes, stepped units which reduce the sight size slightly, see figure 47, are available.

Glazing should be performed in strict accordance with the manufacturer's instructions. For example, all units must be glazed with a non-setting compound, but some may be fronted with metal casement putty or polysulphide sealant. Beads with a minimum of 3 mm non-setting com-

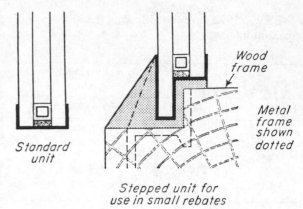

Standard unit

Wood frame

Metal frame shown dotted

Stepped unit for use in small rebates

47 *'Insulight'* units

pound at front and back are advisable for larger windows, and resilient, non-absorbent distance pieces must be used opposite each other on both faces of the units.

Edge clearances between unit and frame must be at least 3 mm all round for units up to 2.8 m and not less than 5 mm for larger units.

Units should be set on two non-absorbent resilient blocks each about 40 mm to 150 mm long and 3 mm wider than the thickness of the unit.

(m) Hollow glass blocks Glass blocks comprising two 'trays' of glass fused together are used to construct non-loadbearing walls or screens. Blocks made in clear glass, colours and various patterns are now imported.

Panels of glass blocks provide:

Thermal transmittance U-2.5 W/m^2 deg C average for a north wall which is approximately equal to the value for a 215 mm solid plastered brick wall.

Solar heat transmission Hollow glass blocks provide a form of solar heat rejecting glazing – see page 264. Transmission is negligible where the angle of incidence of the sun's rays is less than 30° and increases to approximately 60 per cent maximum where the sun's rays are normal to the panel and where there are no shading devices.

Visible light transmission of panels of white blocks is about 50 per cent.

Light diffusion Various patterns provide a degree of privacy with much greater depth of light penetration into rooms and more evenly distributed illumination than normal transparent glazing.

Fire resistance Half hour (BS 476:Part 8:1972 *Fire tests on building materials and structures*) in panels built in mortar.

Sound insulation 35–40 decibels over the frequency range 100 to 3 150 Hz.

Ventilation can be provided by special blocks of all glass construction.

Appearance Blocks are made in white, red, amber, green and blue glass, in various patterns:

Sizes
    240 × 240 × 80 mm
    240 × 115 × 80 mm
    115 × 115 × 80 mm
    190 × 190 × 80 mm
    146 × 146 × 98 mm
    190 × 190 × 100 mm (for glass/concrete roofs)

Installation Glass blocks are non-loadbearing and must be built independent of the main structure. The manufacturers' instructions should be closely followed. There are four methods:
1 In situ traditional – using dry 'fatty' mortar
2 In situ supported – using 'stiff' mortar and steel reinforcement
3 Pre cast panels
3 Dry – for internal use only – using plastics spacer strips.

(n) Channel glass This imported product consists of rectangular troughs 262 mm wide × 41 mm

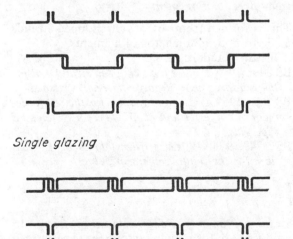

Single glazing

Double glazing

deep, in lengths up to 6 m in 6 mm rough cast glass with or without eight embedded longitudinal wires. Uses include translucent walls, screens, lay lights and roofing. Large areas of glazing, fixed only at the ends of the channels, are possible. Size depends upon prevailing wind pressures and whether units are used either in single thickness, or in double thickness so as to form thermal insulating cavities.

Panels must be firmly held at all edges but have clearances at top and ends to avoid any load being transferred from the structure. Internally, adhesive PVC extrusions can be used to seal the joints between units. Externally, in sheltered conditions special compressible *Neoprene* gaskets can be used to give a 4 mm joint. In severe exposures and in greater lengths preformed mastic strips with an external capping of butyl or polysulphide sealant are recommended. Where mastics or sealants are used spring clips are needed to control the width of the joints.

(o) Corrugated glass Corrugated glass produced in 6 mm rough cast glass, with and without wire in its thickness can be used for glazed walls and screens and for roof glazing. Profiles match 76 and 152 mm corrugations in metal and asbestos-cement sheets.

(p) Glass domes Glass domes are made for use as roof lights in rough cast glass in sizes from 437 mm to 1 829 mm diameter in 50.8 mm increments and in rectangles up to 2.438 × 1.219 m. Certain sizes are available in wired glass.

(q) Glass lenses and pavement lights Glass having a coefficient of thermal expansion similar to that of concrete, *lenses* can be incorporated in concrete floors and roof slabs. Lenses are pressed in moulds in 95 to 299 mm squares and in a diameter of 206 mm. Toughened lenses 152 × 152 × 51 mm are suitable for heavy duty. Resistance to impact loads is five times greater than that of annealed lenses and should they be broken they expand and the fragments remain in position, as they do also in fire thereby preventing the passage of flame and burning material. Toughened lenses withstand a thermal change of 180°C.

Pavement lights are made in clear glass 101.6 × 101.6 × 22.2 mm for incorporation in precast or in situ reinforced concrete frames at the following centres to carry loads up to 1953 kg/m³.

Centres of glass mm	Depth of construction mm
127.0	63.5
152.4	101.6
165.1	121.0

(r) Bullion glass Machine-made simulations of crown glass – see page 246 are rolled by Pilkington Brothers in five standard sizes in clear and amber tinted glass. Unlike genuine crown glass the reproductions have one flat side. Other manufacturers press mock bullion glasses.

(s) Copper lights (electro-copper glazing)
Copper light units consist of glass held in very narrow electrically welded or specially interlocking copper cames. A copper strip is laid between the glass squares and metal is deposited electrolytically to retain the glass. They combine very neat appearances with good vision (wire is not required in the glass), and fire resistance. The GLC (Amending) By-laws 1979 allowed a period of $\frac{1}{2}$ hour fire resistance for 6 mm glass 'in direct combination with metal the melting point of which is not lower than 900°C, in squares not exceeding 0.015 m² in area', and panels not exceeding 0.4 m² in area. BS 6262 describes composite panels with dividing bars to give 1 hour fire resistance.

(t) Leaded lights Lead cames soldered at their intersections have been used for many centuries to support small pieces of glass, often irregular shapes and of varying thickness. Today, large units of glass are sometimes used and without recourse to painting or firing. A range of cames of varying weights is available, in *round, rounded, beaded* and *flat* sections, including those with steel cores. Cames are wired to saddle bars which should be hot-dip galvanized steel, or preferably bronze, or stainless steel.

4 Glass fibre products

Glass fibres (or *filaments*) were made in ancient Egypt. Today, fibres about 0.006 mm diameter are formed from molten glass which is either thrown out from apertures in a rotating dish or dropped through apertures.

The fibres are used as *wool* and bonded with size to form *strands*. *Rovings* are a number of parallel strands twisted one full turn in about 250 mm.

Glass fibres are strong, and fungus, bacteria, insect and damp 'proof'. They are non-combustible, but binders and coverings to quilts can render products combustible (BS 476).

Glass fibres are used loose, and in quilts and batts as thermal insulation.

Batts in 50 mm cavities achieve 0.6 W/m²k in brick/block walls. The layered fibres minimise the risk of water penetrating to the inner leaf.

Glass fibres are also woven into non-combustible (but not *fire resisting*) textiles, and they are used as reinforcement in: plastics (GRP – see chapter 13); plaster (GRG) and in GRC – see chapter 10); and in bituminous roofing felts (see *MBS: Finishes*, chapter 7).

Alkali-resistant Cem-FIL® glass fibres are necessary in cement (GRC products – see chapter 10).

Glass wool

Fibres are sprayed with a binder and formed into random masses. Glass wool is supplied in rolls.

Glass fibre reinforcement

This is formed from rovings or chopped strands made from 51 mm lengths of filaments.
British Standards are:
BS 3496:1973 *E glass fibre chopped strand mat for the reinforcement of polyester resin systems;*
BS 3691:1969, *Glass fibre rovings for the reinforcement of polyester and expoxide resin systems (E type glass);*
BS 3749:1974, *Woven roving fabrics of E glass fibre for the reinforcement of polyester resin.*

Glass fibre textiles

Strands of up to 816 filaments are processed to produce yarn for weaving into tapes for electrical insulation and into fabrics, including window curtains.

Staple tissue

Made from 762 mm long strands staple tissue is used for filters, for reinforcing bituminous bonded roofing felt and for battery separators.

5 Foamed glass

Foamglas[1] available in slabs 300 and $600 \times 450 \times 30$–130 mm thick is pure glass containing discontinuous gas-filled cells. The subject of an Agrément Board certificate dated January 1976, it provides good thermal insulation combined with non-combustibility and immunity from attack by fungi and insects. Being impervious to water and vapour it is very suitable for *inverted roof construction* where roof insulation on flat roofs is above the water-proof covering. Tapered slabs are available. Deflection of roof decks must not exceed 1/240 of the span, but *Foamglas* type T2 is adequate for normal car parking, and type S3 for heavy lorry parking, where a superimposed rigid concrete slab limits compressive loads to 167 kN/m² for type T2 and 216 kN/m² for type S3 material.

Properties are as follows:

	T2	S3	
Density	125	135	kg/m³
Compressive strength (average)	490	640	kN/m²
Thermal conductivity at 20°C(K)	0.046	0.049	W/m°C
Flexural modulus of elasticity	980	1 180	MN/m²
Coefficient of thermal expansion	8.5	8.5×10^{-6}/°C	
Sound transmission (100 mm thickness)	28	28	dB

Work on glass

Toughening has been considered on page 258. Glass can be modified in many other ways for utilitarian or decorative reasons involving some highly skilled techniques. These are often costly but products have the unique qualities of glass. Processes include the following:

(a) **Glass appliqué** Rich decorative effects

[1] Pittsburg Corning UK Ltd.

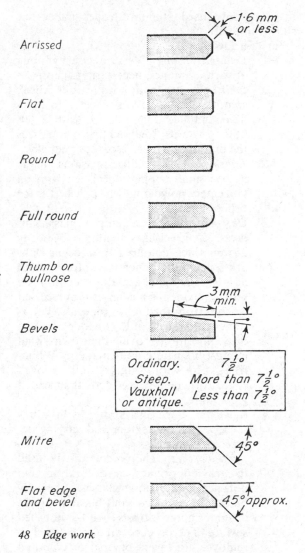

48 *Edge work*

are obtained inexpensively by sticking together with transparent adhesive, one or more coloured pieces of glass. The spaces between the pieces of applied glass may be filled with pigment.

(b) **Fused glass** A wide range of interesting rich visual effects are obtained by melting pieces of coloured glass together.

(c) **Bending** After being heated, most glass can be bent around formers for curved windows, non-reflecting windows, showcases and domes.

(d) **Edge work** Cut edges may be ground and smoothed and if required polished. Examples

267

of commonly used standard edge shapes are shown in figure 48.

(e) Surface work

1 *Brilliant cutting* Very precise incisions of various shapes, necessarily tapering at their ends, are made with a stone wheel, smoothed by a willow wheel with powdered pumice and polished with a felt buffing wheel. Edge lighting emphasizes the brilliance and iridescence of cut glass.

2 *Engraving* Very fine incisions made with a small copper wheel, fed with oil and emery powder may be polished as for brilliant cutting.

3 *Chipping* Hand chipping of glass slabs gives a jewel-like quality. Typically 25 mm thick coloured translucent slabs are set flush in concrete reinforced with non-ferrous metal armatures or they are set in epoxide resin composition recessed 6.5 mm from the face of the glass. Variations in the width of the opaque concrete or composition provide their own visual interest. Concrete units made by Whitefriars Ltd with tongued and grooved edges are generally less costly than leaded glass units.

4 *Grinding* Surfaces ground with abrasives give a fine texture and varying degrees of coarseness.

5 *Sand-blasting* The process can be used to provide privacy or decoration. On polished pot opaque glasses, especially black, the textured sand-blasted surface is lighter and contrasts with the polished surface. The process can also be used to remove thin layers of opal or coloured flashed glass. Hard edges and various depths, including graded effects, can be obtained by skilful manipulation of stencils. Shaped holes such as key holes can be cut only by sandblasting. There are five main finishes:

  (i) *Surface or matt* A flat obscured finish.

  (ii) *Peppering* A very light sandblast which does not entirely remove the polished surface, giving a mottled effect.

  (iii) *Deep or gravé* A more deeply bitten surface, parts of which can be filled with colour.

  (iv) *Modelled* Deeply bitten, but in varying depths giving a mottled appearance.

  (v) *Shaded* Graded from clear to full obscuration.

6 *Acid etching* A mixture of hydrofluoric acid and an alkali (usually sodium bifluoride) etches glass leaving a white frosted surface, smoother than a sand blasted one. There are four main finishes:

  (i) *White acid* from one application of acid only

  (ii) *Satin or velvet* from one application of acid followed by a neutralizing alkali

  (iii) *Embossing* Designs in relief are referred to as 'embossed' because they appear so when viewed from the reverse side of clear glass. Embossing may be *single, double,* or *triple* giving full obscuration, according to the number of applications of acid.

  (iv) *Stippling* results by strewing grains of mica over glass before it is flooded with acid.

7 *Silvering* Silver or other metals can be deposited on any glass, flat or textured, for decoration, to create the illusion of greater width, height or length in rooms, or to reflect light into interiors, to provide vision around blind corners or to create one-way vision – see *One-way glasses*, page 259.

High quality mirrors have a coating of copper deposited on the silver, followed by under-coat paint and a layer of enamel which is then stoved. In damp atmospheres or where there is danger of attack by sulphur eg in the atmosphere, in clinker blocks or fixing media the back must be further protected by at least 0.075 mm lead foil. If attack by alkalis could arise, an alkali-resistant coating is necessary. In special atmospheres such as hairdressing salons aluminium foil may be preferable to lead foil.

For screw fixing, holes in glass should

accommodate resilient sleeves on the screws. In high humidities there should be ventilation gaps behind mirrors.

8 *Gilding* Both gold and glass are extremely durable and many ancient gilded glass mosaics survive. Nevertheless, the back of gilded glass should be kept dry to prevent deterioration of the bond. Gold leaf, available in various qualities and thicknesses and in colours from pale lemon to deep gold, is applied on *isinglass* (a transparent fish glue). For ordinary work, the first layer of gold is *faulted* by applying leaf to any blemishes but for the best work two layers are applied.

Edges of the gold are trimmed and sealed with clear gold size, the back is polished, to give added lustre the gold may be scalded. Backs are then protected with paint.

Other metals which can be applied in the leaf form include silver, platinium, copper, aluminium and tin.

9 *Painting* Coloured ceramic enamels, metal oxides and stains painted and fused on glass are equally durable as the glass. Examples of stained glass by John Piper are to be seen in Coventry Cathedral and the Liverpool Roman Catholic Cathedral.

Both plastics and rubbers are polymers consisting of large chain-like molecules. Plastics are dealt with here and rubbers are dealt with on page 293.

Plastics

Forms and uses of plastics in buildings include:
Sections, including tubes and mouldings
Sheets and films
Cellular products –
   thermal insulants
   wood substitutes
   foamed fillers
Glazing – including safety and anti-vandal products
Light fittings
Adhesives
Sealants
Coatings – including paints
Floorings

Plastics[1] are by common consent taken to be mainly organic materials derived from petroleum and, to a small extent from coal, which at some stage in processing are plastic when heated.

Like paints, rubbers and sealants, plastics consist of molecules joined together to form chain molecules, or polymers which are normally randomly coiled. Identical molecules or monomers, when polymerized, produce homopolymers. Two or more different monomers produce copolymers (eg ethylene/vinyl acetate copolymers).

Monomer molecules   *Homopolymer chains*   *A copolymer chain*

[1]In this context, 'plastics' is a singular noun, not to be confused with the adjective 'plastic'.

Two distinct types of plastics can be distinguished: thermoplastics and thermosetting plastics. The *thermoplastics* always soften when heated and harden again on cooling, provided they are not overheated – page 276.

Thermosetting plastics undergo an irreversible chemical change in which the molecular chains cross-link so they cannot subsequently be appreciably softened by heat. Excessive heating causes charring.

Thermosetting plastics are described on page 284.

The range of properties in plastics materials is so great that generalizations are difficult to make. Generally, however, plastics are light in weight and have good strength:weight ratios, but rigidity is lower than that of virtually all other building materials, and creep is high.

Plastics have low thermal conductivity and thermal capacity, but thermal movement is high. They resist a wide range of chemicals and do not corrode.

Like all other organic materials, plastics are *combustible* as defined by BS 476. Some are highly flammable, while others such as polyvinyl chloride, are difficult to burn.

Plastics lend themselves to a wide range of manufacturing techniques, and products are available in a wide range of forms.

Various textures are available, and all colours, many of which fade externally. The harder surfaces such as those of typical decorative laminates are easy to keep clean, but dirt becomes engrained in the surfaces of many products.

Plastics for building are dealt with in the following publications:
BS 1755: *Glossary of terms used in the plastics industry:* 1974 and 1980
BS 4901:1976 *Specification for plastics colours for building purposes*
Production of plastics products is dealt with in *Manufacturing Technology*, M Haselhurst, Hodder and Stoughton.

Material	Density kg/m[1]	Coefficient of linear expansion per deg C × 10⁻³ and mm/m/°C	Max. temperature recommended for continuous operation °C	Short term Tensile strength N/mm²	Behaviour in fire
Thermoplastics Polythene:[1]					
Low denisty	900	0.2	80	7–16	Melts and burns like paraffin wax
High density	945	0.14	104	20–38	
Polypropylene	900	0.11	120	34	
Polystyrene	1 040	0.07	70	41	Melts and burns readily
Polymethyl metacrylate (acrylic)	1 185	0.07	80	70	
Rigid PVC² (UPVC)³	1 395	0.05	65	55	Melts but burns only with great difficulty
Plasticized PVC²	1 280	0.07	40–65	10–20	Melts, may burn– depending on plasticizer used
Nylon	1 120	0.08	80–120	50–80	Melts, burns only with difficulty
Thermosetting Phenolic laminates	1 410	0.03	110	80	Highly resistant to ignition
Melamine laminates	1 440	0.03	120	96	
GRP⁴ laminates	1 600	0.02	90–150	100	Usually flammable, but relatively flame-retardant grades are available

[1] High density and low density polythene differ in their basic physical properties, the former being harder and more rigid than the latter. The values shown are for typical materials but may vary considerably, depending on composition and method of manufacture
[2] PVC   Polyvinyl chloride
[3] UPVC Unplasticized polyvinyl chloride
[4] GRP   Glass-reinforced polyester

Table 95   Typical properties of plastics used in building – based on BRE Digest 69

Information about Plastics can be obtained from the Information Department of the British Plastics Federation, 5 Belgrave Square, London SW1 8PH.

PROPERTIES OF PLASTICS

Properties vary widely depending upon the basic formulation and the presence of additives such as stabilizers, ultra-violet absorbers, fire retardants, reinforcements, fillers and pigments. Table 95 gives typical properties of some plastics used in building.

The present lack of interrelation between test methods and of a co-ordinated and agreed system for comparing the properties of different plastics materials is being rectified by the BSI. However, it cannot be overemphasized that test results can rarely relate directly to the conditions in which materials will be used and, in the absence of expert advice, they should be taken only as a general guide.

Specific gravity

This is similar to that of wood, about 0.9 to 2.2.

Strength

Plastics have tensile strength:weight ratios which are much more favourable than those of most metals but the modulus of elasticity (E) is low, even compared to aluminium and glass, and rules them out, for load-bearing beams. Thermoplastics are also precluded from such use by their tendency to creep under sustained loading coupled with a serious loss of strength at elevated temperatures.

However, the low modulus of elasticity ceases to be a disadvantage in continuous surface structures in which the load is uniformly distributed over the whole surface and these structures are also better adapted to accommodate the characteristically high thermal movement of plastics materials. They would also accommodate creep although most surface structures are likely to be formed in a thermosetting material in which creep is very small.

BS 6206:1981 specifies *Impact performance requirements for flat safety glass and safety plastics for use in buildings*

Behaviour at elevated temperatures

Creep and degradation are more rapid at high temperatures and strength properties are reduced. Thermoplastics soften at points which are not well defined, in most cases between 60 and 110°C.

Behaviour in fire

General considerations in respect of fire and the BS 476 fire tests are dealt with in chapter 1, page 35. There are references to plastics in fire on page 40 and in table 95, page 271.

Thermoplastics soften at high temperatures see page 276 and all plastics are *combustible. Spread of flame*[1] over the surfaces of some plastics is high and burning plastics generally produce large volumes of smoke – the worst menace to life in fire, and toxic gases, mainly carbon monoxide are produced. Ignition of a plastics oil storage tank by radiation from a nearby fire is an example of a possible hazard. Similarly, plastics hinges are unsuitable for fire resisting doors. Hazards can result from collapse of plastics ventilation ducts and from soil stacks and from flaming droplets from roof lights. On the other hand, the opening of roof vents can, with advantage, allow smoke to escape.

No treatment can make any plastics material *non-combustible*. However, by reason of the inherent molecular characteristics of plastics such as PVC, or by the inclusion of fire retardant additives, many plastics are very difficult to ignite, and some are self-extinguishing.

An element comprising two skins of glass fibre reinforced polyester with a 19 mm cellular phenolic core, had a *fire resistance*[1] of 15 minutes when tested by the Fire Research Station, but prolonged fire resistance is not possible for glazing.

Thermal conductivity and capacity

For 'solid' plastics this property is similar to that of wood, but thermal capacity is higher, and the visual 'warmth' of wood is often preferred, eg in wc seats. Expanded plastics have considerably lower values – see page 291.

Transparency

See page 286.

Electrical properties

Plastics are excellent insulators but electrostatic charges attract dust and the sparking can be a hazard where flammable vapours are present. Polyester floorings are less likely to cause problems than nylons and acrylics, but conductive ingredients must be included where an anti-static surface is necessary – see *MBS: Finishes*, chapter 1.

[1]As defined by BS 476:Part 7:1971 – see page 37.

Thermal movement

Thermal expansion of plastics materials is generally several times, and in some cases as much as ten times, that of steel. For example, a 3.7 m long rigid PVC gutter expands about 6 mm in response to normal annual temperature variations in this country.

Moisture movement

Most plastics absorb very little water; the chief exceptions, cellulose acetate and certain types of nylon, swell with moisture absorption.

Durability

Plastics do not rot or corrode and in general they have extremely good resistance to the chemicals normally encountered in building work.

The performances of plastics products vary widely with the type and grade of polymer and with any compounding ingredients used, the control exercised in manufacture and the dirt deposits, temperature and weather peculiar to the situation in which they are exposed.

BRE Digest 69 *Durability and application of plastics* points out that few plastics have been available long enough to assess their durability in use. For the time being predictions must rely largely upon knowledge of composition and manufacture, and upon accelerated laboratory ageing tests.

The Digest states: 'Sunlight, particularly its ultra-violet component, is the most significant single factor responsible for the breakdown of plastics and for producing changes of colour. Ultra-violet radiation initiates many of the chemical reactions by which plastics are oxidized and degraded. These are often chain reactions which are accelerated by favourable conditions of warmth, oxygen and moisure.' An increase in temperature of about 10 deg C doubles the rate of chemical reactions which lead to breakdown. BS 5255:1976 for polyethylene pipes states that where suitable pigments and/or stabilisers are not used in pipes as protection from degradation by UV light, they should be painted with a paint which has no deleterious effects upon the plastics material. Moisture is needed for many reactions initiated by ultra-violet light to continue and often contributes to loss of bond between resins and fillers or reinforcement.

Moist salt-laden air combined with a high ultra-violet light intensity in coastal exposures present severe conditions, while the combined 'effects of warmth, moisture and sunlight in the tropics are much more drastic. However, pigments and also dirt deposits, filter ultra-violet light and reduce the rate of breakdown, and some plastics materials, in particular acrylics and polyvinyl chloride, have retained their physical properties well externally for a number of years. Products such as wc pan/soil pipe, and clay drain pipe connectors are expected to remain watertight for sixty years.

Colour retention, as in paints, depends upon the correct choice of pigments.

COSTS

Compared with other building materials the cost per kilogram of unformed plastics materials is in many cases high and machines and moulds often involve high capital investment. However, manufacture is economical in labour, and complex products which in other materials would have to be assembled from many separate parts, can often be formed in one operation. Thus the cost per unit volume of products is generally favourable, and further savings may accrue from ease in handling, fixing and in maintenance, if not always in visual satisfaction.

FORMING METHODS

Plastics articles can be formed by many diverse techniques, some of which are not applicable to other materials.

Table 96 lists the main processes for forming building products, of which hand lay-up spraying and some foaming techniques are feasible on the building site.

Continuous processes

Extrusion

Tubes and complex sections can be extruded in

Process	Suitable[1] Material		Products	
			Precision[2]	Economical volume of production[3]
CONTINUOUS				
Extrusion	TP	(TS)	1–2	A
Film blowing	TP		2	A
Film and sheet casting	TP	(TS)	2	A
Calendering	TP		1–2	A
Paste spreading	TP		2	B
Spraying (with or without fibre reinforcement)	(TP)	TS	2	C
DISCONTINUOUS				
Moulding:				
compression	(TP)	TS	1	A
transfer	TP	TS	1	A
injection	TP	(TS)	1	A
blow	TP		2–3	A
slush	TP		3	C
rotational	TP		2	C
Thermal-forming from sheet				
simple pressing	TP		3	C
blow forming	TP		2	B
deep drawing	TP		2	B
punching die process	TP	TS	2	B
vacuum forming	TP		2–3	B–C
drape forming	TP		2–3	B–C
Laminating				
fibre reinforced products (GRP)				
hand lay-up		TS	3	C
mechanical lay-up		TP	3	B
sheet forming		TS	3	A
Coatings				
dip coating	TP		2	C
sinter coating	TP	TS	2	C
Casting	TP	TS	2	B–C
Foaming	TP	TS	2	A–C

[1] TP thermoplastics; TS thermosetting plastics (less commonly employed materials are in parentheses).
[2] 1 high; 2 medium; 3 low   [3] A large; B medium; C small

Table 96   Methods of forming plastics products

thermoplastics by forcing the *melt* through a suitably shaped die and cooling it as it emerges.

Film blowing

Air is blown into an extruded thermoplastics tube to form a cylindrical balloon of film in thicknesses from 0.01 to 0.25 mm.

Film and sheet casting

Hot melt or solution is fed over a chilled roll or on to a chilled moving band.

Calendering

This is the forming of sheet by passing plastics

melt between hot rolls followed by cooled rolls. By this means two or more sheets can be laminated together or bonded to a hessian or felt backing as in some flooring materials.

Other continuous processes are: *Paste spreading* see *MBS: Finishes*, chapter 6.

Discontinuous processes

Moulding methods

1 *Compression moulding* Powder, or preformed pellets of thermosetting resins are placed in a mould which is then closed, heated and subjected to pressure. The charge is plastic for sufficient time to fill the mould before the molecules irreversibly cross-link and the charge sets.

Apart from practical limitations such as shapes which present difficulties in filling moulds or which cannot be extracted from moulds, very intricate mouldings are possible. Metal and other inserts can be incorporated in injection and compression mouldings.

2 *Transfer moulding* This is a form of compression moulding in which plastics material in a *shot cavity* is preheated before being forced through channels into a mould. The process can be used for thermosetting as well as thermoplastic materials and is suitable for delicate or complicated articles.

3 *Injection moulding* Thermoplastic granules or powder are made plastic by heat and injected into a mould in which they cool and harden. The process is rapid, several small products are made together in a single mould while mouldings as large as a 273 litre cistern can be formed. Thermosetting plastics can now be formed by injection moulding. Incidentally, the mould for a plastics manhole base costs many thousands of pounds. However, in reaction injection moulding (RIM), tooling costs are low. Complex shapes are possible in deeply dished products of varying thickness. The method is suitable for small production runs for which pressed steel is not economical.

4 *Blow moulding* A soft pliant tube is extruded and the halves of a mould close round it nipping it together at the top and bottom. At the same time a probe punctures the tube, air is blown in and the plastics material takes the shape of the mould.

Cold water cisterns are being made by blow moulding an enclosed 'tank' and cutting it to form two units each of which has 227 litre (actual) capacity.

5 *Slush moulding* A simple method whereby PVC paste is poured into a heated mould, the excess is removed and the moulding is gelled in an oven.

6 *Rotational moulding* This is used to form large hollow articles such as cold water cisterns. A measured quantity of thermoplastic powder, usually low density polythene, is placed in a female mould which is rotated about two axes and heated so that the plastic sinters over the internal surface of the mould forming a wall of uniform thickness.

Thermo-forming from sheet

In these processes thermoplastics sheets are heated and made to conform with a single sided mould.

1 *Simple pressing* Heated sheet is clamped on a die plate and pressed into shape by a former. The process may be assisted in the early stages by a vacuum on the reverse side of the sheet.

2 *Blow forming* Heated sheet is formed by clamping it at the edges and blowing a bubble, with or without a female mould.

3 *Deep drawing* A sheet is lightly clamped at the edges so it can slip and pull out in order to conform to the shape of a plunger which is pressed onto it.

4 *Punching die process* This is similar to deep drawing but both a die and a roughly matched punch are used.

5 *Vacuum forming* of preheated thermoplastic sheets is done into a mould which is perforated with small holes through which air is extracted sucking the sheet in so it assumes the shape of the mould. To form local deep draws the process is *plug assisted*. The size of mouldings is limited by the width of available sheets which is at present 1500 mm.

Vacuum formed articles display the interesting phenomenon known as *plastic memory*, whereby when it cools, because heating of the flat sheet before forming was insufficient to reorientate the molecules, the product tends to revert to its original flat condition.

Casting Casting in open moulds is possible in the case of polymethyl methacrylate (eg *Perspex*), epoxides, polyesters and even PVC.

Discontinuous processes described elsewhere are:

Laminating: GRP lay up – page 287
Sheet laminates – pages 271 and 296
Coatings – see *MBS: Finishes*, chapter 6
Foaming – page 288.

JOINING

Plastics components generally can be joined by screws, bolts, snap action and by adhesives, and thermoplastics can be joined by:

Heat welding

The materials to be joined are pressed together and momentarily heated by: *high frequency heating, hot-knife, ultra-sonic* or by *friction* methods. Welding by *hot gas* and *filler rod* can be done on the site by skilled operators.

BS 5955, Part 6:1980 gives *Recommendations for methods of thermal fusion welding*

Solvent cementing

A very easy and convenient method of jointing materials such as PVC, *Perspex*, ABS and polystyrene, but not polyethylene or polypropylene.

THERMOPLASTICS

When thermoplastics are heated the molecular chains are able to move relative to one another. In most cases softening points, which are not generally well defined occur between 60 and 110°C and up to about 171°C for some nylons and polyacetals.

On cooling, thermoplastics regain their original degree of stiffness, a process which can be reversed repeatedly, provided the material is not heated so much as to cause degradation. Clearly, thermoplastic products cannot be used near to sources of heat (table 95 recommends maximum temperatures for continuous operation), but thermoplasticity makes for versatility in forming methods. With a few exceptions, thermoplastics unlike thermosetting plastics, are softened and swell in various organic solvents and may be dissolved by them.

Thermoplastics vary from hard and rigid to soft and pliable and have wide ranges of other properties when they are modified by additives. Under prolonged and constant stress they exhibit increasing deformation with time (ie *creep* or *cold flow*) and permissible working stresses must take this phenomenon into account.

Apart from the natural polymer cellulose, nearly all thermoplastics are prefixed 'poly'.

The properties relevant to the main building uses of the following thermoplastics are now considered:

Polyethylene (PE)
Polyvinyl chloride (PVC)
Polyvinyl fluoride (PVF)
Polyvinyl acetate (PVAC)
Polypropylene (PP)
Polymethyl methacrylate (PMMA)
(eg *Perspex* (ICI))
Polystyrene (PS)
Polytetrafluorethylene (PTFE)
Acrylonitrile butadiene styrene (ABS)
Cellulose nitrate (CN)
Cellulose acetate (CA)
Cellulose acetate butyrate (CAB)
Coumarone indene resin
Nylon (PA – polyamides)
Polyacetals (POM)
Polycarbonates (PC)

Polyethylene (PE)

Polyethylene, often called *polythene* is one of the polyolefin group. It is made in high density (*HD*) and low density (*LD*) forms. Both forms have high thermal movement, are extremely resistant to many chemicals at room temperature and are good electrical insulators. They burn without external aid. They have high impermeability to water and to water vapour. A degree of permeability to gas does not preclude the safe use of polyethylene tubes for conveying drinking water near to gas pipes below ground (see *Model Water Byelaws* 1966 edition, HMSO).

Low density (specific gravity 0.91 to 0.93) is resilient and very tough, particularly at low temperatures. It softens at about 80 to 105°C.

High density polyethylene (Specific gravity 0.94 to 0.97) is less resilient and tough and softens at a higher temperature. Natural polyethylene is colourless and translucent, embrittles and loses strength in sunlight within about three years. However, carbon black pigmentation increases the serviceable life considerably. Even items such as cisterns because they are exposed to sunlight before they are installed are required to include carbon black in their composition.

Uses include:

Cold water cisterns and floats Polyethylene cisterns which are designed so they will not be overstressed, are 'expected to last for the life of a building' and polythene floats 'appear to be very suitable'. (BRE Digest 69.) BS 4213:1975 describes *Cold water storage cisterns and covers.*

Water pipes Water is less liable to freeze, but if it does so, less damage will result in polyethylene pipes than in metal pipes. In other respects it is expected that pipes of the correct gauge and density to suit the service pressure, properly supported and fixed with slight *snaking* to take up thermal movement in long runs, should carry cold water for at least 50 years. However, the high thermal movement of polyethylene precludes its use for conveying hot water in buildings.

The narrow temperature range suitable for fusion welding necessitates close control of the operation in a factory or a workshop on the building site. If this is not possible, compression joints must be used in spite of their clumsy appearance. British Standards for polyethylene pipes include:

BS 1972:1967 *Polythene pipe (Type 32) Small diameter for cold water services* Max 32 kg force/cm^2 (black)

BS 1973 1970 *Polythene pipe (Type 32) for general purposes* Max 32 kg force/cm^2

BS 6437:1984 *Polyethylene* pipes (Type 50) *for general purposes* Max 50 kg force/cm^2 (black for non-potable water)

BS 6572:1985 *Specification for blue PE pipes up to 63 mm nominal size below ground for cold potable water*

BS 864:Part 3:1975 *Compression fittings for polythene pipes*

Bath, basin and sink wastes Polyethylene is not normally attacked by ordinary effluents including strong detergents and fats, and waste pipes are expected to remain serviceable for thirty years at least. However, paint strippers, cleaning fluids and similar solvents can cause deterioration and polypropylene pipes are more suitable in such circumstances, particularly where very hot water is discharged.

BS 5255:1976 deals with *Plastics, waste pipes and fittings.*

Damp-proof courses and membranes In gauges suitable to resist building loads, polythene damp-proof courses can be expected to last indefinitely. BS 743:1970 *Materials for damp-proof courses* requires the materials to be 95 per cent by weight pure, with mineral filters and at least 2 per cent by weight carbon black, the sheet to be not less than 0.46 mm thick and to weigh about 0.48 kg/m^2. Again, the black rather than the translucent form prevents degradation by sunshine before installation. The minimum recommended thickness for dpms is 0.127 mm.

Separating membranes Polyethylene film can be used as a separating membrane between screeds or floor tiles and concrete slabs. Tests have shown that a '250 gauge' film 0.064 thick separating a lean concrete base from a thick concrete slab offered only about one tenth of the resistance to relative movement of that offered by a bitumen emulsion membrane.

Chemical resisting membranes Polyethylene sheeting is permeable to mineral oils, and petrol causes swelling but a *1000 gauge*, 0.254 mm film can be used to prevent certain chemicals penetrating downwards into a reinforced concrete floor, or as an isolating membrane where it is desired to lay a screed on certain contaminated bases.

Temporary glazing and protection 'Natural' polyethylene film and sheet last for a year or two in direct sunlight or for three or four years in shaded positions if they are not unduly stressed by wind and are of the appropriate gauge. Temporary glazing should be not less than *250 gauge*, 0.064 mm, and preferably *500 gauge*, 0.127 mm, while temporary roofing in exposed positions may require to be *1000 gauge*, 0.254 mm.

Film containing carbon black, however, is

expected to last for at least ten years even in direct sunlight.

Concrete curing Polyethylene film coverings reduce evaporation from concrete surfaces and where they are laid on hardcore they prevent loss of moisture downwards.

Polyvinyl chloride (PVC)

Polyvinyl chloride in either the 'natural' rigid form or made flexible by the addition of plasticizers, is versatile and low in cost.

PVC can be recognized by a greenish tinge to the flame when it burns and an acrid odour when the flame is extinguished. It has very good weathering properties and is unaffected by dilute or concentrated acids and alkalis. PVC is attacked by aromatics and is soluble in ketones and esters. It begins to soften at about 70°C and creep is rather high. At very low temperatures PVC has reduced impact strength. Joints can be formed by solvent 'cementing', which is particularly useful for joining pipes and sheet floorings.

Ordinary PVC is not suitable for hot water pipes. *Chlorinated PVC* (CPVC) shows promise here, but the problem of accommodating thermal movement, although less than that of polyethylene and polypropylene, remains.

As in all plastics and paints, colour stability depends mainly on the pigment system. Generally colour, in particular reds and yellows, fade when they are exposed externally.

Unplasticized ('Rigid') PVC (UPVC)

In this form PVC burns only with great difficulty and is self-extinguishing.

Uses include:
Ventilation ducts These should perform well indefinitely if allowance is made for thermal movement and if they are not heated by very hot air or an external source of heat.

Soil and waste pipes

Soil and waste systems BS 4514:1983 covers *Unplasticized PVC soil and ventilating pipes, fittings and accessories* and BS 5255:1976 deals with

Modified unplasticized PVC waste pipe and fittings to convey normal domestic effluents. The material used in the latter specification has a higher softening point than that used for soil pipes and fittings, to allow the use of thinner walls while carrying hot wastes flowing full bore. Manufacturers should be consulted as to the suitability of pipes for non-domestic hot and continuous effluents.

Certain waste systems in recently developed chlorinated PVC (CPVC), are claimed to be capable of taking boiling water but *GLC Design and Materials Bulletin 17*, allows the use of PVC wastes only up to two storeys and states that sink wastes from washing machines must not discharge into stacks.

The low softening point of rigid PVC may necessitate protecting pipes larger than 38 mm in diameter against fire where they pass through separating floors.

Rainwater goods Unplasticised PVC gutters and downpipes are cheaper than cast iron goods and the BRE state that although there is some loss of resistance to impact, good service may be expected for twenty years or more. Owing to the difficulty in obtaining light-fast pigments, they are usually made in black, grey and white material.

BS 4576 *Unplasticized PVC rainwater goods*, Part 1:1970 covers: *Half-round gutters and circular pipes*.

Drains and sewers BS 5955:Part 6:1980 is *Installation of PVC pipework for gravity drains and sewers*.

BS 4660:1973 deals with *Unplasticized PVC underground drain pipes and fittings*.

BS 5481:1977 *Specification for unplasticized PVC pipes and fittings for gravity sewers* covers nominal sizes 200 to 630 mm.

Water mains Unplasticized PVC is stronger in tension and more rigid than polyethylenes. Pipes are made in diameters of 50.8 to 152.4 mm and much larger to order. Pipes to BS 3505:1968, *unplasticized PVC pipe for cold water services*, laid in trenches, or by mole plough, should have a life of fifty years.

BS 4346 describes *Joints and fittings for use with unplasticized PVC pressure pipes*. Part 3 describes *Solvent cement*.

Transparent and translucent sheets The natural material is virtually colourless but darkens with prolonged exposure to light. Coloured sheets are available. Corrugated sheets, usually 1.6 mm thick, are cheaper than polyester resin glass-fibre reinforced sheets, but fail more rapidly in fires. See also page 286.

The Building Regulations 1976 permitted the use of rigid PVC sheeting which is self-extinguishing, for roofs more than 6 m from any boundary. Nearer to boundaries it could be used for roofs to garages, conservatories and outbuildings not exceeding 40 m^2 in floor area, and for roofs or canopies over balconies, open car ports, covered ways and detached swimming pools. PVC sheeting has high resistance to breakage by impact. Flat sheet can be used for 'safety glazing' in doors and windows. It is available with wire-mesh reinforcement.

Sheet BS 3757:1978 describes: *Rigid PVC sheet 0.25 mm thick and above* (6 types).

Corrugated opaque sheets These should remain mechanically sound for at least 20 years. However, colour is likely to change within a few years and they may become rather liable to impact damage. If backed with thermal insulation and exposed to the sun, degradation will be accelerated, and dark sheets may distort.

BS 4203:1980 describes: *Extruded rigid PVC corrugated sheeting.*

Electrical conduits and accessories Rigid PVC conduits are available in *light* and *standard* gauges. Flexible PVC conduits are also available. Compared to steel conduits they are light and easy to bend and fix. Mechanical joints, which avoid weakening the tube by threading, and the saddles accommodate the relatively high thermal movement. Where necessary, joints can be cemented to exclude air, dust and water. Although a separate earth wire is required, it is claimed that the fixed cost of PVC systems is lower than that of steel systems.

BS 4607 Part 1: describes *Rigid PVC conduits and conduit fittings* and Part 5:1973 describes *Rigid PVC conduit fittings and components.*

Windows Lack of rigidity and high thermal movement has discouraged the use of plastics by themselves, and they have been used mainly as a protection on wood and metal frames and sashes. Nevertheless, windows having sashes formed with unplasticized PVC hollow extrusions have been in use in Europe for over twenty five years. They do not require painting and avoid the 'cold bridging'

caused by ordinary metal windows. Window boards and reveal liners are also available.

Fencing This competes in durability and cost with 'non-durable' species of timber which are not treated with preservative, and with steel which is not galvanized. Sections can be strengthened with wood or metal inserts.

Floor tiles BS 3260:1969 describes *PVC (vinyl) asbestos floor tiles.*

Other uses of unplasticized ('rigid') PVC include expanded rigid PVC – see page 291.

Plasticized (flexible) PVC

This is very suitable for extrusion, injection moulding, calendering and blowing into film. The ease of ignition and rate of burning depend upon the type of plasticizer which is used.

Uses include:

Floor coverings The most important use of flexible PVC in building is in vinyl sheet and tile floor coverings – see *MBS: Finishes*, chapter 1.

Sarking This must resist high temperature and wind pressure without draping excessively between the rafters.

Water stops The BRE state that 'provided they are of sufficiently heavy section to permit site handling and that differential movements between concrete sections are not excessive, there is every indication that they will perform satisfactorily indefinitely'.

Preformed joint seals PVC extrusions have been used as loose baffles and as tubular or cruciform gaskets to be held in compression. See chapter 16.

'Clip-on' extrusions Sections of PVC are available to fit over standard handrail cores with a minimum radius of 76 mm. After heating, preferably with a hot air blower, the extrusion is fixed from the top downwards, care being taken to avoid stretching and to allow an overlap for shrinkage. Joints are butt welded by pressing the contacting surfaces together after softening them with a hot plate or knife.

Bright colours are available but where rails are exposed to sunlight, black PVC is recommended. Reinforcing bridges underneath the core are advisable at bends.

Electrical cable insulation PVC has good electrical insulation and it is water resisting, flexible and sef-extinguishing.

BS 5803 for roof insulation draws attention to embrittlement due to the migration of plasticizer, which may be caused by the insulation.

BS 6004:1974 deals with *PVC insulated cables*, BS 6500:1975 *Insulation flexible cords* and BS 6746:1984 *PVC insulation and sheathing of electric cables*.

Transparent swing doors These heavy duty, self-closing doors allow safe 'push-through' passage for trollies and vehicles.

Sheet BS 2739:1975 describes: *Thick PVC sheeting (calendered, flexible, unsupported, 250–900µm)*.

Flat roof coverings A proprietary sheet consisting of PVC laminated to impregnated asbestos is intended as a single layer flat roof covering, the laid cost of which is stated to be comparable with that of three layer built-up bituminous felt. The manufacturer suggests that the covering will remain serviceable for over 20 years.

Coatings

These include: dipped and sprayed melt coatings, *organosol* and *plastisol* paste coatings, film and sheet – see *MBS: Finishes*, chapter 6.

Expanded PVC – see page 291.

Cast plasticized PVC may be used as moulds for concrete.

Polyvinyl fluoride (PVF)

Polyvinyl fluoride has recently been used as a surface film bonded to asbestos cement, metal and plywood sheets. If it is not damaged, the BRE expect a decorative and protective life of at least twenty years.

Polyvinyl acetate (PVAC)

The uses of PVAC are limited by its low softening point to:

Adhesives for joinery – see chapter 14
Emulsion paints
Plaster bonding agents ⎫
Screed bonding agents ⎬ see *MBS: Finishes*
In situ floor coverings ⎭

Polypropylene (PP)

This relatively new polyolefin material softens as a higher temperature than most common thermoplastics, and like nylon and polytetrafluorethylene (PTFE) a very costly material, it can be sterilized. However, polypropylene pipe systems for conveying hot liquids must be very carefully designed to accommodate its very high thermal movement. Polypropylene homopolymers have low impact strength in cold conditions but copolymers are suitable in this respect for all normal conditions found in building in this country. Polypropylene is attacked by chlorinated solvents. Sheet is not so easily vacuum formed at PVC or ABS. Specific gravity is low, about 0.90.

Uses include:

Fittings for clay and pitch fibre drain pipes.
Drain inspection chambers
Waster pipes and fittings BSs 5254:1976 and 5255:1976
Road gullies
Expansion tanks
WC cisterns, siphons, etc
WC seats
Dpcs and cavity trays
Chair shells
Reference:
BS 4991:1982 *Propylene copolymer pressure pipe*

Polymethyl methacrylate (PMMA)

This clear *acrylic* resin is best known in the sheet form as *Perspex* (ICI) and is also available as a powder for injection moulding (*Diakon*, ICI). A related form of resin is used in acrylic paints.

Polymethyl methacrylate begins to soften at about 90°C and can be moulded at about 140°C. It burns with a yellow flame, like paraffin wax, and has a sweet odour when the flame is extinguished. Large areas of glazing burn rapidly, and because burning PMMA drops from roofs, large areas must be avoided unless special precautions are taken in design, eg sprinkler installations.

It is not attacked by strong solutions of alkalis and it resists most dilute and many concentrated acids, fats and mineral oils. It is dissolved by many organic solvents, and can be cemented by solvents. The BS for domestic baths requires them to be labelled with a warning to the effect that some dry cleaning agents and paint strippers, and burning cigarettes, cause damage.

Sheet material

Sheet material is cast in transparent, translucent and opaque forms in a range of bright colours and textured surfaces are available. Resistance to impact is very much better than that of glass although inferior to that of glass-fibre polyester resin products.

Unlike ceramics, glass and vitreous enamel, polymethyl methacrylate surfaces may soon lose their initial gloss but although easily scratched, they can be restored with metal polish.

Light transmission of clear sheet is about 92 per cent (glass about 90 per cent). Its ability to *pipe* light can be exploited in internally illuminated signs. See also page 286.

Crazing has occasionally resulted from poor annealing during manufacture, but when exposed to the weather sheet has good resistance to ultra-violet radiation and a life of at least forty years may be expected.

Uses for transparent and translucent sheet include:

Corrugated sheeting
Roof lights including domes
Lighting fittings
Illuminated signs.

Uses for opaque coloured sheets include:

Baths (BS 4305:1972 *Baths for domestic purposes made from acrylic sheet*)
Basins
Shower and bath enclosures
Urinals
Shop fascias and signs.

Polystyrene (PS)

A low-cost thermoplastic, in its unmodified form (BS 1493:1967) is crystal clear but inclined to be brittle and can be recognized by a metallic ring. It is also available in *high-impact, medium impact* and other grades and in the expanded form. It is attacked by certain organic solvents such as white spirit, softens in boiling water and burns readily with a sooty flame. Transparent polystyrene yellows and weakens on exposure to ultra-violet light. See also page 286.

BS 2552: 1955 *Polystyrene tiles for walls and ceilings* specifies materials, dimensions, opacity, colour fastness and finish of injection moulded tiles. Fronts are glossy or matt and backs are recessed to facilitate fixing with adhesive.

Sizes are:

101.6 and 152.4 mm square	2.54 mm
202.6 × 50.8 and 25.4 mm	overall
152.4 × 76.2, 38.1 and	1.575 mm
25.4 mm	minimum

Other uses include:

WC cisterns
Lighting fittings
Concrete formwork
Paint (mainly of copolymer form – see *MBS: Finishes*, chapter 6.
Expanded polystyrene – see *Cellular plastics*, page 290.

Polytetrafluorethylene (PTFE)

This thermoplastic is highly resistant to heat and many chemicals and solvents and has a very low co-efficient of friction. It is, however, very costly indeed and is used only for special applications such as sliding expansions joints in heavy structures and for wrapping as a film around pipe threads to lubricate them so that joints can be tightened by hand. See BS 3784:1973 *PTFE sheet* and BS 4375:1968 *Unsintered PTFE tape for thread sealing applications*.

The resistance of PTFE to heat is demonstrated by non-stick cooking utensils.

Acrylonitrile butadiene styrene (ABS)

ABS distorts about 85°C, a higher temperature

than PVC, but it deforms more than is desirable in an ordinary domestic hot water system and the cost is higher.

ABS is extremely tough and strong and retains good impact strength even at low temperatures. It supports combustion, evolving black smoke in burning and is recognizable by a bitter smell as well as those of styrene and rubber. Specific gravity is low, ie 1.02.

Uses include:

ABS is very suitable for vacuum forming garage doors, small boats and taxi-cab roofs. ABS waste pipes and fittings and drain inspection chambers are available.

BS 5255:1976 describes *Plastics waste pipe and fittings*, BS 5391:Part 1:1976 describes *Pressure pipe for industrial uses.*

Cellulose nitrate (CN)

Cellulose nitrate, first developed in the form of celluloid in 1862, is water-white, easy to shape and has good water resistance. However, it is highly inflammable and in buildings it is used only for certain paint finishes – see *MBS: Finishes.*

Cellulose acetate (CA)

This, although similar to celluloid is very tough and burns much less readily. In burning it can be recognized by a yellow flame followed by a smell of burning paper and vinegar.

It has high moisture movement, poor water resistance and embrittles at high temperatures and is used only to a limited extent as:

    Binder in emulsion paints, see *MBS: Finishes*
    Lighting fittings
    Three-layer corrugated sheet thermal insulating inner glazing
    Door furntiure
    Coverings for handrails

CA is also extruded as a tape with interlocking edges for wrapping spirally around circular cores to provide non-slip hand-grip.

Cellulose acetate butyrate (CAB)

This transparent material is tough and has lower moisture absorption than cellulose acetate.

It has been used for conveying natural gas and for illuminated signs and is also used for coatings – see *MBS: Finishes*, chapter 6.

Casein (CS)

Casein is made from milk whey reacted with formaldehyde. It has a high moisture movement. It is used to a very limited extent for small items such as drawer pulls and as an adhesive – see page 297.

Coumarone indene

These resins are used as a medium in paints; as an alkali-resisting finish, as an electrical insulating varnish and as a binder in *thermoplastic floor tiles* – see *MBS: Finishes*, chapter 1.

Nylons

There are many forms of nylon (polyamide resins) the most important in building being *Nylon 6, Nylon 66* and *Nylon 11*. Nylons are off-white. They are outstanding among thermoplastics for their resistance to organic solvents, oils and fuels and resistance to caustic alkalis up to 20 per cent concentration, is good at room temperature. Among the few chemicals which attack nylons are mineral acids, phenols and cresols, although dilute aqueous solutions of these chemicals have little effect.

Nylons are tough, have high strength, excellent wear resistance, a low coefficient of friction and ability to absorb any particles which would score shafts making it a very suitable material for gears Nylons damp vibrations and noise, and bearings can be run with water or even without lubrication. They can be machined by normal methods. They are good electrical insulators at normal temperature and humidity and have better resistance to high temperatures than many other thermoplastics. Unlike other plastics, nylons absorb up to 2 per cent moisture with some swelling and loss of strength. If a cold metal point is pressed onto a heated surface and drawn away, threads form fairly easily. Nylons burn with difficulty with a yellow flame and when extinquished a smell similar to that of burning hair remains.

Apart from its use as fibre it is used for nuts and

bolts, castors, curtain rail and sliding door fittings and ball valve assemblies. The BRE states that: 'Nylon has been used in door and window furniture and cold water fittings and seems durable enough for these applications. It has also been used with encouraging success as coatings for railings and outdoor furniture' – see *MBS: Finishes*, chapter 6. It is not, of course, suitable for hinges to fire resisting doors.

Polyacetates (POM)

These have a dense crystalline structure and resemble metals in many respects. They are 'heavy' (SG = 1.42), strong and rigid but they are resilient while having very little creep under continuous stress. They have high resistance to heat, abrasion and organic chemicals and dimensional stability is good. However, polyacetals are costly and their use is mainly confined to plumbing components such as taps, gear wheels, etc.

Polycarbonates (PC)

These also have properties which may justify their high cost in uses such as vandal-resistant and burglar deterrent 'glazing'. Polycarbonates are dense, hard and tough. They are the strongest transparent materials, with tensile strength and ductility rather like metals. They are transparent with a slight amber tint but 86 per cent light transmission. The softening point of polycarbonates is high, about 130°C, and they are virtually self-extinguishing. See also page 286.

Vandal-resisting 'glazing'

'Lexan' polycarbonate (a Dutch product), in thicknesses from 1 to 13 mm, is vandal-resistant. It is claimed to be 250 times stronger and 50 per cent lighter (1.2 kg/m² per 1 mm thickness) than 'safety glass'. It is available transparent, transparent/tinted, and obscured and can be surface coated. *Lexan* can be safely bent cold to a radius not less than 100 times the sheet thickness, or it can be thermoformed. It has low flammability, is self extinguishing (BS 2782:1970 Method 508 D) and is the BS 476 Part 7:1971 *Spread of flame* rating is *Class 1*.

Maximum water absorption 0.35 per cent by weight (after 24 hours immersion at 20°C).
Moisture movement is neglible.
Thermal movement: 0.67×10^{-6} per deg C.

An expected product life of ten years may be reduced by incorrect fixing or by the use of abrasive cleaning materials.

It is not adversely affected by service temperatures from – 20°C to 120°C. Five years exposure of the UV stabilised *Lexan* sheet to sunlight in Florida, USA caused no loss of impact strength. It is not damaged by insects or fungi or by most chemicals, but strong alkalis affect it.

Light transmission of clear *Lexan* sheet is 84.0 – 89.7 per cent varying with thickness, and of tinted sheet 50 per cent.

Solar energy transmission of tinted sheet is 57 per cent.

UV transmission of 3 mm sheet
0 per cent	at 385 nonometres
50 per cent	400 nonometres
virtually 100 per cent	385 nonometres

Thermal resistance is three times that of glass and conductivity is 4.748 k cal/m²h°C (glass is 5.6 k cal/m²h°C).

Strength in tension is 3.16 kN/m².

Examples of recommended thicknesses of sheet and fixing dimensions for 200 kg/m² maximum static load are:

| Size of pane | Thickness | Glazing rebate | |
| | | Minimum width | Minimum 'edge engagement' of sheet* |
mm	mm	mm	mm
300 × 300	3	8	6
450 × 600	3	10	6
600 × 600	4	15	11
600 × 1200	4	25	20
1200 × 1200	6	27	22
1200 × 2400	6	28	22
1500 × 2500	9.5	31	25
1830 × 2500	13	34	28

*width of cover provided to edge of sheet by sash

The sheet, cut with a fine-toothed saw, must be held by beads, with elastomeric sealant on four edges. Putties which harden must not be used.

Windows should be washed only with mild soap or detergents and not in direct sunlight. Abrasives, even paper towels, should not be used.

Hair-line scratches can be removed with polish after cleaning the surface.

Cellular glazing

Polygal is cellular acrylic coated 4 – 10 mm double wall and 16 mm triple wall sheet in clear, bronze and opal finishes. 6 mm and thicker sheets are *Class 1* BS 476 (*Spread of flame*).

Bullet-resistant 'glazing'

'*Lexgard*' 27 and 33 mm thick laminates consist of *Lexan* sheets bonded with a patented interlayer film and with MR 4000 mar-resistant outer sheets. It is 60 per cent lighter than equivalent glass. Uses include protection of bank and post office counters, valuable exhibits and merchandise.

THERMOSETTING PLASTICS

Cross-linking, which produces a characteristically rigid structure, is brought about by a chemical curing agent (*catalyst* or *hardener*). In resins such as epoxides and polyesters the reaction occurs at room temperature, while in resins such as phenol formaldehyde (*Bakelite*) the catalyst becomes active only when subjected to heat and pressure. Generally, the higher the temperature the more rapid is the cure.

Subsequently, thermosetting plastics cannot be appreciably softened under the influence of heat which is in many cases an advantage, and essential for structural adhesives.

The properties and costs of *thermoset* products are influenced very much by fillers such as wood flour, asbestos fibre, cotton flock, silica and metallic powders and reinforcement such as glass or asbestos fibres.

Thermosetting plastics are generally relatively rigid and hard and resist scratching to varying degrees. Compared with thermoplastics creep is very small, although greater than that of most metals.

The main use of thermosetting plastics in build-ing are as impregnants for paper fabrics, adhesives, binders for glass-fibre reinforced plastics and binders in paints and clear finishes.

The main thermosetting plastics are:

Phenol formaldehyde (PF)
Urea formaldehyde (UF)
Melamine formaldehyde (MF)
Resorcinol formaldehyde (RF)
Polyesters (UP)
Polyurethanes (PU)
Epoxide resins (EP)
Silicones (SI).

The main properties and uses of these plastics are as follows:

Phenol formaldehyde (PF)

Phenol formaldehyde or *Bakelite*[1], first produced commercially in 1910, is the cheapest thermosetting resin.

Various properties result from differing resin formulations and wood flour, cotton flock or macerated fabric and asbestos give brittle, tough and heat resistant mouldings respectively. Products are usually black or brown in colour. When a lighted match is held to the corner of a moulding the vapour has a characteristic odour of phenol.

Uses include:

Mouldings, eg electrical accessories, door furniture, WC seats (dark colours) (BS 1254: 1971).
Impregnants for paper and fabric laminates – see page 288.
Paints – see *MBS: Finishes*, chapter 6.
Adhesives – see page 299.
Cellular or foamed products – see *Cellular Plastics*, page 292.

Urea-formaldehyde (UF)

A clear thermosetting resin; products are usually white or brightly coloured. It is self-extinguishing and has a fishy smell when it burns.

---

[1] *Bakelite* is the trade name for phenolic materials manufactured by Bakelite Zylonite Ltd.

Uses include:

Mouldings, eg electrical accessories, WC seats (BS 1254: 1971).
Paints, including stoving enamels – see *MBS: Finishes*, chapter 6.
Adhesives – page 299, eg for particle board manufacture.
Cellular or foamed products – see *Cellular Plastics*, page 292.
Paper and textile treatments.

Melamine formaldehyde (MF)

Melamine resins are clear but can be made in a wide range of bright and lightfast colours. They resist hot and hold water better than urea formaldehydes and modern resins are claimed to give good durability when they are exposed externally. Being hard and very resistant to cigarette burns they are used to surface *decorative paper laminates* – page 288.

Like urea formaldehydes, melamine formaldehyde has a fishy smell when it burns and is self-extinguishing.

Other uses include:

Mouldings, eg door handles.
Clear finishes – see *MBS: Finishes*, chapter 6.
Surfaces to hardboards and plywood.
Adhesives – page 299.

Resorcinol formaldehyde (RF)

The chief use for this type of resin which is dark red in colour is as a water and boil-proof adhesive for wood – page 299.

Polyester resins (UP) (unsaturated)

The available resins have a wide range of properties. Some can withstand temperatures over 230°C for short periods without degradation.

Polyester resins harden without heat or pressure and are used mainly with glass fibre reinforcement in GRP (see page 287) and in polymer concrete (see page 172). Different types of polyester are made as films, those stuck on glass to improve shatter resistance or to provide solar control and another film coated with aluminium is stretched over a frame to provide a light-weight condensation free and 'safe' mirror. Polyester resins are also used in Paints and Clear finishes – see *MBS: Finishes*, chapter 6.

In situ floor coverings with aggregate – see *MBS: Finishes*, chapter 1.

GRPC is glass fibre reinforced polymer cement-free 'concrete' see page 237, not to be confused with polymer impregnated conventional concrete – see page 172.

Polyurethanes (PU)

This group has an even wider range of properties than polyesters.

Uses include:

Paints
Clear finishes } *MBS: Finishes*, chapter 6.
Sealants, chapter 16.
Foams – see *Cellular plastics*, page 292.

Epoxide resins (EP)

For most uses, epoxide systems are provided in two parts, as a resin and curing agent or hardener and the properties of the latter have considerable effect on the physical properties of the hardened product as also have inclusions such as glass fibres, mineral fillers and aggregates, fabrics and metallic powders.

Epoxide resins are extremely tough and stable, have excellent electrical properties and very good resistance to chemicals, especially to acid and alkaline solutions. They adhere well to most materials, including impervious ones, partly because no volatiles are released during hardening. For the same reason hardening shrinkage is very small. Adhesion to timber in wet conditions, however, is not good.

Uses include

In situ floorings – see *MBS: Finishes*, chapter 1
Concrete repair compositions
Paints and
Clear finishes – see *MBS: Finishes*, chapter 6
Glass-fibre reinforced plastics – page 287
Adhesives – page 299

Silicone resins (SI)

Uses include:

1 Transparent waterproofers and water repellents for the surfaces of stonework and brickwork – see *MBS: Finishes*, chapter 6.
2 Chemicals for injecting into walls to prevent damp rising where conventional damp proof courses are not present, or are defective. BS 6576:1985 is a *CP for installation of chemical dpcs against rising damp in existing buildings* BRED 245 *Rising damp in walls diagnosis and treatment* does not recommend these systems in highly alkaline conditions such as obtain in new walls containing Portland cement mortar.
3 Sealants – see page 310.
4 Paints – see *MBS: Finishes*, chapter 6.
5 A lubricant in floor and furniture polishes.

OTHER PLASTICS PRODUCTS

Floorings – see *MBS: Finishes*, chapter 1.
Products considered here are:
    Pipes
    Transparent and translucent plastics
    Fibre-reinforced plastics
    Sheet laminates
    'Improved wood'.

Pipes

British Standards include:
BS 5955 (two parts) *CP for plastics pipework, (Thermoplastics materials)*
BE 5255:1976 *Plastics waste pipes and fittings made from ABS, MUPVC, PP and PE*
BS 3943:1979 *Plastics waste traps*
BS 4962:1982 *Plastics pipes for use as subsoil drains.*

Tests carried out at the BRE on plastics pipes for hot water services[1] show that ordinary PVC and high density polythene are unsuitable, and ABS and polypropylene are likely to be suitable only in closely controlled and relatively low temperature systems. Recently developed chlorinated PVC (CPVC) has relatively good high temperature characteristics and 'shows promise'. It is interest-

[1]BRE current paper (CP 7/78) *Trial of plastics pipes for hot water services.* J R Crowder and A. Rixon.

ing to note that even here, although thermal movement is reduced, success must await the development of an expansion joint which will accommodate movement which is three times greater than that of copper pipes.

Transparent and translucent plastics products

Transparent plastics are lighter and tougher than glass, do not break into dangerous fragments and some of them, eg *Perspex*, transmit more light than glass. Unlike glass, thermoplastics can be bent cold to gentle curves and they can be thermoformed to complex shapes such as pyramidical roof lights. On the other hand, plastics are more costly. They are more easily scratched and like all plastics tend to acquire an electrostatic charge which attracts dust. Also, their weathering properties vary but in all cases are inferior to those of the equivalent pigmented plastics and glass.

BS 6262:1982 a *CP for glazing for buildings* relates area and thickness of panes to wind loadings, and recommends methods of fixing. Generous allowances must be made for thermal movement which is much greater than that of glass. Mirrors and frameless doors are referred to. Toughness and flexibility make some plastics suitable for 'see through' heat-saving crash doors in factories.

The softening point of the thermoplastics is much below that of glass and all plastics burn, although flame retardant grades are available.

Transparent forms of the following plastics are used in building:

polythene (films only)	
polyvinyl chloride	
polymethyl methacrylate (*Perspex*)	
cellulose acetate	Thermoplastic
cellulose acetate butyrate	
polystyrene	
polycarbonate	
melamine formaldehyde (clear finishes including surface on decorative laminates)	
	Thermosetting
polyurethane (clear finishes)	
epoxides	

Glass-fibre reinforced polyester resin products, which are translucent, are described below.

Fibre-reinforced plastics

These products are formed from fibres arranged in uni-directional, multi-directional or random patterns and impregnated with synthetic resin. Some complex shapes can be formed and very large mouldings are possible. The fibres are usually glass, in mat, roving or fabric form, hence *glass fibre reinforced plastics* (GRP), but cellulose, asbestos and other fibres are sometimes used.

The resin is usually polyester, but here again, epoxide, phenolic and other resins which can be cured at atmospheric pressures are used.

Hand *lay-up* of the resin and fibres has a high *labour content* but the process is simple requiring only a rigid mould. Surfaces can be improved by withdrawing air from the underside of the *lay-up* or by pressing it against the mould by means of an inflated bag.

For mass production, mechanized spray lay-up is employed and hollow articles can be formed by winding resin-coated fibres around a former (filament winding). Articles such as water cisterns and tanks are pressed between metal dies, giving water resistance superior to that obtained by cold contact moulding, greater accuracy and a good finish on both sides. Dies are often heated to give more complete curing.

Glass-fibre reinforced polyester resin products, are dealt with in *GRP and Buildings*, A J Leggatt, Butterworths. GRP products, especially those with unidirectional reinforcement, have high strength and resistance to impact, and a strength: weight ratio superior to that of mild steel. Stiffness is greater than many plastics although much less than that of mild steel. The best products are likely to remain structurally sound for considerably more than 30 years.

Externally, exposure of the fibres in poor products can slowly lead to the breakdown of their bond with the resin, and surfaces weather better if they are protected with a *gel-coat* of resin. Difficulty in repairing scratched surfaces makes GRP less suitable for urinals, baths and basins than PMMA. Some GRP sheet is now protected with a PVF film.

Typical unpigmented products have initial light transmissions up to 70 per cent, but inferior ones especially in self-extinguishing grades, may suffer serious loss of transparency in 10 years. Colours exposed to the weather may fade in about 5 years.

Ordinary glass fibre reinforced polyester plastics have *Class 3 medium spread of flame* (BS 476:Part 7:1971) but with flame retardant additives they can be raised to *Class 2* or even to *Class 1*.

British Standards for glass fibre reinforcement are listed on page 266.

Uses include:

Translucent sheets (flat and profiled, clear or pigmented) BS 4154: *Corrugated plastics translucent sheets made from thermosetting polyester resins (glass fibre reinforced): Parts 1 and 2 1985.*

Slates Redlands make an interlocking roofing slate comprising crushed natural slate in a resin binder reinforced with glass fibre.

The interlocks allow single laps so that the weight of the man-made slating is reduced.

Rooflights (50.8 mm thick with honeycomb core)

Cold water cisterns and hot water cylinders BS 4994:1973 describes *Vessels and tanks in reinforced plastics*

Cladding panels – see BRE Digest 161

Architectural features, eg church spires, pinnacles, dormer windows and door hoods.

Formwork for concrete – temporary and permanent.

Pipes (BS 5480 describes: *GRP pipes and fittings for water supply or sewerage*) Part 1:1977 *Dimensions, and materials classification*, Part 2:1982, *Design and performance requirements*.

Cesspools

Drain inspection chambers

Baths and shower cubicles

Portable cubicles

Urinals

Copings

Roof edge-trim – for mastic asphalt and felts.

Hollow sections can be used for a variety of purposes such as: guard rails, fencing, cable troughing and conduits for light structural work.

Windows with hollow frames jointed with aluminium spigots

Flagpoles up to 18 m high, are designed to resist winds up to 161 km/h

Nuclear fall-out shelters

Security glazing reinforced with expanded steel mesh.

Sheet laminates

Laminates of paper, wood, asbestos, glass fibre and fabrics, impregnated with phenolic, epoxy, melamine, polyester and silicone resin binders are hot pressed into sheets.

Some products which are extremely tough have strengths approaching that of some metals and are suitable for mechanical parts such as gear wheels. Their appearance is however dark and unattractive and in time they lose gloss, and fade.

Decorative paper laminates have a pattern printed on the top sheet of paper and this is surfaced with clear melamine formaldehyde with a matt, satin or gloss finish. In addition to a very wide range of standard patterns, artists' originals can be incorporated to order. The surface is impervious and highly resistant to organic and dilute mineral acids, alcohols, oils and in some cases to alkalis. It is easy to keep clean, does not taint foodstuffs and is resistant to burns by cigarettes so that the cigarette resistant grade incorporating an aluminium foil lamina is not usually required. However, the surface is scratched by coins passed over counters, and is not suitable as a cutting board in kitchens.

Special grades are now deemed to be sufficiently durable for use externally.

BS 3794:Part 1:1982 describes *Decorative laminated plastics sheets based on thermosetting resins* mainly by performance and as follows:

Class 1 1.6 mm nominal thickness, with one decorative surface and the reverse roughened or treated to aid adhesion to a substrate

Class 2 3.2 mm nominal thickness, with a decorative surface both sides

Classes 1A and 2A include a metal foil lamina.

The BS recognizes, *general purpose, horizontal* and *vertical* categories – the latter for the least demanding uses. *Post-formable* and *fire-retardant* grades are also described.

A typical product consists of ten sheets of kraft paper, impregnated and bonded with phenol formaldehyde. The laminate is 1.6 mm thick in sizes up to 2743 × 1219 mm and the back is usually sanded to give a key for adhesive.

Sheets can be sprung to a radius of 150 mm below which a *post-forming grade* is required. They must be adhered overall to a rigid background (they are available ready-mounted on plywood).

The edges of laminates are included to delaminate if they are knocked and should be protected.

Other *British Standards* for plastics laminates are:

BS 4965:1983 *Decorative laminated plastics sheet veneered boards and panels.*

BS 5102:1974 *Phenolic resin bonded paper sheets for electrical purposes.*

BS 2572:1976 *Phenolic laminated sheets and epoxide cotton fabric sheets.*

BS 3953:1976 *Synthetic resin bonded woven glass fabric laminated sheets.* 0.4–50 mm thick. These sheets, made with low-alkali (*E*) glass, are generally more heat resistant and have greater impact strength than laminates having paper or cotton fabric bases.

Improved wood

These products are timber, impregnated with resin, and subjected to heat and pressure.

Whereas the specific gravity of plywood is about 0.5, improved wood, by reason of the high resin content and the pressure to which it is subjected, has a specific gravity up to 1.35.

An important use in building is as flooring.

Particle boards

These consist of wood or other cellulosic particles bonded under pressure usually with urea formaldehyde – see page 100.

CELLULAR PLASTICS (FOAMED AND EXPANDED PLASTICS)

Some plastics can be formed into rigid or flexible cellular materials either by a chemical change which causes an additive to evolve gas or by aeration of the soft plastic.

BRE Digest 224 *Cellular plastics for building* should be studied. Table 97 gives properties of

Material	Density	Compressive stress at yield point	Coefficient of linear expansion	Thermal conductivity at 10°C	Thermal resistivity (Reciprocal of thermal conductivity value) l/k	Maximum temperature recommended for continuous operation	7 days' water absorption	Water vapour diffusance[2] 25 mm thick board at 18°C	Behaviour in fire
	kg/m^3	$N/m^2 \times 10^4$	per °C $\times 10^{-5}$	W/m °C	m °C/W	°C	Volume per cent	MN s/gm	
Expanded polystyrene bead boards extruded extruded, with surface skin	{16 24 32 40 (mean)	7 12 27 27	5–7 7 7 7	0.035 0.033 0.035 0.032	29 30 29 ~31	80[1] 80[1] 75[1] 75[1]	3.0 2.5 1.5 1.0	270 420 1 300 1 300	Burns fairly rapidly often softens and collapses Flame-retardant grades available
Expanded polyvinyl chloride (PVC)	40 72	27 90	3.5 5	0.035 0.043	29 23	50 50	3.0 3.8	800 1 300	Collapses but burns with difficulty
Foamed urea-formaldehyde (UF)	8	Negligible	9	0.038	26	100	10	22	Resistant to ignition
Foamed phenol-formaldehyde (PF)	48	14	2–4	0.036	28	150	*	35 240	Highly resistant to ignition
Foamed rigid polyure-thane (fluorinated hydrocarbon blown)	32	17	2–7	0.01–0.03	100–33	100	2.5	360	Generally burns rapidly, with thick smoke. Flame-retardant grades available

[1]These temperatures may be slightly lower for flame-retardant grades. Manufacturers' advice should be sought in cases of doubt
[2]15MN s/g is considered a suitable value for a vapour barrier in building applications, but because of the risk of interstitial condensation within a cellular material, an additional vapour-sealing skin may be required
*High – open cell
Fairly low – 95 per cent closed cell

Table 97 Typical properties of cellular plastics used in building. Based on BRE Digest 224:1979, Cellular plastics for building

289

various cellular plastics and table 98 gives suitable uses.

Densities commonly vary from 16 to 72 kg/m^3 so that although they are combustible, in the lower densities, contribution to the *fire load* is negligible and 'k' values as low as 0.020 W/m deg C are obtainable.

Cellular plastics have either *closed* or *open* cells. Expanded plastics have a substantially closed-cell structure and foamed plastics can have either closed cells, or cells which are mostly interconnected. Closed cells provide the best thermal insulation and products float on water, but absorb very little. Those having a water vapour diffusance of 0.067 g/MNs are considered to be suitable as vapour barriers in building applications, although joints must be sealed, and an additional vapour sealing skin may be desirable. *Open* cell materials provide flexible material for upholstery, and air filters.

Foamed plastics contribute less to fire load than related dense products, but they burn more rapidly and smoke and toxic gases are serious hazards.

BRED 294 states that combustible cavity insulation is relatively safe if cavities are sealed and filled, and less so with board insulants which allow a vertical air gap.

Polystyrene

Expanded or foamed polystyrene (EPS) is resistant to fresh and sea waters, acids other than concentrated nitric acid, alkalis, alcohols and to animal and vegetable oils. It is very readily attacked by ketones, esters, chlorinated hydrocarbons, benzene, turpentine and ether. Adhesives and paints must not contain solvents. Ultra-violet light gradually embrittles surfaces but they can be decorated with water-based and emulsion paints, or with wallpaper. They can be plastered, preferably on a pva emulsion bonding agent.

BS 6203:1982 is a *Guide to the fire characteristics and performances of expanded polystyrene (EPS) used in building applications.*

EPS is combustible in all its forms. Carbon monoxide and of heavy dense black smoke –

Cellular plastics	Form	Blocks	Boards	Sheets	Mouldings	Loose fill	Liquid(s) for in situ foaming[1]
Expanded polystyrene { bead board extruded	rigid semi-rigid rigid	✓ ✓	✓ ✓	✓ ✓	✓	✓	
Expanded PVC	rigid flexible	✓ ✓	✓ ✓	✓	✓		
Foamed urea formaldehyde	rigid and friable	✓					✓
Foamed phenol formaldehyde	rigid and friable	✓					✓
Foamed polyurethane	rigid semi-rigid flexible	✓ ✓	✓	✓ ✓ ✓	✓ ✓ ✓		✓ ✓ ✓
Expanded polythene	semi-rigid			✓			
Expanded ebonite	rigid		✓				

[1]For filling cavities or coating surfaces

Table 98   Types, forms and uses of cellular plastics

greater than for equal masses of other materials – are produced. Flame retardant additives only delay ignition and generally increase the smoke produced.

The calorific value is about twice that of an equal mass of timber, but as EPS is 98 per cent voids the calorific value by volume is a small fraction of that of timber. 16 kg/m^3 density EPS requires 150 times its volume of air for combustion.

Surface spread of flame is *Rapid – Class 4*, BS 476 – more so when coated with oil gloss paint, which should never occur. Plaster skim coats, aluminium films, flame retardant paints, intumescent coatings and water-based paints with a high inorganic content.

Securely fixed facings, such as 9 mm plasterboard and 10 mm gypsum plaster can delay ignition, but flaming occurs where molten polystyrene or gas escapes through joints or fissures.

BS 6203 recommends methods of installing linings and incorporating them in constructions without reducing their overall fire resistance.

Expanded polystyrene boards are manufactured either: (a) from beads or (b) by extrusion.

(a) Beads of expandable polystyrene are heated to form closed cell products as light as 16 kg/m^3 with a low k value of 0.035 W/m deg C and very low compression strength, 7×10^4 N/m^2. Bead-boards cut from blocks by an electrically heated wire have a smoother surface than those cut by a band saw.

(b) Extruded boards have a simpler and more regular structure. They are slightly denser but absorb less water and have lower water-vapour diffusance. They retain good thermal resistance in damp positions, eg behind cladding applied to existing external walls, below water-proof membranes to solid ground floors and above water-proof coverings on 'upside down' flat roofs. The latter insulation must be weighted down, and protected by gravel or paving slabs against mechanical damage and against ultra-violet degradation – see *MBS: Finishes*, chapter 7.

BS 6203 grades boards and blocks as:

  EHD   extra high duty
  HD   high duty
  SD   standard duty
  UHD   ultra high density
  ISD   impact sound duty
  SHD   special high duty

They are made in two densities 16 and 24 kg/m^3. *Grade SE* is *self extinguishing* when tested in the manner prescribed, with *spread of flame* equivalent in hazard to *Class 1* BS 476. (As a wall and ceiling lining the boards absorb little sound even in the form of perforated *acoustic* tiles). BS 3932:1965 *Expanded polystyrene tiles and profiles for the building industry* deals with products which are usually made by the bead-fusion process. Precompressed boards sandwiched between a floating screed and a structural base in suitable constructed floors are effective in reducing the transmission of impact and airborne sound – see *MBS: E and S*.

Although water vapour permeability is low, where condensation is likely BRE Digest 224 advises the addition of a polythene or aluminium foil vapour barrier. Similarly, in very wet conditions, a water-proof barrier is recommended to prevent loss of thermal insulation.

Boards are used as preformed roof 'screeds', wall and ceiling linings, as insulation to ground floors and below floor warming installations and as cores for sandwich panels. Boards are also used for lagging cold water cisterns, but they are not suitable for insulating hot water systems, which do not have closely regulated temperature control.

Expanded polystyrene can be formed, or cut from blocks, in special shapes including preformed lagging for cold water pipes. It can be carved and used as formwork to produce relief on concrete surfaces. It is useful for forming holes and channels in concrete where crushing will not occur and subsequent removal is easy by burning, solvent action or by mechanical means.

Hollow concrete wall blocks are available with the voids filled with expanded polystyrene. Unfused expanded beads are being used as a loose fill for cavities and as an aggregate for lightweight concrete.

Polyvinyl chloride

Plasticized PVC is available in the cellular form as *flexible* sheets, and as rigid cellular PVC. The cells are closed and very small and permeability to water vapour is low.

Expanded rigid PVC is more expensive than expanded polystyrene but much less flammable and it is relatively strong. In a density of 72 kg/m³ the crushing strength is 90×10^4 N/m² and it is a useful core for thermal insulating stressed-skin constructions. The surface gives a good key for plaster with uniform suction. BS 3869:1965 deals with *Rigid expanded PVC for thermal insulation purposes and building applications*. It specifies two densities:

(a) 1.5 and up to (but not including) 24.03–32.04 kg/m² and
(b) 48.06 kg/m³ and above

in blocks, boards and sheets not less than 12.7 mm thick for use in temperatures up to 50°C. It describes a self-extinguishing grade (*SE*) and method of testing. Flame-resistant products are usually coloured pink.

Skirting, architrave and similar sections, which can be worked with joiners' tools, are made in cellular PVC with a dense 'ready decorated' skin.

Urea formaldyhyde

The cellular product has poor strength properties but is comparatively inexpensive and has been widely injected into cavity walls with considerable improvement to their thermal insulation. However, foamed urea formaldehyde tends to shrink and fissure so that good control is required to completely fill cavities.

In exposed areas where walls may be subject to severe driving rain, walls should be rendered and the BRE states that: 'the technique should not be used in buildings where rain penetration problems have previously occurred'. Also more than 0.5 ppm UF in air irritate the eyes, nose and respiratory tract. The inner leaves of cavity walls must therefore be soundly constructed.

Wall cavities must not be filled before construction is complete in case mortar droppings collect on the set foam and convey moisture from the outer to the inner leaf.

BS 5617 a specification and BS 5618 a CP deal with *UF foam systems for thermal insulation of cavity walls having masonry or concrete inner and outer leaves.*

Phenol formaldehyde

Phenolic foams from 16.02 to 320.37 kg/m³ can be formed in situ with special equipment by stirring a rapid-acting acid hardener into liquid phenolic resin to which a foaming agent has been added.

The material is self-extinguishing and can be used at a continuing temperature of 130°C or for shorter periods up to 200°C.

BS 3927:1965 describes: *Phenolic foam materials for thermal insulation and building applications.*

Polyurethane

The material is available in flexible grades with an open or closed cell structure, and rigid grades with closed cells.

Density can be controlled down to about 24 kg/m³ with a high strength:weight ratio. Foamed polyurethane provides excellent thermal insulation (k is 0.025 W/m deg C for a density of 32 kg/m³). Although more expensive than foamed polystyrene, it is suitable for use at higher temperatures. There is no odour, transmission of water vapour is low and dimensional stability and resistance to many chemicals and solvents is high. Foam burns rapidly (650°C in 30 seconds and up to 1200°C recorded). It evolves hydrogen cyanide and CO and much more smoke than an equivalent wood fire. Unfortunately, flame resistant foam produces twice as much gas as normal foam. Foamed polyurethane should not be used where it could be exposed directly to flame in the event of fire. An air gap between foamed polyurethane boards and wall or ceilings substrates is a serious spread of flame risk. The GLC banned such uses. It adheres strongly to surfaces and is used as a structural core in stressed skin sandwich construction.

For thermal insulation foamed polyurethane is used to fill cavities in some proprietary lightweight wall blocks and it is injected into cavities in external walls, although in severe exposures shrinkage of the foam has sometimes led to water penetration – see BRE Digest 236, *Cavity insulation.*

Another use is as bitumen-impregnated foamed polyurethane preformed strips for sealing joints – see page 307.

Isocyanurate foams based on polyurethane materials have similar physical properties, but greater thermal stability and resistance to oxidation and therefore less rapid spread of flame. Isocyanurate foams are available as blocks, sheets and laminates.

British Standards include:

BS 3379:1975 *Flexible urethane foam for load-bearing applications.*

BS 5608:1978 *Preformed rigid urethane and isocyanurate foam for the insulation of pipework and equipment.*

BS 4841 *Rigid urethane foam for building applications* comprises:

Part 1:1975 *Laminated board for general purposes* eg as bats in cavity walls and insulation under floor screeds.

Part 2:1975 *Laminated board for use as a wall and ceiling insulation.*

BS 5241:1975 *Rigid urethane foam when dispensed on a construction site*, eg as a cavity fill.

Rubbers

Rubbers are similar to thermosetting plastics but differ in the ease with which the molecular chains are able to uncoil.

In 1844 Charles Goodyear (USA) patented a process for heating natural milky white rubber latex, a natural polymer, with sulphur to make a non-tacky product. In manufacture, acids are added to the latex. The spongy coagulum is first 'masticated', fillers are added to modify the properties (and price), and carbon black to increase strength in tension and improve wearing properties. After forming, the product is vulcanised by heating under pressure, usually with sulphur. The process is analogous to the hardening of thermosetting plastics, in which the individual macro molecules cross-link (or bridge) forming a three dimensional network, although in rubbers cross linking is sufficient only to provide the required degree of resilience. In vulcanization strength and elasticity are increased and sensitivity to changes in temperature is reduced. *Ebonite* is a fully vulcanized hard rubber.

Today, modified-natural and synthetic rubbers, often called *elastomers*, have a very wide range of properties and are used in many building products. Unlike natural rubbers, materials such as *Neoprene*[1] and butadiene-acrylonitrile have good resistance to oils and solvents. *Hypalon*[1] a chlorosulphonated polyethylene is virtually immune to the action of ozone, oxygen and sunlight. Butyl is extremely tough, has excellent resistance to acids, good weather resistance and is unique in its low permeability to air. Nitrile rubber has excellent resistance to oil and aromatic hydrocarbons and is used in adhesives.

Polysulphides have excellent resistance to oils and solvents and low permeability to gas. Silicones are resistant to many oils and chemicals. They have outstanding heat resistance and good electrical properties. Mechanical properties, however, are generally not so good as those of other synthetic rubbers.

Recently developed ethylene-propylene rubbers have extremely good resistance to ozone and to ageing, and look promising for use as roof coverings.

Uses include:

Adhesives – see chapter 14

Adhesive tape

Paints – see *MBS: Finishes*, chapter 6

Anti vibration and sound absorption. Rubber is useful as anti-vibration mountings for machinery (see DD47:1975 *Vibration isolation of structures by elastomeric mountings*) and as a floor covering or underlay to absorb impact noise. Rubber can also be used as resilient mountings for floorings.

Roof coverings. In butyl or composite *Neoprene, Hypalon*[1] sheets are a recent development. It remains to be seen whether cost-in-use is competitive with bituminous felts – see *MBS: Finishes*.

Mastics. Butyl, polysulphide and silicone synthetic rubber mastics are used for sealing joints (see chapter 16), and natural rubber is often included in bitumen-based mastics.

Gaskets, *Neoprene* is particularly able to maintain tight contact with surfaces, but natural rubber butyl and PVC are also used – see chapter 16.

Preformed seals for wall cladding joints. *Neo-*

[1]Trade names.

prene extrusions are suitable as loose baffles and for tubular or cruciform sections to be held in compression.

Electrical insulation.

Thermal insulation. This is provided by *Onazote*[1],

[1] Trade name

an expanded ebonite, manufactured by the Expanded Rubber and Plastics Co Ltd. Density is about 64 kg/m³. It has a closed cell structure and water absorption is very low, ie 1.5 per cent after 7 days. The 'k' value is low, 0.029 W/m deg C. It can be used continuously at temperatures up to 50°C.

Natural adhesives have been in use for thousands of years, but recently the development of synthetic products which give rapid bonds, extremely high strengths and which are durable in damp conditions or where exposed to the weather has brought adhesives into the structural and mass production fields. Most combinations of surfaces can be successfully bonded, a large range of products is available, although some of them are highly specialized. In the absence of standards for quality and for methods of testing, the advice of manufacturers should be sought in the selection of adhesives for particular uses.

The term 'glue' is usually applied to animal and fish adhesives – particularly for wood. The term 'cement' should not be used to described organic adhesives.

Good references are:

Adhesives, Eric W Allen in *Specification*, Architectural Press.

Adhesives Guide, Joyce Hord BSc, British Scientific Instrument Research Association.

Adhesion and Adhesives, R S R Parker and P Taylor, Pergamon Press.

Adhesives Handbook, J Shields, Newnes-Butterworth.

Industrial Adhesives and Sealants, ed B S Jackson, Hutchinson Benham.

BS 6138:1981 *Glossary of terms used in the adhesives industry.*

BS 5442: Classifications of adhesives for construction,

Part 1:1977 – *for use with flooring materials,*

Part 2:1978 – *for use with wall and ceiling coverings*, (excluding rolled, decorative, flexible materials). This part refers to possible effects of wall and ceiling linings on Surface spread of flame ratings. It also refers to flammability of adhesives, and risks of inhalation in unventilated spaces, during and immediately after use.

Part 3:1979 – *for use with wood,*

BS 5980:1980 describes *Adhesives for use with ceramic tiles and mosaics.*

BS 3046:1981 *Adhesives for hanging flexible wall coverings.*

BS 5268 *Structural use of timber*, Part 2:1984.

Preferred adhesives are given in each specification.

BRE Digest 175, *Choice of glues for wood*, HMSO.

BRE Digests 211 and 212 *Site use of adhesives* HMSO.

Adhesion may be derived from *specific adhesion, mechanical adhesion*, or both. Specific adhesion is due to molecular attraction between surfaces which are in very intimate contact. This happens in mica and between smooth sheets of damp glass without any intervening *adhesive*, and where liquid adhesives with good 'wetting properties' make similarly close contact with surfaces. Mechanical adhesion occurs where bonding agents key into porous surfaces.

For maximum bond strength, surfaces should be brought into close contact to give a thin glue line. With wood/wood bonds a gap of 0.076–0.15 mm gives the greatest strength. The term *gap-filling* can be misleading. BS 1204: *Synthetic resin adhesives (phenolic and aminoplastic) for wood:* Part 1:1984. *Gap-filling adhesives* describes them as those which do not craze due to initial shrinkage when they are used in films up to 1.3 mm thick. Part 2:1979 deals with *Close-contact adhesives* with 'gaps' up to 0.1 mm.

Contact adhesives give *instant tack*, but generally surfaces must be clamped together until a bond is achieved, while avoiding excessive pressure which would cause 'glue starvation'. Some modern adhesives give very good bonds even between smooth metal and glass, but roughening or etching of such surfaces is usally desirable.

Surfaces to be joined must be firm, clean, usually dry, and free from grease, so that the glue is not adversely affected either mechanically or chemically.

Glued joints are mechanically efficient. By distributing loads evenly over wide areas they avoid the concentrated local stresses which are the weakness of nails, screws, bolts and rivets. Although all *wood adhesives* are stronger than most timbers, ie those having shear strengths less than about 21–28 N/mm². Consequently durability and convenience in use are usually the criteria in choice of glues for woodwork, rather than strength. Table 99 compares the durabilities of adhesives for wood. Wood preservatives[1] and water-repellants are not compatible with many glues: in some cases only resorcinol formaldehyde is suitable. Specialist advice should be obtained in this respect. At present no glue gives a permanent structural bond with timber that has been treated with fire retardants. Where there is an existing coating such as paint on a surface, the effective bond strength can be no greater than that of the coating to the base.

Joints which are subject to shear stress are stronger than those subject to tension, and peel stress should be avoided. The length/thickness ratio is important and although with overlaps up to about 25 mm the increase in strength is approximately linear, with greater overlaps it does not increase proportionately – see figure 49.

Adhesives set in one of the following ways by:

1 Gelling on cooling, a process which is reversed by reheating, eg animal glues.

[1]See – *Gluing preservative treated wood* (BRE Technical note 13).

2 Loss of moisture or solvent by evaporation and absorption into the materials being joined, eg starch pastes, PVA and rubber-based adhesives.

3 Loss of moisture and some chemical change, eg casein and thermosetting resins such as urea-phenol-, melamine and resorcinol formaldehyde.

4 An irreversible chemical reaction accelerated by a hardener or catalyst, eg epoxides or polyesters, either induced by heat, or at room temperature, by reacting two chemicals together.

5 Hardening on cooling (hot melts).

Shelf life is the time during which an adhesive can be stored without deterioration.

Pot life is the time available for using an adhesive after it has been prepared.

Closed assembly time is the period during which the parts can be moved to obtain the exact position desired.

The durability of adhesives varies widely. Forest Products Research Bulletin 38:1968 edition *The efficiency of adhesives for wood* describes the behaviour of many types. BS 1204 gives classifications for phenolic and amino plastics, summarized here as follows:

Type of adhesive	Full exterior conditions	Semi-exterior and damp interior conditions	Dry interior conditions	Class (BS 1204)
PF, RF and combinations	25 years expected	Indefinitely		WBP
MF, MF:UF, Fortified UF	5–10 years	10–20 years estimated		BR
UF	2–5 years	5–10 years		MR
Casein	1–2 years	2–5 years	Indefinitely	—
PVA	Comparable with casein[1]			—
Animal	Fail in a few months	1 year		—

[1] FPRL Bulletin no. 20

Table 99   Durability of wood joints bonded with various adhesives

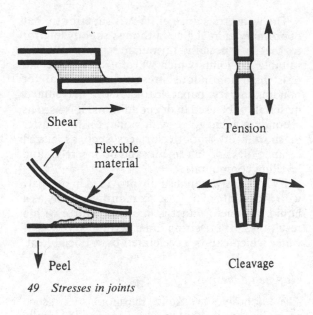

Shear

Tension

Flexible material

Peel

Cleavage

49   Stresses in joints

WBP   *(Weather and boil-proof)*   Highly resist-
        ant to weather, micro organisms, cold and
        boiling water, steam and dry heat.
BR   *(Boil resistant)*   Good resistance to boiling
        water, but fail under prolonged exposure to
        weather. Withstand cold water for many
        years. Have good resistance to micro-organ-
        isms.
MR   *(Moisture resistant)*   Moderately weather
        resistant and will survive full exposure for a
        few years. Withstand prolonged exposure
        to cold water, but only limited exposure to
        hot water. Resistant to micro-organisms.
INT   *(Interior)*   Withstand cold water for a
        limited time only and only suitable for
        occasional damp conditions. Not necessar-
        ily resistant to micro-organisms.

The requirements of BS 656 Part 8:1985
*Specification for bond performance of adhe-
sives for plywoods* are similar but the BR
designation is replaced by CBR which is
assessed by a *cyclical boil resistant test*.

Broadly, adhesives can be classified by:
1 properties:
   close contact or *gap-filling*,
   contact or *requiring sustained pressure –
   internal only* or *external, rigid* or *flexible*
2 materials to be joined:
   eg flooring adhesives and ceramic tile

adhesives. *MBS: Finishes*, chapters 1
and 4
3 composition of adhesive.

Adhesives are considered here under the
latter heading with particular regard to
their use for wood/wood, concrete/con-
crete bonds and for general purposes.

TYPES OF ADHESIVES

Adhesives derived from natural products

Starch

Flour with water or chemicals such as sodium
hydroxide which must be able to evaporate, has
been used for plywood suitable only for internal
use.

Cellulose derivatives

This group of adhesives is available as solid
'melts' solvent solutions or as water soluble
powders. Methyl cellulose, which is suitable for
joining porous materials is well known as wallpaper paste applied with water.

Animal glues

Animal glues made from hides, skins, bones and
sinews set slowly, are sensitive to moisture and are
now rarely used. See BS 745: 1969 *Animal glue for
wood* (joiner's glue). They are supplied in the solid
form or as a jelly for hot application, and as a
prepared liquid for cold application. In dry conditions animal glue gives strong wood/wood
bonds but above about 80 per cent RH it is
attacked by micro-organisms, and even where
fungicides are added animal glue and other natural glues are suitable only for interior uses.

Casein glues

Casein glue is made from soured mild curds dried
and ground to a fine powder and mixed with an
alkali and fillers. The *pot life* of powder dissolved
in cold water is 5 to 24 hours and hot-setting
caseins have a *pot life* of 12 to 24 hours after
liquid stabilizers or hardeners are added.

297

Casein is easy to use at ordinary temperatures. It sets by a chemical action which is not unduly sensitive to temperature, and by evaporation of water. In dry conditions, the hardening glue has good resistance to heat and develops high strength. However, when wet it loses strength and is not durable externally, Casein is suitable for bonding wood to linoleum, plasterboard, asbestos-cement and to decorative laminated plastics sheets and for general joinery – although the alkaline constituent causes dark red or purple stains with timbers such as oak which have a high tannin content.

Bituminous adhesives

BS 3940:1965 (withdrawn 1984) *Adhesives based on bitumen or coal tar* described types for: (i) laying wood flooring blocks and mosaic, linoleum sheets and tiles, thermoplastics and vinyl-asbestos tiles, ceramic floor tiles and quarries: (ii) bonding roofing felt; (iii) bonding paper/paper and paper/fabric laminates.

Bituminous adhesives are available for use hot and as emulsions and solvent solutions for use cold. They have good resistance to water and many chemicals, but tend to flow at higher temperatures and staining can result from migration of the bitumen. Coal tar emulsion adhesives have less tendency to flow at higher temperatures, and the inclusion of rubber enables them to accept small movements.

Adhesives based on synthetic thermoplastics

Polyvinyl acetate (PVA)

BRE Digest 209 describes: *Polyvinyl acetate adhesives for wood.*

These adhesives described by BS 4071:1966 *Polyvinyl acetate (PVA) emulsion adhesives for wood* are white liquids, transparent when set, which generally do not discolour other materials although some grades stain in contact with ferrous metals. They are easy to use, they set at room temperature and do not blunt cutting tools.

PVA emulsion adhesives soften at about 70°C. Copolymers of vinyl acetate adhere to a wider range of substrates.

The water resistance of PVA is superior to that of ordinary animal glues although slightly inferior to that of specially formulated types. PVA is suitable for joints which will not be required to resist high continuous stresses, in particular for bonding joinery, paper, leather, cloth and similar materials to be used in dry conditions. It is used as a bonding agent to provide adhesion for plaster or mortar on smooth surfaces such as glazed ceramic tiles and in cement:sand floor screeds and levelling compounds.

PVA adhesives are usually provided in one part with a shelf life of at least six months, usually as a liquid but one product is in a gelled form with a resulting shorter setting time. Setting is by loss of water which can be accelerated by artificial heat.

Hot-melt adhesives

These adhesives are 'solid' thermoplastics – such as sealing wax – but modern types are usually based on ethylene-vinylacetate (EVA) copolymers.

Adherands may require to be heated. The molten adhesive ('melt') is then applied to one of the surfaces, adequate *wetting* of which is ensured by a sufficient temperature during application. The other surface is immediately applied and held in contact until the adhesive cools.

Alternatively, precoated materials, such as veneers, can be hot-ironed onto substrates.

Hot-melt adhesives are very suitable for flow line production – there are no flammable solvents and the bond is formed in seconds. Adhesion is good to wood, metals and plastics.

Synthetic resin thermosetting adhesives

These adhesives are capable of extremely high strength, in some cases even when joining metals and although they are combustible their performance at high temperature is superior to the natural and thermoplastic types. They are in some cases extremely resistant to moisture and microorganisms, but they tend to blunt cutting tools. Hardening is essentially by heat action which is usually accelerated by either a catalyst or hardener enabling practicable curing times to be achieved at room temperature, or very fast times at elevated temperatures. The resin and hardener may be mixed together immediately before use or alternatively dry powdered resin and hardener supplied ready mixed are activated by water. Where fast setting is required the resin is applied to one surface and the hardener to the other, and hardening begins when the surfaces are brought together. In each case special techniques and close control of gluing operations are necessary and BS 4169 states that bonds of the requisite standard for laminated timber structures are rarely possible on the building site.

British Standards are:

BS 1203:1979 Synthetic resin adhesives (phenolic and aminoplastic) for plywood

BS 1204 Synthetic resin adhesives (phenolic and aminoplastic) for wood

Part 1:1979 Gap filling adhesives

Part 2:1979 Close contact adhesives

BS 4169:1970 Glue laminated timber structural members

Urea formaldehyde

These colourless adhesives are inexpensive and are widely used in building where they are not exposed to the weather. Uses include: laminated timber construction, joinery (UF adhesives are available with exceptionally long closed assembly times) and in the manufacture of plywood and particle boards.

Urea formaldehyde is available as a one-part adhesive which includes an acid hardener, requiring dissolution in water, or with a separate hardener as a two-part adhesive. Water must be able to evaporate. Setting is quicker at higher temperatures, taking about 6 hours at room temperature, but only a few minutes at 80°C.

The inclusion of up to 20 per cent wood flour as an extender reduces cost, and by reducing brittleness it makes thicker gap-filling films possible.

Normal UF glues are classified as MR but inclusion of resorcinol or melamine in the hardener (fortified UF glues) brings the durability classification up to BR.

Phenol formaldehyde

Phenol formaldehyde adhesives are classed as WBP (not affected by weather or boiling). They are provided in an alkaline solution, or film suitable only for weather-proof or marine plywood manufacture, which set at temperatures from about 115°C.

A cold setting glue for timber assembly and constructional work is provided in an acid-hardened non water-soluble form. Durability classification is WBP.

Melamine formaldehyde

Although comparatively expensive, these adhesives, like UF adhesives, are colourless and can be used for similar purposes and particularly for veneering where increased durability or heat-resistance are required.

They are provided as a film for setting at 100°C or as a powder to be mixed with water and hot pressed at 95–130°C.

The durability classification of melamine-formaldehyde adhesives in BR.

Cold-setting reactive adhesives

These adhesives set at room temperature by chemical change.

Resorcinol formaldehyde

This reddish-purple adhesive is more costly than phenol formaldehyde but easier to use at low curing temperatures, and more rapidly up to about 70°C. It has a long shelf life and the pot life after adding a hardener, usually in powder form, is from 1 to 5 hours at 20°C.

Although water soluble until cured, when hardened the durability classification is WBP.

Resorcinol formaldehyde is used to form extremely strong and durable joints in timber structures with a moisture content not exceeding 15 per cent at the time of bonding. For less exacting applications even higher levels can be tolerated. It is also suitable for bonding plastics, rubber and alkaline materials such as asbestos-cement sheets.

Epoxide resins

These two-part resin adhesives, eg Araldite (CIBA [ARL] Ltd) are costly but possess remarkable properties, adhesion being excellent to almost all surfaces, in most cases both internally and externally. Epoxide adhesives can be resilient, electrical resistance is high and they are water and water vapour proof. They are highly resistant to most acids, alkaline and solvents.

Epoxide resins are transparent and ideal for glass showcases and for appliqué glass techniques (see page 267), but colour can be introduced in manufacture. Most grades cure in 24 hours at room temperature, in 1 hour at 80°C or in ½ hour at 100°C. Since they harden without giving off any volatile matter, shrinkage is negligible and they can be used for joining non-porous materials such as metals and glass. Metal/metal bonds can have a shear strength of $14 \, N/mm^2$ for cold cured systems and $21 \, N/mm^2$ or more for hot cured systems. Concrete/concrete bonds which can be stronger than concrete are being used increasingly, eg for bonding the precast segments which form the shell roof of the Sydney opera house.

Epoxide resins are used, with sand or powdered stone, for repairing concrete and masonry. Mortars with a 2:1 filler:binder ratio have a crushing strength of about $103.5 \, N/mm^2$.

Dry bonds with timber are excellent but long term weathering or prolonged soaking in water can produce breakdown of the joint more rapidly than with conventional *WBP, BR* or *MR* adhesives. Also because epoxide adhesives soften at temperatures ranging from 60°C to well over 120°C according to formulation, they do not qualify for the WBP classification of BS 1204.

Cyanoacrylates

Although costly, these one-part adhesives set rapidly without the need for heating and pressing equipment. However, instant adhesion to skin is a very serious hazard.

Bonds formed with glass, rubber and wood are usually stronger than the materials bonded, but gap-filling properties are poor.

Rubber-based adhesives

These natural or synthetic rubber-based adhesives are provided in a volatile solvent or as emulsions ready for use, either for application to one surface only, or as *contact* adhesives to be applied to both surfaces for a *two-way dry stick*. After allowing 10 to 30 minutes for the solvent to evaporate the surfaces are brought together and an immediate contact bond is obtained. There is little or no opportunity for adjustment, but contact adhesives are very suitable for bonding plastics laminates, fabrics, aluminium, and sheet floor coverings. Wood joints are not of great strength, and the tendency to flow under constant load renders them unsuitable for structural work. In general, these types of rubber adhesives are not durable in external exposures.

For fixing boards to walls, gap-filling conditions are obtained by bonding onto walls horizontal strips of adhesive-impregnated foam pads (*Bostik pads*)[1] and bonding the boards to them. This system has the advantage of accommodating any movements of the board and of the background.

[1]Bostik Ltd, Leicester.

15 Mortars for jointing

Dry mortarless walling has a commendable history and continues to be servicable in many ancient structures.

Today, mortar is usually required to bond and distribute loads between separate walling units so they act structurally as brick, block, or stone masonry. Until about 1900 the binding agent for most mortars in UK was lime, the softness of which makes it easy to remove it from bricks and stones so they can be reused. Portland cement then replaced limes as the principal binder, and a proportion of fat lime provided workability in mixes. More recently, chemical plasticisers often provide the latter function.

References include:

BDA publications.

BRE Digest 160 *Mortars for bricklaying.*

PD 6472 *Guide to specifying the quality of building mortars.*

BS 5628, Part 3:1985 *CP for the use of masonry materials etc*, relates masonry units and mortars to situations.

BS 5390:1976 *CP for stone masonry.*

BS 6270 *CP for cleaning and surface repair of buildings.*

BS 4551:1970 comprises: *Methods of testing mortars and specification for mortar testing sand.*

BS 187:1978 *Calcium silicate bricks* deals with mortars – see page 141.

Mortar must cohere so that it does not slip off the trowel too readily, but it should spread easily and remain plastic so that the units can be adjusted for line and level. It should also retain water so that it does not stiffen in contact with absorptive bricks. All this makes for a good brick to mortar bond, which is an important factor in preventing rain penetrating a wall. Typically, rain enters at the brick–mortar interface, rather than through either the mortar or the units.

BS 743 for *Damp proof courses* points out that mortar by itself cannot be a dpc.

Once the units have been laid, mortar should develop strength quickly to avoid squeezing out by the superimposed load and in order to resist frost during its early life. In service, mortar should bond well to the units, and provide frost and salt resistance, without cracking or other deterioration.

'Final' strength must be adequate but it is important to note that the strength of mortar is not directly related to the strength of brickwork, and for any brick strength there is an optimum mortar strength, above which further increases in mortar strength do not increase that of the brickwork.[1]

In drying, Portland cement and lime shrink, and in particular Portland cement tends to crack. With relatively weak mortar any cracking occurs as hair cracks distributed thoughout the joints. Excessively strong mortar with a high Portland cement content, however, tends to concentrate the effects of any movement into fewer but relatively wider cracks. These may extend through both bricks and mortar, are unsightly and may admit water. Flexibility is particularly necessary in mortars used with calcium silicate and concrete bricks and lightweight concrete units which have high drying shrinkage. A relatively weak mortar is also better able to absorb expansion of new clay bricks.

Where medium strength bricks are used 1 Portland cement : 2 non-hydraulic mortar gives brickwork 20 per cent weaker than 1 Portland cement : 3 sand mortar.

New mortar used in restoration work should be of similar strength to the old mortar. Sadly, all too often, such mortar is too hard, dense, grey, and otherwise visually unsympathetic to old masonry. BS 6270 refers to Mortars.

TYPES OF MORTAR

Today, Portland cement is usually the principal binding agent, often with the addition of *fat* (high calcium) lime or semi-hydraulic lime to provide

[1]See *MBS: S and F Part 2.*

Type of construction	Weather Exposure	Type of bricks or blocks[2,3]	Mortar group (table 101)	
			No risk of frost during construction	Frost may occur during construction
Load bearing, engineering	any	Clay (class A or B) calcium silicate (class 5) or concrete	1	1
Earth (drained at rear) retaining walls	any	Clay class (A or B), Calcium silicate (class 4) or concrete	1 — 2[1]	1 — 2[1]
Sills and brick copings	any	Clay (generally class A or B) Calcium silicate (class 4) or concrete	1 — 2	1 — 2
Parapet walls (rendered)	any	Clay (with low sulphate content) Calcium silicate (class 3) or concrete	3[4] — 4	3[4] — 3
Parapets (not rendered) external free standing walls, work below dpc	any	Clay (with low sulphate content) Calcium silicate (class 3) or concrete	2 or 3 — 3[4]	— 3[4]
External walls between eaves and dpc	severe[1]	Clay Calcium silicate (class 2) or concrete	3[4,5] or 4[4] — 4	3[4] — 3
	sheltered or moderate	Clay Calcium silicate (class 2) or concrete	4 — 4	3 — 3
Backing to external solid walls	—	Clay Calcium silicate (class 1) or concrete	4 — 4	3 or 5 — 3 or 5
Inner leaf of cavity walls	—	Clay Calcium silicate (class 1) or concrete	6 — 6	3 or 5 — 3 or 5
Inner walls	—	Clay Calcium silicate (class 1) or concrete	5	3 or 5 — 3 or 5

[1]Severe exposure applies to external walls in districts with a driving rain index of $7\,m^2$/sec or more, and to high walls in districts with an index of $6\,m^2$/sec or more. For more detailed information see BRE Digest 127
[2]Clay bricks BS 3921. (Classes are given only where strength is particularly important)
[3]Calcium silicate bricks BS 187. Concrete bricks BS 1180
[4]Sulphate-resisting cement should be used with high sulphate content clay bricks or below ground
[5]Rendered walls

Table 100   Mortars for uses

plasticity and improve adhesion to absorptive units or with an air-entraining agent.

High calcium limes stiffen only as they dry out, and strength development by absorption of carbon-dioxide is too slow for modern building. Magnesian limes which develop more strength are used to some extent in the Midlands and North of England. Eminently hydraulic limes develop strength reasonably quickly and can be used where limes, and operatives skilled in slaking and handling them are available. To avoid expansion of mature mortar, slaking of lime must be complete before it sets and loses its plasticity. Because hydraulic limes do not possess the same degree of workability (*fatness*) and ability to retain water of non-hydraulic limes, they are less suitable for use with Portland cement. There is no British Standard for hydraulic limes: information on the subject is contained in BS 5390 and BS 6270.

Portland cement mortars

Portland cement contributes high early and final strength, and rapid hardening Portland cement gives even greater early strength – see page 155.

Sulphate resisting Portland cement can be required to provide superior resistance to attack by sulphates, eg those in sulphate-bearing soils, or in clay bricks, although the added sulphate resistance afforded to lean mortars may be negligible.

BS 5838: *Dry packaged cementitious mixes* Part 2: *Prepacked mortar mixes*, requires cement-sand mortar to have a minimum crushing strength of $6 N/mm^2$ at 28 days.

Portland cement, lime mortars

The leanest workable Portland cement mix is 1 volume Portland cement to about 3 volumes of clean sand but such a mix is stronger than required for most purposes. By substituting non-hydraulic or semi-hydraulic lime to BS 890:1972 for part of the cement in a 1:3 mix, strength and the tendency to crack can be reduced, while improving workability, water retention and bonding properties.

1 Portland cement: 2 lime: 9 sand mixes are about 10 per cent weaker, and 1:3:12 mixes about 20 per cent weaker than 1 Portland cement: 3 sand.

BS 4721:1981 describes *Ready-mixed lime-sand for mortar* – mainly for gauging with cement.

BS 5838 requires bricklaying and masonry mortar to have a minimum crushing strength of $3 N/mm^2$ at 28 days.

Portland cement-plasticizer mortars

An alternative, and very common means of improving the workability of cement-sand mixes and thereby reducing their cement content, is by the use of plasticizers which entrain air into the mortar. The pores allow for expansion so that plasticized mortars have greater resistance to

frost both before and after hardening, than cement-lime-sand mixes.

Aerated cement-sand mortars are also more resistant to sulphate attack than cement-lime-sand mortars of the same strength. On the other hand, plasticized mortars have poor water retention and achieve weaker bond with absorptive bricks than cement-lime-sand mortars.

It will be seen from tables 100 and 101 that aerated cement-sand mixes leaner than 1:8 or richer than 1:3 are not recommened.

BS 4887:1973 describes *Mortar plasticizers*.

Masonry cement mortars

These mortars have properties intermediate between cement-lime-sand and cement-plasticizer mortars. They are based on Portland cement, together with a very fine mineral filler and an air-entraining agent which provide workability – see page 154 and BS 5224:1976. The manufacturers' instructions should be carefully followed.

COLOUR

The apparent colour of brickwork depends as much on the colour of the mortar as on that of the bricks.

To some extent colour of mortar results from that of the sand, but it is mainly due to that of the cement, whether grey, white or coloured. Pigments added to cement should comply with BS 1014:1975 *Pigments for cement* or carbon black can be used. If proportions of these pigments exceed ten and three per cent by weight of cement, respectively, the mortar may be seriously reduced in strength.

SAND

Sand should comply with BS 1200:1976 *Sands for mortar for plain and reinforced brickwork, block-walling and masonry*, It should be free from clay, salts, including calcium chloride, and gypsum and other contaminants. Correct grading is important. Both coarse and fine gradings require more water to achieve equal workability, with consequent reduction in the mortar strength.

BATCHING

Measurement of materials should be done in gauge boxes, and *bulking of sand* (up to 40 per cent for damp sand, see page 174), should be allowed for. In the case of cement-lime-sand mixes the effect of bulking is reduced if cement is added to lime and sand *coarse stuff* just before use. Dry hydrate should preferably be soaked at least overnight to improve its plasticity, but if this is not done the proportion of lime should be increased.

Proportioning by mass can be more accurate, but only if variations in bulk densities of materials are checked regularly and allowed for.

MIXING

Mixing can be done by hand, or by machine. Prolonged mixing of cement-lime mortars improves their plasticity and water retention, particularly in high speed mixers or activators. Prolonged mixing of aerated mortars however may cause excessive air-entrainment and loss of strength. Aerated mortars cannot be mixed effectively in mortar mills. Portland cement mixes should be discarded if they have not been used within two hours after adding the cement – since their strength cannot be recovered by *knocking up*.

BRICKLAYING IN COLD WEATHER

Bricks, and also sand and mixed mortar, should be kept dry before laying, and mortar should not contain a lower proportion of cement than:

(i) 1 cement:1 lime:5–6 sand,
(ii) 1 cement:2 lime:8–9 sand with an air-entraining plasticizer,
(iii) 1 cement:5–6 sand with an air-entraining plasticizer, or
(iv) 1 masonry cement:4½ sand.

Bricks should never be laid in freezing conditions, unless a constantly heated enclosure is used. The temperature of the brickwork is critical and work should stop when it falls below 3°C or when frost is imminent.

Work should not recommence until the temperature of the bnrickwork or concrete on which the new work is to be laid is above freezing point. A 'spear' thermometer can be used to check the temperature of mortar.

Frozen mortar will not develop strength and brickwork should be taken down.

The water should be heated and tops of walls

Mortar group	Hydraulic lime: sand	Cement[1]: lime[2]: sand[3]	Masonry cement: sand	Cement[1]: sand[3] with plasticizer
1	—	1:0–¼:3	—	—
2	—	1:½:4–4½	1:2½–3½	1:3–4
3	—	1:1:5–6	1:4–5	1:5–6
4	1:2–3	1:2:8–9	1:5½–6½	1:7–8
5	1:3	1:3:10–12	1:6½–7	1:8

Increasing strength and durability.

Decreasing ability to accommodate movement

←————— Approximately equal strength —————→
————— Increasing resistance to damage by freezing —————→
←——— Improving bond and consequent resistance to rain penetration ———

[1] Normal Portland cement, or sulphate-resisting Portland cement where sulphates are present in soil, or where clay brickwork is likely to remain wet — see page 155

[2] Non-hydraulic or semi-hydraulic lime putty. It should preferably be soaked at least overnight before use. If lime is measured as dry hydrate the amount can be increased up to 1.5 volumes for each volume of lime putty

[3] Where a range of sand contents is given the larger quantity should be used for sand that is well graded and the smaller quantity for coarse or uniformly fine sand

Table 101 Mortar mixes by volume

protected against heat loss with substantial insulation – The BDA points out that wet hessian is useless!

Anti-freeze admixtures should not be used in mortars even if they are satisfactory in concrete.

The water should be heated and tops of walls should be covered against heat loss during breaks.

BS 5628 states that admixtures reduce the strength and adhesion of mortar.

BS 4887 describes *Mortar admixtures*.
If they are used the manufacturers' instructions should be complied with.

Calcium chloride cannot be relied upon to prevent frost damage and, being a hygroscopic salt, it may cause dampness in walls and corrode metals.

POINTING

Integral *jointing* is preferable but if a separate *pointing* mix is desired, perhaps containing coloured cement or a special sand, it should not be appreciably stronger than the bedding mortar.

Repointing of weak bricks or stones should be done with a mortar containing only sufficient cement to resist weathering.

SELECTION OF MORTAR MIXES

The mixes recommended in table 101 show that the proportion of cementitious binder (cement or cement with lime) required to fill the voids in a well graded sand is about 1/3 by volume. Coarse and uniformly fine sands require a higher proportion of binder to fill the voids. Tables 100 and 101 recommend mortars for various conditions which take into account:

Type of unit and its strength, moisture movement and density

Conditions of exposure – internal, sheltered external or severe external exposures

Frost hazard during construction

Likelihood of attack by sulphates in soils or clay bricks which may be accentuated by retention of water in brickwork by renderings.

Sealants are used to seal joints against rain, air, dust, odours and sound. They are required to adhere to surface and may be required to accommodate the full range of combined moisture and thermal movements three or four hundred times a year.

Gaskets – see *MBS: S* and *F* Part 1 chapter 2, must move similarly while maintaining seals against surfaces by a spring action.

References include:
BRE Digests 227–229, Parts 1–3: Estimation of thermal and moisture movements and stresses.
BRE Digest 137 Principles of joint design.
BS 4643:1970 Glossary of terms relating to joints and jointing in buildings.
BS 6093:1981 Code of practice for the design of joints and jointing in building construction.
BS 6213:1982 Guide to the selection of constructional sealants.[1]
BS 5390:1976 Code of practice for stone masonry.
BS 5628 CP for the use of masonry. Part 3:1985 refers to *Materials and components*, including movement joints.
CP 298:1972 Natural stone cladding (non-load-bearing).
BS 8110 Structural use of concrete Part 2:1985 deals with movement joints.
Sealing compounds, K A Nimmo, Specification 1982, Architectural Press Ltd.

Materials which can accommodate a very limited amount of movement, such as *linseed oil putty* for glazing wood sashes (BS 544), and some materials, for fixing tiles, glass and sheets are sometimes called *mastics*.

Movements are usually small and fairly slow, but where they are restrained by friction, as happens in spigot and socket joints in aluminium certain walling, a *slip-stick* action occurs and the rapid movement can cause sealants to fail.

Joints must be sufficiently wide to accommodate the expected movement but excessive width

[1] This BS is being amended to include guidance on sealants for stonework

wastes sealant, which is costly, and it may slump. Consequently the width of filled joints must be related to the properties of the sealant to be used, the expected movement and also inaccuracies in dimensions of units and in construction.

If movement at a joint is greater than can be accepted by a sealant it may be necessary to either increase the frequency of joints or to adopt open drained joints – see BRE Digest 85.

TYPICAL JOINTS

In the absence of calculations, BS 5628 recommends spacings and widths of joints for accommodating thermal and moisture movements during the lives of buildings.

Movement joints are not normally needed in internal walls. Recommendations for typical external unreinforced walls are:

Clay brickwork

Vertical joints in unrestrained walls should be at, say 12 m centres. Expansion decreases with restraint but the spacing of vertical joints should never exceed 15 m, or possibly less in parapets.

Width of joints in mm should be 30 per cent more than their spacing in metres. For example, joints at 12 m centres should be 16 mm wide.

The BS states that vertical movement appears to be of the same order as horizontal movement.

Calcium silicate brickwork

Generally, vertical joints should be at 7.5–9 m centres and not more than 10 mm wide. Panels of brickwork should not exceed 3 length:1height between joints.

Concrete blockwork

Vertical joints should not exceed 6 m spacing and panels should not exceed 2 length:1 height

FORMS AND TYPES OF SEALANTS

The forms and types of available sealants are described in table 102 and BS 6213:1982.

FORMS

Sealants are made in consistencies suitable for application by hand, gun, by pouring or as a preformed strip or tape.

Knife or trowel application requires a sealant which is soft but not too tacky.

Gun application requires a soft sealant which can be extruded through a nozzle by hand or air pressure.

Sealants for *pouring* into joints are based on either bitumen which requires heating before use, eg BS 2499:1973 *Hot applied joint sealant for concrete pavements*, or for cold pouring polysulphide rubber or polyurethane, and BS 5212:1975 *Cold poured joint sealants for concrete pavements – ordinary and fuel resistant types*.

Tapes or strips are available in a variety of shapes, sizes and hardnesses. They are generally only suitable where they will be held in compression without being squeezed out, or where movement is small and here it may be necessary to prime the surfaces to be joined.

Hard sealants are available for flooring joints and to take light loads, but usually 'spacers' or mortar must be provided to relieve mastics of all loads. See CP 202:1972 *Tile floorings and slab flooring*.

TYPES

Sealants are based on viscous media with inert fillers and in some cases with solvents which facilitate application and then evaporate. The properties of the main types are given in table 102.

It is important to differentiate between non-setting *plastic* and *elastic sealants*.

Plastic sealants

Most oil, bitumen, polyisobutylene and some butyl rubber based materials have *plastic flow*. (They are not plastics materials as defined on page 270. They show limited recovery after deformation and are termed *plasto-elastic*. They are squeezed out by constant pressure, or by the action of high winds on large panels and tend to slump in vertical joints more than about 13 mm wide.

Plastic sealants form a tough and protective skin by oxidation. Paint prolongs the life, but eventually these sealants harden and tend to crack.

The maximum recommended elongation for plastic sealants in butt joints is only 5 to 15 per cent, although the rate, rather than the extent of movement, determines their ability to retain adhesion and cohesion. Where movement is small and slow, good quality oil-based (*oleo-resinous*) mastics are likely to be satisfactory in the more sheltered positions.

Butyl sealants are suitable for bedding window and door frames and the better grades with low slump characteristics are usually satisfactory for glazing with beads. Acrylic sealants are used for sealing gaps between baths, lavatory basins and wall surfaces.

Dark surfaces exposed to sunlight in the UK may rise to 70°C. Such temperatures with associated thermal movements, together with UV radiation, accelerate deterioration of sealants.

Elastic sealants

Elastomeric sealants include polysulphide rubber, silicone, polyurethane and some of the butyl rubber materials. Polyurethane and polysulphide rubber sealants are supplied in two parts, the base and a curing or vulcanizing agent, which must be very thoroughly mixed together immediately before use. The resulting product is sticky and is best applied by a gun. It is important to realise that it remains plastic for a few days, during which it may be unable to accept loads such as the weight of glass or sashes. However, when thoroughly cured, the sealant is able to accept compression without extruding, and to withstand elongation in some cases up to 75 per cent in shear. Polysulphide and silicone sealants may have some plastic flow but polyurethane is truly elastic.

Greater stresses are imposed on adhesion to joint surfaces by elastic sealants than by plastic

307

materials and to avoid adhesion failure, soft plastic sealants are preferred.

Elastic sealants are durable but costly, and in order to take full advantage of their superior flexibility, joints must be as far as possible, of uniformly optimum width and depth. These conditions of accuracy are possible in metal assemblies such as curtain walling, and here elastic sealants may be essential to accommodate *stick-slip* movements with only short periods spent at full elongation. High grade sealants such as poly-sulphide or silicone are also very suitable for capping two-stage joints; sealed inside with mastic strips or with preformed synthetic rubber gaskets, and are advisable for pointing around windows and doors in the more exposed weather conditions.

Two-part polysulphide sealants are probably best for glazing with beads, and for narrow joints in external tiling and mosaics (see *MBS: Finishes*), where movement is often relatively considerable. Joints in flooring should be filled with a hard grade of sealant to minimize dirtying and treading in of grit.

Silicone products, which are obtainable in a translucent form, a very durable white and a wide range of colours, are very suitable for sealing around baths, sinks and basins and for joints in wall tiling.

SELECTION

BRE Information Paper IP 25/81, *The selection and performance of sealants* is a useful reference.

Manufacturers should be consulted at an early stage in the design of joints and their written assurance obtained as to the suitability of the materials they recommend.

Not all of the very wide range of types and formulations with widely varying properties are provided by all manufacturers.

Sealants, especially those which satisfy the more exacting requirements, are costly both to buy and to apply, so that a long life is very desirable. However, although the effects of fourteen days' storage at 70°C, usually approximate to those of a year's natural weathering, laboratory

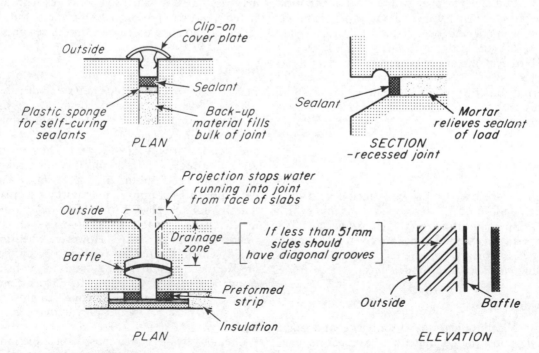

50   *Sealants in protected positions*

ageing tests on new materials require many years of correlation with field experience.

Factors which must be taken into account in the selection of sealants include:

Function of joint Weather seal in cladding; flooring movement joints; glazing.

Exposure of joint

Maximum total extension of joint and toleration of movement by sealant

Compatibility with adjacent materials Other surfaces, alkalis in concrete; primers; back-up materials; setting blocks; other sealants.

Durability Freedom from: adhesion failure; cracking; crazing; cohesive tearing; slump in open joints; hardening.

Resistance to: weather; traffic.

Position of joint and ease of application of sealant Vertical: horizontal; upward or downward facing; accessibility.

Appearance Tendency to stain adjacent materials; colour; suitability for painting.

Cost Sealant, fillers; primers, etc; labour; 'cost-in-use'.

DESIGN OF JOINTS

This subject is dealt with in *MBS: S. and F., Part 2*, but some considerations relevant to the use of sealants are discussed here.

Because the life of sealants is much shorter where they are exposed to the action of light and air, and where they are subject to the movements which are greater on exposed faces 'where possible, complete reliance against rain and wind penetration should not be placed on the sealant'. Thus provided they are accessible for renewal, sealants are best protected from the weather and used mainly as air and dust seals, see figure 46 and the lap joints in figures 47 and 48.

Butt and lap joints compared

Butt joints are easier to form, but for equal movement in joints, the extension of sealant in a lap joint is only about half that in a butt joint. Also, in lap joints, because the sealant is usually better protected from the weather, should it fail in adhesion to one surface, serious leakage is less likely than with similar failures in a butt joint, as shown below:

Geometry of joints The width and depth of sealant in a joint must be carefully considered. Table 102 shows maximum widths and minimum depths. The width/depth ratio is also important – see footnotes to the table.

Width of joint For given overall movement in a butt joint, the movement per unit width of sealant increases as the width of the joint decreases and the joint must be sufficiently wide to avoid overstressing the sealant. Joints should never be narrower than 3 mm and preferably at least 6 mm wide. Also sealants must be free to move over the full width of a joint, unlike the examples shown in figure 52 where adhesion to panels limits their effective widths. Particularly, where elastic sealants, are used, a flexible backing such as polythene or polyurethane sponge is required to ensure that they are effective over their full width, and therefore subject to a much lower percentage movement.

The maximum extensions which various sealants will tolerate in service are given as percentages in table 102 and, if the expected movement in a joint is known, the minimum width in butt and shear joints can be calculated.

Summarized, the design steps are:

(a) estimate movement due to moisture and thermal changes, deflection, etc. BS 5628 *CP for the use of masonry* Part 3:1985 refers to *Materials and components*. BS 5628, Part 3 includes ways of assessing the magnitude of individual movements.

(b) determine minimum width for the sealant intended to be used

(c) allow for manufacturing and assembly tolerances

(d) determine design width of joint.

Depth of sealant

Where necessary, *back-up materials* such as fibreboard, rope or expanded plastics, should be provided to limit the depth of sealant, and to give a stop against which the sealant can be forced to ensure good adhesion with the surfaces to be joined. In closed joints back-up materials must compress easily.

Oleoresinous sealants can be allowed to adhere

Medium	Nature	Available forms	Skinning setting or tack-free time hours	Maximum width of joints[1] mm	Maximum total extension in service		Minimum depth of mastic in joint[6] mm
					Butt joints per cent	Shear joints per cent	
Oleo-resinous one part	Plastic form skin	Strip	24	51	2	5	6
		Knife grades	24	25	5–15	20	9.5
		Gun grades	24	13	15	30	13
				25–31 Special grades		(Better grades 40	Up to 25; for very wide joints
Bitumen-rubber one part (BS 2499:1973 Hot applied joint sealing compounds for concrete pavements)	Mainly plastic some form a skin	Strip	—	51	10	20	6
		Knife grades	—	25 or more	10	20	13
		Gun grades (solvent type)	—	13	10	20	19
		pourable grades (for hot application in horizontal joints in paving)	When cool	up to 51 (horizontal joints)	10–15	20–30	19
Butyl (and related polymers) one part	Plastic some form a skin and Elastic types	Strip	—	51	2–10	5–25	6
		Knife grades	24	25	2–15	5–25	6[5]
		Gun grades	24	13	5–15	10–40	6[5]
				25 (Special grades and techniques)			
Acrylic one part	Plastic with slight elastic 'memory'	Gun grades	12	13 (or wider by use of special techniques)	15 25 (terpolymer type)	30	6[5]
Polysulphide one part Gun grade (BS 5215:1975)	Elastomeric may have some plastic flow	Gun grade	24	19	15[3]	30[4]	6[5]
two part (BS 4254:1983 Two-part poly-sulphate sealants[2] (not heavy duty floors)		Gun grades	48	13	15–33[3]	40–75[4]	6[5]
		Pourable grades	24	25 or wider by use of special techniques	15–33	40–75[4]	6[5]
Silicone one-part (BS 5889:1980 Silicone based building sealants)	Elastomeric may have some plastic flow	Gun grades	2 (full cure takes 7–14 days)	25 (or wider by use of special techniques)	Probably similar to polysulphide sealants		6[5]
Polyurethane two-part	Elastomeric	Knife grades Gun grades Pourable grades	24	51	Probably 20	Probably 50	6[5]

[1]The minimum width should be preferably 6 mm and never less than 3 mm
[2]For general building and glazing – not heavy duty floors

Table 102   Sealants for movement joints (the information is intended as a guide only and the advice of manufacturers should be obtained in the design of joints and selection of sealants, primers and back-up materials)

Comments		Colour	Life expectancy in external movement joints (years)
…rface oxidizes rapidly, but in particular, gun grade mastic remains soft below …otection, if only by paint, prolongs life …dhesion: fair-good. Strip mastics may require primers. …ery slow movements can cause cohesive failure.		Grey and wide range of colours	2–10
…urable grades generally more flexible than the solvent type but overheating …estroys their properties. Very suitable for joints in mastic asphalt and tarmacadam …aving …rface tends to oxidize, craze and discolour …ifficult to paint …dhesion: fair		Black	2–3 (Solvent type)
			5–10 (Hot applied)
…ide range of properties …asy to place and recent formulations retain softness well …urability is affected by nature of movement. Some butyl mastics split through the centre …hen subjected to repeated slow movements. Dark colours have superior durability …dhesion: fair. Some strip mastics require primers …ood resistance to alkalis. Most collect dirt		Grey and black (standard)	10
…e have similarly to many butyl mastics …ust be applied 'hot' 50°C …ight dirt pick-up …able to breakdown if subjected to prolonged wetting		White and wide range of colours	15–25
…imers are required to provide adhesion to …orous or friable surfaces and on the inside …f glass to prevent loss of adhesion caused …y sunlight passing through …kaline surfaces may require to be sealed …o prevent staining during the curing period …esistance to acids, alkalis, petroleum and …olvents is generally good, but not to chlor-…nated hydrocarbons, eg trichloroethylene …ifficult to replace	Cure by absorption of moisture. A skin forms rapidly but cure of interior may be very slow, particularly in large joints and in cold dry weather. Not suitable where appreciable movement is likely before curing is complete. It is recommended that the two-part type should be use where the calculated movement exceeds 10 per cent	Aluminium and black also available in dark greys and colours (not all colours are light fast)	15–20
	Cure by chemical reaction rapid at normal temperatures. Less risk of failure due to movement during cure than with one-part type. Extremely flexible		20
…ure by absorption of moisture from air, slowly especially in cold or dry conditions and …n thick sections although some types cure faster than one-part polysulphide sealants …urable under water. Good adhesion, but elasticity places adhesion under stress and …reparation of surfaces is extremely important. Primers are required on most porous …urfaces and on some non-porous surfaces		Translucent, white and wide range of colours but pick up dust which is difficult to remove	Possibly 20
…uly elastic, therefore adhesion under stress is only fair-poor …eparation of surfaces is extremely important …imers may be required		Cream grey, black	Possibly 20

…he maximum tolerance for movement is obtained where the width of sealant is twice the depth
…he maximum tolerance for movement is obtained where the depth is not less than the width
…reater depths may be necessary to allow for inaccuracies in joints in masonry, brickwork and concrete
…epth of sealant is measured perpendicular to the weather surface for shear (lap) or butt joints

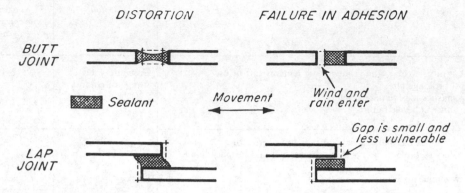

51 Butt and lap joints compared

to back-up materials, but with rubbery products adhesion to inflexible back-up materials can seriously affect their performance. With these it is essential either to use a suitable *parting agent* or to interpose a flexible back-up material such as polyurethane or polythene sponges, as may be recommended by the supplier of the sealant – see figure 52.

APPLICATION

Good workmanship is essential for forming satisfactory joints with sealants and specially trained operatives should be employed.

As far as possible, application of sealants is best delayed until drying shrinkage of concretes, and calcium silicate bricks, expansion of new clay bricks and settlements have taken place. Surfaces to be joined must be free from loose matter and clean. Aluminium, especially extrusions, must be cleaned with solvents, using frequently changed rags to remove grease. Surfaces must be dry, a gas torch or hot air blower may be required.

A primer supplied by the maker of a sealant should be applied to surfaces:

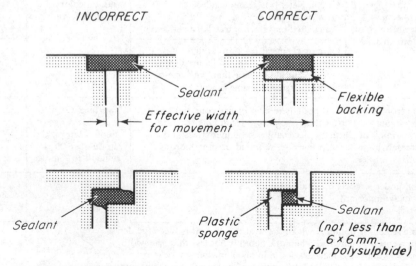

52 Effective widths of joints

1 To reduce absorption of the medium from a sealant by porous surfaces, or staining of the surrounding material.
2 To prevent attack on oleoresinous sealants by alkaline surfaces.
3 To improve adhesion.
4 To prevent loss of adhesion by action of water, eg in new concrete.

Joints in ceramic tiling, and particularly in ceramic floor tiling, should be finished concave rather than flush, to avoid bulging of the sealant later due to expansion of the tiles.

Index

INDEX